THE TRUNKED RADIO AND ENHANCED PMR RADIO HANDBOOK

THE TRUNKED RADIO AND ENHANCED PMR RADIO HANDBOOK

NEIL J. BOUCHER

A WILEY-INTERSCIENCE PUBLICATION
JOHN WILEY & SONS, INC.
New York / Chichester / Weinheim / Brisbane / Singapore / Toronto

This book is printed on acid-free paper.[∞]

For ordering and customer service, call 1-800-CALL-WILEY.

Library of Congress Cataloging-in-Publication Data:

Boucher, Neil J.
  The trunked radio and enhanced PMR handbook / Neil J. Boucher.
    p. cm.
  Includes index.
  ISBN 0-471-35289-6 (alk. paper)
    1. Trunked radio.   2. Mobile communication systems.
3. Radiotelephone.   I. Title.
TK5103.488.B68   1999
621.3845—dc21   99-34062
                                              CIP

Printed in the United States of America.

10 9 8 7 6 5 4 3 2 1

CONTENTS

PREFACE

This book has been written for engineers and other professionals working in the field of trunk radio and enhanced public mobile radio (PMR). While the book is primarily aimed at operators of large systems (nationwide or statewide), the principles outlined and the methods developed are equally applicable to the smallest of systems. This book covers all systems, whether analog or digital, and has seven chapters dedicated to describing the many systems available on the market today.

Because this book is meant to be used by practising engineers, the use of mathematics has been limited to what is necessary for explanatory purposes only. It is the experience of the author that school and university are the only places where practitioners can define their problems in ways that have neat mathematical solutions. The real world is much messier, and most of the mathematics that is needed for trunked radio design is too complex for a simple mathematical treatment, with numerical iterative techniques being necessary for all but the most basic calculations.

Some subjects can only be adequately treated mathematically and those have, as far as is possible, been placed in chapters that can be skimmed through by those who are daunted by the mathematics, without too much loss of overall continuity.

However, while some mathematics can clarify, much of the mathematics that is necessary for the successful design of a modern trunked or enhanced PMR network is neither elegant nor readily comprehended. To this end, you will find that the hard work has been done by the available software which contains 20,000 lines of code and most of the calculation and number crunching power that you will need for the successful implementation of networks large and small. This software is menu driven, and after a short familiarity period, is intuitive and straight-forward to use. See page 373 for instructions about how to obtain the software.

The diversity of trunked radio systems is as much a product of the governing regulations that apply in the respective countries as it is a market-driven phenomenon. The push for digital systems follows the cellular trends, but with much less enthusiasm. Details on all the major systems have been elaborated upon, and it is left to the reader to do the comparisons.

It has been surprisingly difficult to obtain detailed information about the various trunked radio systems that are available, and the limited treatment of some of them is in part due to the paucity of information. I would be delighted to correct this in a future edition if the information is forthcoming. I can be contacted at my E-mail address nboucher@ozemail.com.au

NEIL J. BOUCHER

Maleny, Australia
August 1999

ABOUT THE AUTHOR

Neil Boucher is a communications engineer with more than 20 years experience in large enhanced PMR and trunked networks. Having designed and operated systems on both a statewide and nationwide scale, he has developed a deep understanding of the industry.

On occasions he has been called upon to design and implement small enhanced PMR and trunked radio systems for specialized services. These systems have been used in a number of different countries around the world, and you will find his treatment of the topic reflects that international experience.

He is fluent in a number of computer languages and has written the accompanying software package TRUNK (Ver. 4.0) over a period of many years, to address the design and operational problems that have arisen from time to time.

In addition to network design, Mr. Boucher has been involved in hardware design, particularly of telephone interconnects, PSTN interfacing, and off-air interfacing of incompatible systems.

He is the author of a number of other technical books on mobile communications including *The Cellular Radio Handbook* and the *Paging Technology Handbook*, both published by John Wiley & Sons, Inc. He has written dozens of technical papers for various trade publications.

Currently a free-lance mobiles technology consultant, Mr. Boucher's onside interests include flying, sailing, classic cars (and classic car rallys), and astronomy.

THE TRUNKED RADIO AND ENHANCED PMR RADIO HANDBOOK

CHAPTER 1

A BACKGROUND ON TRUNKED AND ENHANCED PMR RADIO

HISTORY

Conventional mobile radio for decades relied simply on all-stations broadcast calls, followed by either a general message or a voice page to a particular user. Originally, mobiles were expensive (and so rare) and spectrum was readily available. Ordinarily, two radios in communication would use a common repeater channel, with the repeater relaying the message to the other mobile as in Figure 1.1. This is done by using two channels, one for transmit and one for receive, with a suitable spacing in frequency to enable both transmitter and receiver to work at the same time at the base. The mobiles ordinarily have a press-to-talk action that switches the receiver off when the transmitter is on. Figure 1.2 shows how this is done. The combiner (also known as a duplexer) is a device that allows both the transmitter and receiver to share a common antenna by isolating the transmitter power from the receiver, discriminating on frequency. The base station receiver then picks up the transmitted signal from mobile A, amplifies it, and then couples it back (at audio level) to the transmitter, which sends it on to mobile B on another frequency.

This mode of transmission is known as either "two-channel simplex" or "half duplex" and is characterized by the press-to-talk operation. Full duplex, which requires a duplexer in the mobile station as well as in the base station, is feasible so long as the frequency of the transmit and receive are sufficiently far apart that a reasonably sized duplexer can be built (this is not necessary for digital radios, which generally do not require a duplexer). This will allow the parties to communicate without press-to-talk keying, giving a cellular-like voice path.

The disadvantage of this full duplex, from the network point of view (and the reason that it is not too often done in conventional mobile or trunk networks), is that full duplex *requires* that each of the mobiles uses a separate pair of frequencies. If they were to share the same repeater, with both of them transmitting for the duration of the

1

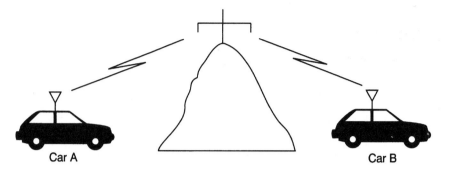

Figure 1.1   The basic concept of a mobile repeater station.

call, then the mobile transmitter signals would interfere, sometimes destructively, at the base station receive antenna, causing the transmission to break up from time to time. Using two transmitters for the call ties up twice as many resources as would a half-duplex call. In addition to this, the duplexers in the mobiles, which must necessarily be small, will have significant losses. This will result in reduced coverage from any given site.

There are, however, times when full duplex can conveniently be used. When a mobile calls a land line (telephone number) and there is only one mobile path, full duplex can be used without network resource liabilities. Also in the case of intersite calls the mobiles will be using separate base transceivers regardless of whether full duplex is used or not. For very large networks the number of calls that use a common repeater can be expected to be small, and so full duplex may be permitted on request.

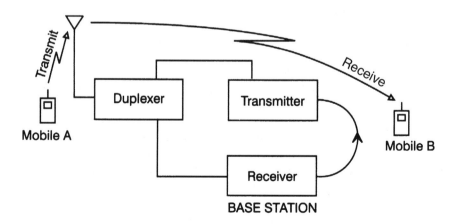

Figure 1.2   A basic duplex repeater.

Invariably however, half duplex will still be available because in fringe areas it may be the only way to connect and also because half-duplex mobiles are cheaper and therefore in demand.

Simplex transmission is possible on most mobiles and is used when the mobiles need to communicate directly without going through the repeater. Typically, this is done when the mobiles are out of range of the repeater(s) or when for other reasons, like security or even to save air-time charges, it is desired to operate independently of the repeater. In simplex transmission the *same* frequency is used for transmit and receive.

As mobile systems began to mature in the late 1970s and 1980s, it was noted that during these high inflation years the prices of mobiles remained roughly constant, or in some cases even came down. This new affordability put many more mobiles on air, resulting in spectrum congestion in many cases.

Although mobiles became increasingly affordable, the cost of a repeater site, which ranged from $10,000 upwards (depending on the associated infrastructure such as towers, buildings, and sometimes even roadways), was still an obstacle to many small users. Also with cheaper mobiles came a demand from the more cost-conscious, and the shared (or community) repeater became common.

Early shared repeaters simply allowed a number of often unrelated users to share a common channel. Often the repeater was owned by a third party whose business was solely to provide mobile repeater services to the public.

To overcome the inconvenience of having third parties listening into company business, the subaudible tone repeater was used. This involved allocating to each user group a subaudible tone (that is, a tone below 300 Hz) and fitting each mobile with a tone-driven squelch. Each member of a group would automatically generate the company's subaudible tone (also known as CTCSS) and thereby broadcast the call only to mobiles in the same company group. A further minor refinement was to arrange for the mobiles to be wired so that the presence of a foreign CTCSS tone would disable the PTT. This prevented third parties from breaking into an existing call that they were not meant to be involved in. Effectively, this allowed a degree of privacy on a shared repeater, and at the same time reduced the incidence of false triggering of the repeater by distant users on the same frequency.

An alternative to CTCSS signaling is DTMF (dual-tone multifrequency). Here the same tone signals are used as for tone dialing in the wireline telephone network. Being inband (that is, you can hear these tones), the signaling is potentially faster, and with a suitable decoder the number of users with individual IDs (identities) can be quite large. DTMF was designed originally to work on the relatively "clean" wireline medium, and the noise inherent in radio can cause decoding problems. Also like CTCSS (but for different reasons), DTMF decoding is a relatively slow affair, so that anything using this system has to be economical with the number of instructions sent using this code. Because DTMF is inband, it is often used in conjunction with CTCSS, the latter being used to send signals on the speech path when necessary. DTMF seems to be the code of choice for the simpler telephone patching techniques.

While both DTMF and CTCSS are widely used, they are far from secure. Decoders for both are freely available, and a number of up-market signal monitors include decoders as standard equipment. Mobiles can easily be reprogrammed to "*look*" like one another, and listening in on calls or even cloning a mobile with the same identity can readily be done.

The mobile identity is used by the network operator to allow (or disallow) calls. In many instances, calls will be time-charged, this being particularly so if telephone access is permitted. The mobile ID is used by the billing system to generate the bills. The security of the mobile identity is hence of considerable importance to both the user and the network operator.

SINGLE SITES

The problems associated with being limited to a single-site repeater soon became evident. The coverage from any one site is typically a 25–60 km radius, and many users have an area of interest that will be wider than the area that can be covered from any one site. To make matters worse, handhelds have a much lower transmit power than mobiles, and they often have to be operated from inside buildings where the building itself presents a significant obstacle to good radio wave propagation. The effective range for a handheld operating from a typical base station can be from a few kilometers in the CBD (Central Business District) to tens of kilometers in open country.

Despite the limitations, there are a lot of instances when a single site will give effective coverage, and a number of trunking systems have evolved to cater for this demand. When demand at a single site exceeds the capacity of a one-channel repeater, additional ones can be added. A large number of individual channels can present a maintenance liability and do not use the potential traffic capacity efficiently. Most trunked radio systems and many enhanced PMR systems permit the pooling of channels on demand, with the main advantage of this being increased traffic efficiency.

Figure 1.3 shows the comparative traffic capacity of a number of single-channel repeaters compared to a system of independent channels. Based on the assumption of a calling rate per subscriber that allows 26 users on one channel, it is seen that while the conventional mobile system capacity grows linearly with the number of channels, the trunked radio systems begin with the same efficiency as conventional (26 users on the first channel), but by the time the system has 20 channels the trunked system is carrying 6.6 times as much traffic as the 20 conventional channels. So for large single-site systems, the trunked radio is not only a more efficient way to carry the traffic, it also should cost less per user because the RF channels are the biggest component of cost in the network.

Multiple-site operations are often called for when the service area is greater than that which can be covered by a single site. Wide area systems are not necessarily traffic efficient, particularly when there is a significant proportion of group calls.

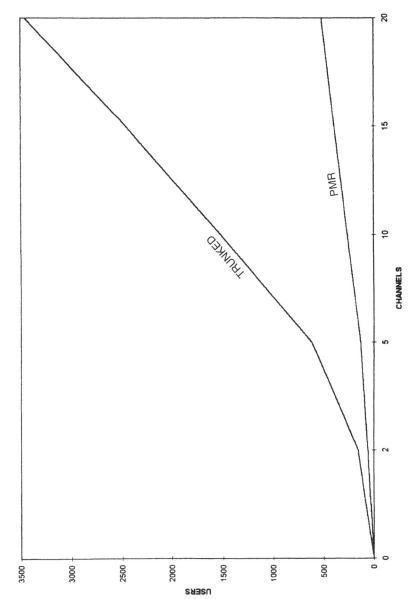

Figure 1.3   The relative traffic efficiencies of trunked and conventional radio.

5

Trunked radio systems are well-suited to wide area operations, but this can be achieved with conventional mobile systems also.

For conventional mobiles, this wide coverage area can be done in a number of ways. The simplest to understand (but not necessarily the simplest to implement) is called *simulcast*. This is implemented by having a number of base sites operating in parallel with the frequency of each site being the same. This system is very effective for the users because they do not have to change channels as they roam from site to site. From a network viewpoint it is less simple.

When two sites transmit a signal of the same frequency, there will be some interference between the transmissions at points of overlap. This interference may be constructive (the two signals add) or destructive (the signals subtract and may even cancel each other). At locations where the signal strength from two sites of the same frequency are similar in level, complete cancellation is possible and local "holes" in the coverage appear. These areas of effective total cancellation can be kilometers across, and their actual location and size can drift as the relative phase of the interfering transmitters drift. It is important to realize here that the interference occurs both at the radio-frequency (RF) level (that is, interference between the two carriers) and at the modulated level; that is, the signals impressed on the carriers (often audio) can also interfere.

Everyone who has used a mobile phone has experienced carrier interference. In areas of low signal strength, it is often the case that a multipath signal will be similar in strength to the main path signal. When this happens—for example, in an office—there will be some parts of the office in an area of poor reception where the phone works fine, but there will be other parts of the same office where it may not work at all. Carrier-to-carrier interference is pervasive, and there is not much that can be done about it in a simulcast situation. This is mainly due to multipathing, which makes the phase of the carrier at any given point not only virtually impossible to estimate, but also highly variable over time.

The modulation frequency interference is, however, a different problem. While the wavelengths of the carrier frequency will be of the order a meter or less (and so any interference pattern will also be of this order), an audio signal traveling on a radio carrier at the speed of light will have a very large wavelength. Take the case of a 1-kHz signal, traveling at 300,000 km/s, and you will see that it has a wavelength of 300 km. However, in the time domain a wavelength is 1/1,000 of a second. So any processing delays of the order of a fraction of a millisecond will cause big phase shifts. Virtually all digital systems (such as microwave links) will introduce some processing delays, additional to the delay caused by the time it takes the signal to traverse the route.

To avoid this problem the interfering transmitters can be locked together in phase to ensure that they are in phase in the overlapping areas of nearly equal field strength. This is what is done in quasi-synchronous systems. Although it is necessary to lock the transmitters together in phase, the transmissions will *only* be in phase in the areas of nearly equal field strength (where total signal cancellation could occur if the phases were random), hence the prefix "quasi."

DIFFERENT FREQUENCY WIDE-AREA

Another way to provide conventional wide area coverage is to repeat the site but use a different frequency at each repeater site. This avoids the need for synchronization (and therefore is more attractive to the service provider) but means that the mobile user will have to change channels when roaming between base sites. This can be done manually; alternatively, smart mobiles with scanning capabilities can do this job automatically.

The common problem with these types of wide area systems is that they are not frequency efficient, nor do they make efficient use of the channel resources. The paralleling of the wide area channels is done in such a way that one channel at each site repeats all the traffic. This is done even if no mobile in the region is there to hear the traffic. Once there are more than three sites involved, this will mean that there will always be channels tied up in mobile-to-mobile calls, which are not actually carrying any traffic.

CONVENTIONAL VERSUS TRUNK

It has already been shown that the pooled channels of a trunked radio system are more efficient than the same number of individual channels on a conventional system.

It is evident that a three-lane highway can carry considerably more traffic than the total of three single-lane carriageways, and this is the same advantage that trunked radio systems offer.

Thus the expected limitations of conventional systems are as follows:

- Coverage dictated by geographical constraints
- Inefficient use of channel capacity
- Inefficient use of spectrum
- Frequent traffic contention
- Voice traffic security (unless encryption is added)

Trunking systems differ from conventional mobile systems in that they permit a pool of channels to be shared by all users. This requires intelligent switching by both the mobiles and the network; therefore, trunking was not really practical until the time when microprocessors and frequency synthesized mobiles under processor control became commonly available.

Microprocessors are needed to perform the complex task of controlling a mobile signaling and responding to signaling to the network and delivering responses, such as ring tone, short messages, channel hunting, and other mobile functions that in conventional mobiles are handled manually by the user.

Synthesized mobiles permitted rapid frequency changes under the control of software to replace the manual switching (which generally meant physically switching in to the circuit the crystal that was needed to select a particular frequency) that was required in earlier conventional radios. In contrast, synthesized radios use one reference crystal to generate a clock pulse that can be used to set the mobile to the desired frequency simply by performing mathematical operations that derive a set of clock frequencies from the reference. Because software is cheap to manufacture and crystals are not (relatively at least), today even most conventional mobiles will have synthesizers if they need to cover more than a few different frequencies. In fact, many single-channel radios have synthesizers that are set up once only, to the single-channel frequency.

WORLD SYSTEMS

By mid-1997 there were an estimated 4.3 million trunked radio users around the world. Although a rapid growth is being experienced, the number of conventional users is estimated to be an order of magnitude greater.

Trunking worldwide is essentially divided into two groups. The US-based trunking systems are almost exclusively 800- and 900-MHz systems, while the rest of the world generally uses the bands 66–520 MHz (but may use any band including the US bands). In recent times the United States has begun to use some spectrum below 500 MHz for trunking.

North and South America, along with India and Sri Lanka, use the 800- to 900-MHz bands following (more or less) the FCC band plans.

In the United Kingdom a number of TV channels were recovered to make what is known as the Band III frequencies (174–225 MHz) available for both nationwide and regional trunking. Generally, across Europe the 410- to 430-MHz band is widely used.

In the Asia/Pacific region, it is more general that no strong distinction is made between trunking and conventional mobile bands, and frequencies are allocated spectrum where available. Eastern Europe sometimes uses frequencies that are outside the "regular" mobile bands.

For a trunked radio supplier to offer a universally useful product, it is nearly mandatory that the RF equipment be capable of continuous coverage from 66 MHz to 520 MHz. Only a few can do this.

There are many different ways of achieving the functionality of a trunked system, and thus many different standards have evolved over the years. A major distinction between the various trunked radio systems are those that are essentially single-site systems and that provide wider area coverage by paralleling sites in a manner similar to conventional systems and those that are true wide-area ones. A wide-area system can be described as any system with more than one base site location as seen in Figure 1.4. True wide-area systems can establish which sites both the called party and the calling party need to use. The system resources are then allocated so that only the channels that are required to make the call are utilized.

Parallel operation of either trunked or conventional systems is very wasteful of resources, and the cost per subscriber of a network increases in direct proportion to the number of sites operating in parallel for such systems. They typically have an economic limit of two to five sites, although many of them can physically handle more.

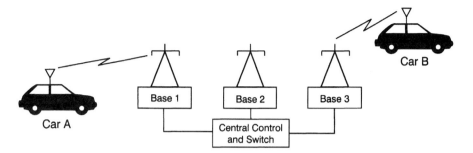

Figure 1.4 A wide-area system works transparently with more than one base station site.

A CONTROL CHANNEL

Some trunked systems have control channels (which operate much like cellular control channels). These are generally dedicated to calling mobiles, receiving call requests from mobiles, and issuing instructions. While some can be arranged to default to operate as voice channels when no other channels are available for connection, they do so only at the expense of losing most of the systems control functionality. For example, if a control channel is operating as a voice channel, it will fail to receive incoming call requests and therefore cannot queue them. Most large system operators prefer to use the control channel dedicated only to that function.

The alternative approach used on some systems is to have no control channel and rather send control information over idle voice channels. This requires the mobiles to scan for free voice channels carrying control information. For very small systems, this has the advantage of making all channels available for voice calls, but it does mean that things like queuing cannot be properly regulated by the network. Also, because any voice channel can potentially become the control channel at any time, calls can be missed during the relatively long scanning time that may be necessary on large systems. This can be countered by repeating the message (s) for a period of time or by holding data until all mobiles have had time to scan, but this will slow up the total throughput of the control data.

What needs to be understood about control channel functionality is that for small systems any mode of operation that is selected will to some extent be a compromise, but for large systems a control channel is a definite advantage.

DIGITAL TRUNKED RADIO

Like all other mobile systems, trunk radio has gone digital. Recently, two main standards have emerged: (1) the European TETRA (which is TDMA and very GSM-like) and (2) APCO-25, a US system that has been designed to allow a smooth transition from conventional networks to digital. Both these standards are open and nonproprietary. Time division multiplex (TDMA) systems send messages out in bursts, so that over one channel a number of messages can be sent, with each of them being allocated a time slot. Most TDMA systems also use frequency division multiplexing (FDMA).

In the background, others are developing CDMA-based systems, but no firm information is yet available on these. Code division multiplex (CDMA) systems transmit a number of messages simultaneously over the same channel and use a decoding key at the data level to separate them at the receiver. The key works by making the unwanted signals appear as low-level background noise.

Digital offers improved data capabilities, and perhaps a moderate improvement in spectrum efficiency. Both TETRA and APCO-25 systems come with handover capability, and while this increases the network design options available, it increases complexity and switching costs.

Digital trunked radio with handover leaves little left to distinguish it from cellular except for the ability *designed into* cellular for operations in a frequency reuse environment. Cellular networks expect interference from other base station sites and come with the ability to adapt to it. Cellular base sites in fact transmit an ID that identifies the source and thereby indicates interference if the ID is foreign. Trunk radio systems (except iDEN, a Motorola trunked system based on GSM) have no such capabilities, and a frequency reuse pattern of $N = 9$ (nine cells with unique frequency grouping) is needed to ensure good operations.

TRUNK OR CELLULAR

System designers need to be reminded of the *real* difference between cellular and trunked radio, particularly as the hardware capabilities merge. Cellular was meant for one-to-one calls, while trunked radio is meant for one-to-many (group) calls as well as one-to-one calls. For large and widely dispersed groups a cellular-type design for a trunked network can make group calling prohibitively expensive. The smaller the coverage from one site, the more sites you will need to actively handle a single group call. Because a channel at one site typically costs around $10,000 to $20,000, a group call involving 15 sites can tie up $150,000 to $300,000 worth of resources. In other words, a group call utilizing 15 sites will cost the operator 15 times as much as a purely local call.

While one response may simply be to charge the user for the sites used, significant resistance can be expected from users who are asked to pay 15 times as much for a group call as for a one-to-one local call. With cellular systems, increased traffic leading to more cell sites usually (at least initially) results in improved service quality for all, because the sites are increasingly placed closer to the users. In trunked systems, while the same would be true for the voice quality, especially on one-to-one calls, the ability of the operator to offer cost effective wide-area coverage decreases with every new site.

Digital systems like TETRA, which use TDMA and sacrifice path budget for spectrum, may prove to be *less* spectrally efficient than the analog counterparts that they are intended to replace, simply because the extra sites needed for adequate coverage will consume more spectrum than the narrow-band operations will save.

Modern trunked systems are very cellular like. Large ones will almost certainly use cellular-type frequency reuse plans ($N = 9$), the hardware and software will be very similar, full duplex is featured on many trunked systems, and some will even have handoff. Trunked handhelds are getting smaller and in 1999 are about the size of a typical cellular phone circa 1990. The higher transmit power of a trunked handheld probably means that it will always have a bigger battery and a more substantial heat sink than a cellular phone, so size will remain a distinguishing feature (note that lower-powered handhelds are possible but will mean that network repeaters need to be placed closer together, thereby making wide-area group calls problematic). Generally because of the lower market volumes, trunked mobiles are significantly more expensive than cellular mobiles and typically cost twice as much (before operator subsidies).

In the early days of trunking, the classic problem was this. Most systems have the intelligence to connect two parties to one repeater channel, if they are registered on the same site. If the system has a significant overlap, then two mobiles that could work from a common repeater may in fact be logged onto adjacent ones. The classic solution was to minimize overlap. This was fine when all mobiles were equal, but the advent of handhelds mean that there were two different coverage areas and two different regions of overlap. Later came the concept of a "home" site, whereby mobiles would, if possible, log onto their home site even if it was not the best one, but it was at least above a certain threshold.

Thus the trunked radio designer *must* compromise speech path quality for wide-area coverage, and he or she needs to design the impossible—namely, a system with just enough sites, but not even one too many. Because of this there will always be areas of unsatisfactory performance. The trunk radio designer also needs to be more expert with repeater utilization, because these can extend coverage without increasing the number of sites. On the down side, although repeaters are much cheaper than base sites, they offer much less coverage. If this sounds difficult enough, it has become even worse in the last few years with the ready availability of reasonably priced, small handhelds, which look like cellular phones. The real problem is that people expect them to work like cellular phones and to be effective in most buildings. Providing good downtown in-building coverage, while at the same time minimizing the number of cell sites, is not easy.

ENHANCED PMR

There are many systems enhancements that can upgrade the capabilities of conventional public mobile radio (PMR) networks. The most basic improvement, and one that is mandatory in most countries today, is to add selective calling so that only the mobile (or group of mobiles) addressed will participate in a call. This feature has become widespread because of its inherent ability to permit a number of unrelated users to share a repeater and also because it can prevent false triggering of a distant repeater that might happen to share the same frequency.

In the mid-1980s, telephone interconnection became available. The concept was powerful because it enabled the mobile network to be linked to the outside world. Dual-tone multifrequency (DTMF) signaling, which is widely used in the public subscriber telephone network (PSTN), is almost universally used for interconnection.

The demand for wide-area operations can be met with a number of base station sites, operating in parallel. Initially, with the earlier mobiles this mostly meant that the mobile users had to change channels as they moved from one area to another. Smart radios with scanning capability eventually made the channel changing procedure automatic. The scanning capability also led to multichannel operations initially at single sites and ultimately on multisite systems.

With the advent of synthesized tuning came processor control of the mobiles, and this meant that considerable processing power was available. Mobiles could scan, do signal strength comparisons, preferentially seek home channels, and store information about hundreds of different users, channels and groups.

With all this processing power a considerable number of "home-brew" trunking systems evolved to meet local needs. While even today most of these systems would be called enhanced PMR, most were really trunked systems, often with both multichannel, wide area and PSTN access. The line between advanced PMR and trunked systems is blurred, and convention may be the main identifying factor. Trunked systems could be regarded as enhanced PMR systems that have operating systems and protocols that are used widely. Many enhanced PMR systems have evolved slowly over time to meet ever more demanding local needs, and they may have greater capabilities than some of the recognized trunked radio systems. The techniques used in enhanced PMR will be covered more fully in Chapter 21.

CHAPTER 2

PLANNING

Planning is the first step toward setting up the trunked radio network in the most effective manner. This can be most difficult, made more so by the large number of options available. Planning consists of quantifying the network ahead of time so that the required resources can be allocated. It is also essential to identify the type of resources to be used.

For example, should single-channel links be used to connect a site in preference to microwave or wireline? The decision will depend in part on the future growth of the network, the position of the site in question in the overall link hierarchy, and of course the relative costs of the various technologies.

A well-planned system is optimized for the foreseen growth. The short-term optimum configuration may conflict with the long-term interests, and this has to be avoided. For example, it may be cheaper to occupy a small site with limited room for expansion for the initial installation. However, the cost of relocation at a later date may well justify a more substantial site from day one.

There should be a formal plan, which clearly spells out the foreseen area of coverage, the number of sites, the approximate location of those sites, and the link and power requirements. This is best done about three years ahead, with most of the details (such as channels and sites) being held in a computerized format for ease of updating (spreadsheets are ideal for this).

For a new operation it may seem that the first question to be addressed would be which system to use. That may be a good place to start, but maybe it is easier to answer the question, "What frequencies can be used?" Because the frequency assignments are local and probably somewhat inflexible, the frequency constraints may well eliminate some systems. Where it is not possible at first to get a definitive frequency allocation, it may be possible to get the most likely band (or bands) to be allocated.

Next consider the capacity and number of sites envisaged. For small local systems with one or two sites, almost any trunked radio system can do the job. However, for larger wide-area systems with large numbers of linked sites the choice is smaller. MPT

1327, TETRA, APCO25, EDACS, and iDEN in concept are designed for operations in a very wide area. Most other systems are more suitable for regional, single-site applications.

If digital technology appeals to you, then the choice is limited to a few systems. To date, there is little evidence that digital systems offer any real advantages over analog except in security and in some data applications. Digital systems offer more features, but most of these are of little relevance to the average user. Digital mobiles cost about 20% to 50% more, and the network is more expensive to build.

Wide-area calling can be a limiting factor. If there is to be a lot of wide-area calling, then a system that identifies the site where the called party is located, and activates that site only, will be an advantage. Systems that deliver wide-area calls by broadcasting on all sites simultaneously waste a lot of channel resources. Wide-area group calling often necessarily uses a lot of channels, but some systems are better than others at optimizing channels used.

It will not be easy to select the basic system, and probably even harder to choose the ideal vendor. However, once that has been done, it is time to go about planning the basic system.

INITIAL SYSTEM DESIGN

If the system is to replace an existing conventional mobile radio network, then the coverage of the existing network can be used as a guideline for the trunked network. Be careful, however, to make allowances for frequency differences (if they exist) and possible reduced coverage, particularly if a digital system is being considered.

Where a large system is to be built, it will be necessary to estimate the number of sites required. Here, local knowledge can come in handy, because it is likely that conventional systems are in place that cover most of the area required and will give a good first approximation of the number of sites needed. Local consultants should have this information readily available.

If no local systems exist, then the system can be designed in concept by using one of the many propagation software packages that are available to get an estimate of resources. Alternatively, map studies can be undertaken to achieve the same end.

For a rough estimate it can be assumed that most reasonable sites will cover about a 10-km radius to mobiles and a 5-km radius to handhelds (about 50% more in rural areas). The estimates of the number of base station sites can then be used to estimate link, site, and system requirements.

CUSTOMER FORECASTING

In the case of "in-house" systems, the forecasting of the user base can be almost trivially simple. There will probably be an annual budget for mobiles that will dictate the numbers. In the case of a public system, it can be most difficult. However, there are a few simplifying factors. First, the requirement to adequately serve a defined coverage area will, to a large degree, define the number of base sites. Where security is an issue each of these should have at least two voice channels. This effectively defines the smallest practical system.

Calculating the number of customers to be carried requires even more guesswork. It will be necessary to estimate the traffic per customer. In mature systems this figure will be known. There are no clear guidelines to follow here. Some US dispatch-only systems can carry 70 users per channel. This amounts to around 0.005 Erlangs per customer. Some MPT 1327 systems with liberal telephone access may carry three times that amount of traffic per customer.

Where there is no precedent, it is best to err on the side of caution and choose a higher calling rate (fewer customers per channel) than a lower one.

The objective of the planning at this stage is to define the initial number of sites and channels. Once some live traffic is being carried and measurements have been made, this whole process becomes more precise and is done in the manner set out in Chapter 11, on traffic engineering. Remember that, in general, expansion is a relatively easy process, and provided that the initial loading of the system is done in a controlled way, it will not matter too much if the estimate was only fair.

PLANNING A MATURE SYSTEM

Because most trunked radio systems can be expanded one channel at a time, it is economic to expand the system at a pace consistent with the customer base. Marketing trends should be obvious by the end of the first year, and the calling rates should have settled. With regard to this last point, it is interesting to observe calling rates on new systems. The rates determined from small initial customer groups, rarely, in the long term are typical. As the number of users grows, the calling rate generally settles to a steady rate, most often lower than the initial rate.

Armed with good marketing and calling rate data, it is very easy to produce reliable forecasts of system requirements and, in particular, channel and link requirements. The channel data can be obtained directly from a computer calculation that should be done monthly. The included software allows current traffic data to be used to project channel requirements into the future for any given customer base.

In all but the simplest networks, it will be necessary to link sites to each other and to a central controller and maybe a central switch. To achieve this, microwave fiber-optic cable or copper cables may be used. Whatever means is chosen, links will add significant costs to the network and must be accounted for.

The link requirements are a little more complex to estimate, especially because there may be a number of transit sites. It is a good idea to keep a link diagram available for regular updating against channel growth. A typical link diagram is shown in Figure 2.1. Here site 5 is a transit site for a number of others, and the number of circuits required are those as forecast on a yearly basis. For more complex systems a spreadsheet would be an effective way of a keeping track of the circuit requirements, although a sketch of the configuration would still assist in preparing the spreadsheet.

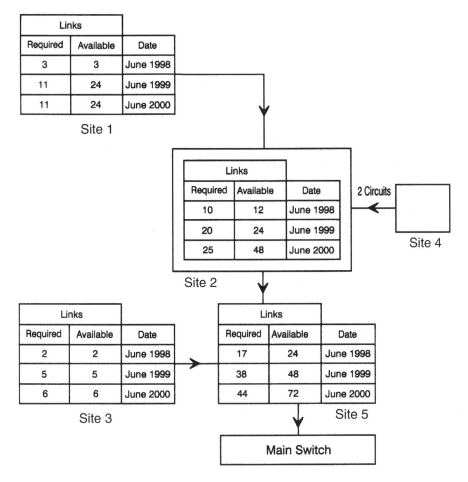

Figure 2.1 A typical link circuit requirement diagram.

SEEKING SUITABLE SITES

It is usual that "existing" conventional sites are considered as candidates in the initial systems planning. Often these are community sites and will have a number of other users. Although conventional single-site systems do not take up too much space or use too much power, it cannot be assumed that the same is true for trunked systems. It must be determined if the sites will be physically able to accommodate a multichannel system. The power requirements should be taken into account, particularly on the more remote sites where this may be limited. Also, due consideration must be taken of the coverage requirements, which today are predominantly dictated by the widespread use of handhelds (which have limited range).

Provision must also be made for links, which can be a considerable part of the total network cost (particularly on small systems). If microwave is to be used, then the rigidity of the tower can be critical to proper operations. All towers sway in the wind, but high-gain microwave systems need to remain pointed toward each other within small tolerances or the paths fail altogether.

TIMING

The timing is an essential element of the planning process. Typical lead times on implementation are as follows:

- Trunked network infrastructure 4–12 weeks (depending on the supplier and often also the frequency chosen)
- Links; single channel 4–8 weeks
- Microwave 6–15 weeks
- Site RF survey 1–3 weeks
- Site acquisition 2–6 months (some sites take a lot longer, particularly if there are zoning problems or objections from nearby property owners)
- Mobiles 4–8 weeks
- Billing system from 4 weeks for a simple system to 25 weeks for an elaborate one

Other factors that will need to be considered are

- Frequency approvals (0–12 months)
- Building permits (2–6 months)
- Tower permits (typically about 2 months, but a lot longer if objections are raised)
- Licensing
- Staff training

LINKS

Most small trunked systems are initially linked by single-channel links. This can be cost effective for sites requiring only a few linked channels, but thin-line microwave is often worth considering. Thin-line microwave is the name given to the low-cost microwave systems, designed for use on the last few hops to rural or low-priority areas. It will rarely have redundancy, and may have a higher MTBF than high-capacity microwave, but the salient feature is that it is inexpensive.

The most basic single-channel links consist of two mobiles at each end, set for low power output (around 1–5 W). When more than a few of these are used, it will be necessary to include combiners so that they can be made to share a single antenna. Where base station sites are used as transit points, it may be that several remote sites

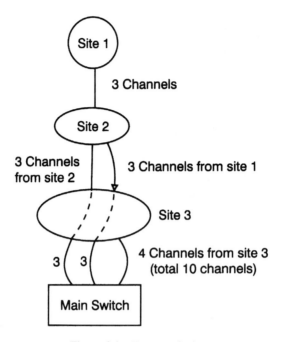

Figure 2.2   Two transit sites.

are linked through a single site (see Fig. 2.2), and considerable capacity will be needed even though each of the sites only has small link requirements.

Thin-line 2-Mbit and 8-Mbit (or equivalently T1 systems) systems can prove to be quite cost effective when compared to the cost of single-channel links and tend to reach the break-even point at four to six channels. Increasingly, the cost differential between 2- and 8-Mbit systems is decreasing (to 10% to 20%), and when in doubt it is best to select the bigger system (provided that a license can be obtained for it). Very big systems might well consider the use of 34-Mbit links and even larger.

SHARED SITES

An attractive proposition is often to share an existing site. However, when this is done, there are a number of factors that must be considered. Ensure the following:

- A sufficiently long-term contract can be obtained with future rentals specified.
- There is adequate room for foreseen expansion.
- Intermodulation/interference from other users is covered in the agreement (e.g., who is responsible, particularly what happens if a new tenant proposes to introduce frequencies that will clash).
- Always do an intermodulation study on the frequencies in use before signing any contract (the included software can be used for this).

- Particularly beware of TDMA-based digital systems (cellular or trunked) at any frequency because they are likely to cause interference at the frame rate, into almost anything, including radio circuits, control systems, and microprocessor-controlled hardware.

BUDGETS

In the early planning stages, it is often important to get a budgetary estimate of a proposed trunked radio network cost. Because this will often need to be done before detailed quotes or design studies are available, some generalized rules will be needed.

BASE SITES

Assuming that the system is a new one, it will be a difficult thing to estimate the number of base stations for all but the smallest of sites. A good estimate can be obtained from a map or software propagation study. If these are not available, however, a few guidelines can be used.

If good handheld coverage is required in the CBD, then it is safe to assume that the maximum range of even a good site is about 3 km (this provides for in-building coverage). Where coverage only to vehicle-mounted mobiles is required, the range can be anything from 15 km to 60 km, and this is best estimated by the techniques described elsewhere in the book. If uncertainty exists, it is best to assume that the maximum range is 10 km, (although for low-priority services, this could be stretched out somewhat).

CHANNELS

It is very difficult to determine both (a) the number of channels that will initially be needed on a new system and (b) their distribution over the service area. Unless there are known indications otherwise, the following can be used. In rural areas provide one channel (or two if redundancy is required or in the cases where a separate control channel and voice channel are to be used). In city areas it is best to provide at least three channels. Because of the modularity of the equipment, channels are easily moved, and this can be done when the traffic patterns become established. A further provision for either two spare channels or 5% of the total should be provided as spares (whichever is the highest).

Remember that channels include the associated combining equipment, power supplies, and other ancillary equipment. This can be a significant part of the cost of a channel and must not be overlooked.

Caution should be exercised with wide-area trunk systems where the wide-area bases are effectively operating in parallel. In this case the number of wide-area channels is determined by the requirements of the largest site in the system, while local traffic channels are provided on a traffic basis. In this case, rural base sites should be provided with only the wide-area channel requirement (because wide-area channels can also function as local).

COSTS

Although costs will depend significantly on the system used, the following can be used for the initial budgetary analysis. Prices here are in $US.

- Cost per channel: $30,000 for the first channel and $15,000 for each additional one
- Cost per controller: $50,000
- Cost per site for buildings: $30,000 (or as determined by local building rules)
- Cost of towers: $1,000 per meter of height
- Installation hours: 80 per base site
- Installation for control room: 100 h
- Test equipment: $2,000 if maintenance is subcontracted, $35,000 if it is done in house
- Billing system basic: $5,000
- Billing system with long-distance calling: $50,000
- Cost of links: see the following section

Note however for digital systems, some of these costs will be much higher.

SINGLE-CHANNEL RADIO LINKS

The cost of hardware for a single-channel radio link is about $6,000, or $8,000 installed. This assumes that there is more than one single channel in the link, and the channels are combined into a single antenna.

MICROWAVE SYSTEMS

Small microwave systems (E1/T1) can be estimated at $25,000 per link, including MUX equipment. Allow another $10,000 for the bigger E1/T1 systems.

CHAPTER 3

DESIGN

The objective of the designer is to provide adequate coverage for both handheld and mobile users within the proposed service area using the least number of base sites. The importance of minimizing the sites cannot be overemphasized, because it will be seen later in traffic studies that the cost per call depends critically on the number of sites. The sites will generally house base stations, but repeaters should also be considered. The sites selected must be accessible for servicing, have adequate space, be secure, and for multisite systems allow for the necessary links.

To a considerable extent, handheld usage will dictate the network configuration and the number of sites. In a typical large city a range of 4–6 km can be expected for reasonable handheld coverage inside buildings. Where the city is extensive or has more than one main business district, it will be necessary to locate stations accordingly.

There are reasons why you should consider limiting the range of the CBD sites. Where they are only providing handheld coverage in a local area, a relatively low site with limited range may both achieve the handheld objective and permit reuse of the same frequency at the edge of the coverage. In systems that operate by only activating sites that are needed for the conversation (such as MPT 1327 and Smartzone), more efficient channel use is made by covering as wide an area as possible from a few sites. Massive overlap, although it will not impair the operations from the point of view of the user, will cause additional multisite calls to be initiated. The reason for this is that the users will log onto the strongest local base and stay there until the signal level drops below a specified level. Within an area of overlap, two handhelds that are quite close to each other may be logged onto different base sites. When one calls the other, the call will be connected through the two sites onto which they are logged. This means that two voice channels and one intersite link are occupied for a call, which could have effectively been connected through a single channel, had they both been logged onto the same site. In terms of network resources the net effect is to have more

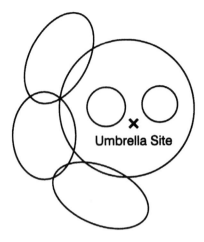

Figure 3.1 A typical coverage plan for a large city.

than doubled the cost of carrying the call. In large systems most calls will be intersite, and the relative effect of overlap is much diminished (but more about this later).

A typical city design will have the "handheld" sites largely overlapped by the umbrella sites as in Figure 3.1.

The selection of CBD sites can be difficult. At first it may seem that the highest site would provide the best coverage and thus should be used. Unfortunately, with the proliferation of cellular technology, the cost of prime sites has risen sharply, with a typical site costing about $10,000 per year (or even more). Some building owners charge per antenna with additional costs for the equipment space. Rental property always incurs the risk that rents will rise to uncomfortable levels, so long-term contracts with specified rent increments may be attractive.

It is often the case that the trunked radio operator will be more comfortable with two lesser CBD sites than a high-priced prime site. In general, when in-building coverage is considered, the two lesser sites will also certainly provide superior coverage. However, the downside of using nonoptimum sites is that more of them may be needed in total, and this becomes *very* expensive when wide-area group calls are carried.

HANDHELD PERFORMANCE

A mobile will always outperform the handheld. Having a higher transmit power, this is perhaps to be expected. There are, however, some other significant factors. One is that the efficiency of the handheld antenna is less that 100%, which means that its gain is less than 0 dB. Also, the vehicle-mounted antenna is usually higher than a handheld, so it will have some relative height gain. The radio-frequency (RF) signal is (approximately) vertically polarized and handheld radios are rarely used vertically.

And finally the handheld radiation pattern will be affected by the body of the user. All of these factors taken together give the mobile an extra 6–10 dB in performance over the handheld, plus any RF power advantage that the mobile may have.

To further penalize the performance of a handheld, it should be noted that handhelds are frequently used inside buildings and vehicles, where the signal strength will be significantly attenuated (typically, attenuation would be 20 dB for a substantial reinforced concrete structure) relative to the outdoors. This increases the subjective evaluation of the relative performance in favor of the mobile.

EXPECTATIONS

It is likely that first-time users will compare the trunked system with a single-site conventional system. Conventional systems nearly always have seriously compromised coverage (something that will not change for single-site trunked systems), frequent-user contention, and no or limited privacy. Even a badly designed trunk system has a lot to offer when compared to conventional two-way radio. It is likely that new users will be impressed.

While the above may lead you to conclude that good design is not important, experience shows that once the users become accustomed to the trunking features and begin to take them for granted, they soon begin to compare the performance with cellular radio.

Cellular radio systems base sites, unlike trunked radio, are mostly installed to provide traffic capacity (not coverage), and so the signal strength will ordinarily be good in most frequently used locations. Trunked radio sites, on the other hand, are designed for acceptable handheld coverage (or in some cases only acceptable mobile coverage). This means that even a well-designed trunked radio system can be expected to be noisier (because part of the design objective is to use a minimum number of sites) and have more coverage difficulties than its cellular counterpart.

This tendency to compare with cellular systems may mean that consideration will have to be given to enhancing performance in some key districts, and this is especially likely in the CBD. Sometimes repeaters can be a cost-effective way of doing this, in a way that minimizes call costs.

PROPAGATION MODELS AND THE REAL WORLD

In order to properly appreciate the limitations of any system that tries to model the RF environment, it is important to understand mobile propagation. In free space the propagation of a radio signal is highly determinant and is easily modeled. The field strength will vary as the inverse square of the distance. For point-to-point systems, like microwave links, this equation can be used to calculate the path loss:

$$Pl = 32.5 + 20 \log F + 20 \log D$$

where F is the frequency in megahertz, D is the distance in kilometers, and Pl is the path loss in decibels. The equation implies that the path loss *increases* with frequency, and of course this is not really the case (at least not to the extent that actual losses increase in direct proportion to frequency, because in free space there is nothing to lose energy to). The path loss equation assumes that the losses are measured between two dipoles. Assuming that the path loss is measured between two dipoles, it will be obvious that a dipole at 1 GHz, which will be about 0.15 m long, would probably capture somewhat less energy than a dipole at 10 MHz, which is 15 m long. Thus, this equation gives the path loss as measured between two dipoles (or any two radiators, assuming that the gain of the radiators is the same for any frequency measured) and will also vary inversely with frequency squared.

The effective capture area of an antenna is referred to as its *aperture* (which accounts for the frequency term in the path loss equation), and it is calculated as

$$A = \lambda^2 G / 4\pi$$

where G is the gain expressed as a ratio [i.e., if the gain is 6 dB, then the value of G is 2 (a power gain of 2)]. The previous expression for path loss has already taken into account the antenna aperture.

The path loss equation is adequate in free space and for distances greater than 1 km. But is it all really this simple? What if we are interested in distances smaller than 1 km? For the case where the distance is zero, the equation above tells us that the path loss is minus infinity. Clearly, this equation is not telling us all there is to know about free-space path loss.

The reason that the equation breaks down is again a real-world reason. As the receive and transmit antennas come closer together, they can no longer be regarded as entirely independent of each other, and the radiation from the transmitter will be, to some extent, dependent on the physical presence of the receive antenna. In this case the receive antenna is said to be in the near field of the transmitter.

In the near field a new and more complex relationship exists between the distance and the path loss. The software package with this book enables you to calculate near-field losses easily and to compare them with the free-space losses.

So if we were to take the near-field effect into account (as necessary), would it be possible to accurately predict the path loss between two antennas? The key word here is *accurate*. The real world intrudes again; and we find that the physical structure of the towers, and even real-world clutter around the antennas, will have an effect on the propagation, and this time it is not easy to account for it.

If we place the link more firmly in the real world and make the path between two points on the earth's surface, then we find that the Earth itself will contribute to the net propagation loss. Of course with even more sophisticated models we can account for some of these extra contributions to the net path loss, but we will never be in a situation where all contributions have been accounted for, and so the calculation will never be entirely accurate.

The path loss varies significantly with atmospheric conditions to the extent that under extremes of atmospheric disruptions, any microwave path can become momentarily unserviceable. This last consideration tells us that there is in fact no such thing as a definite path loss (because it is different at different times of the day), and so the predicted path loss should more properly be called the average path loss.

To account for the vagaries of path losses, microwave systems are designed with safety margins (called fade margins) of around 30–40 dB. These links will also have a calculated probability of outage, which gives an indication of how often the path loss will exceed the design level by even more than the fade margin.

The discrepancy between the predicted path loss and the actual path loss (as measured) is not merely some esoteric one that might amount to an insignificant fraction of a decibel. Over real links, discrepancies of 3–6 dB are regarded as the norm. In fact the path loss between any two links (even those that are line of sight) varies with time and may change a few decibels in a matter of minutes, with occasional drops of 10 dB or more, so that when we talk of path loss we are not really talking about a number of decibels but a statistical average of the path losses taken over a period of time.

The important thing to note here is that in the relatively controlled environment of a point-to-point link the calculated loss is not definitive; we next see how much more so this is the case in the mobile environment.

REAL-WORLD MOBILE PROPAGATION AND PREDICTION

It is nearly impossible to predict mobile coverage with any reasonable degree of accuracy, and yet most trunk radio and virtually all cellular system designs are the result of such predictions. Trunk radio systems fare somewhat better because many of the sites used are tried and proven conventional sites, which may have been in service for decades, and their coverage and performance is well known.

It has already been shown that even for point-to-point operations, link predictions are not precise and a margin of error of the magnitude ±6 dB is to be expected. Mobile propagation is anything but line of sight, and the information that is fed to the computer is far from a complete real-world view of the environment. Real-world clutter includes trees, buildings, people, cars, cranes, aircraft, water (and the waves it generates), grass, rocks, air, and rain. It would be totally impractical to try to quantify all of the environment.

However, it is not just the lack of data that is the problem: Even if all the necessary terrain data were available, it would be well beyond the capabilities of a computer to determine the true coverage. In school most of us were taught that light (and thus by inference radio waves) travel in straight lines. Well it does not!

The concept of light traveling in straight lines comes from some early descriptions of optics, and is a fair approximation for the gross behavior of propagation, in much of classical physics. In fact, light is not constrained by such simple rules and to some extent travels between two points *in all possible ways* (a quantum mechanical description). This can largely be ignored (but not entirely, as the Fresnel zones are a manifestation of this) for point-to-point communications, but for mobile communications it means simply that *real-world mobile propagation is not computable*. Quantum mechanical models of large systems have yet to be reduced to something that a computer can make much sense of. Large systems do tend to behave classically, but because of the complexity of the mobile environment, chaotic behavior applies. Instead of even attempting to definitively predict propagation, computer models use crude algorithms (algorithm is just another word for "computer model" or "approximate equation") to crunch out the approximations that far too many engineers take far too seriously.

But even if we discount quantum effects and go to classical physics for our propagation model, there are problems. Diffraction calculations are routinely done for microwave links, which are usually line of sight (or almost). Even in this instance, for computability it is necessary to make some simplifying assumptions about the obstacles in the path. These obstacles are approximated as regular shapes like "knife edges," or they are "rounded off" to make the calculation easier. These approximations mean that after about three obstacles have been encountered, the accuracy is such that it is pointless to consider additional ones.

A typical mobile radio path will encounter dozens of obstacles along a number of multipaths, which may include trees moving in the wind, moving vehicles, and even the body of the user. There is no way to model such a complex environment with realistic computing resources.

This is not to say that all is lost. Using enough approximating assumptions, some useful calculations can be made. However, it is essential to recognize that any computer model has made a lot of assumptions and that the design subsequently produced should be treated accordingly.

THE MODELS

The simplest approximation of mobile propagation is that the field strength varies as the inverse fourth power of distance (compare this with the inverse square law that applies for point to point links). A slight refinement suggests that an inverse power of 3.8 might be more appropriate. This rule is useful for comparative studies, and is often used in simple models of cellular propagation, but a more complete model is needed for most applications.

A landmark study by Yoshihisa Okumura with Ohmori, Kawano, and Fukuda and published in *Review of the Electrical Communications Laboratory*, Volume 16, Numbers 9–10, September 1968, was one of the first attempts to quantify, in a statistical way, propagation in the mobile environment for frequencies from 150 MHz to 2000 MHz. The study was done in Japan and was conducted from a van, moving at 30 km/h; the results were measured on recording paper. The equipment had a high sensitivity of −125 dBm and a dynamic range of 50 dB. The computer algorithm for this model is known as the Hata model.

Referring to some early attempts to obtain something definitive from their studies, Okumura et al. reported on page 831: "Some propagation curves hitherto published

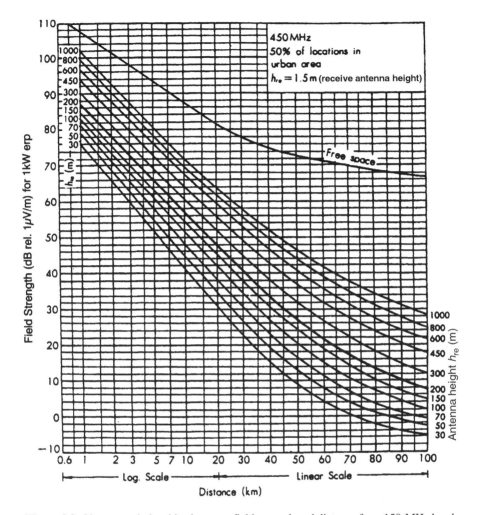

Figure 3.2 Okumura relationships between field strength and distance for a 150-MHz band.

were not definitive, because they took in all sorts of terrain irregularities and environmental clutter haphazardly so were not very useful for estimating the field strength or service area adapted to the real situation." What this is saying is that the real world is a mess and too hard to generalize. To get around this they went on to define types of terrain and compare like with like. While this is consistent with the reductionist scientific method, the engineers and computer modelers who came along afterwards to use these results generally ignored the context in which they were derived. It would be fairly rare that the service area of a trunked radio site would be entirely one terrain type; more likely it would be a mixture of types, like rolling hills, flat terrain, urban terrain, and plains with the population density and foliage varying considerably. Despite the shape of the real world, many computer algorithms accept only one terrain description for the service area of a site. The consequence of using the information out of context is wrong answers!

Used intelligently (more of this later) and not blindly in a computer model, the results of the Okumura study can give some good insights into RF propagation. Figures 3.2 to 3.5 give the Okumura relationships between field strength (referenced to a transmitter with a 1 kW ERP) and distance in kilometers.

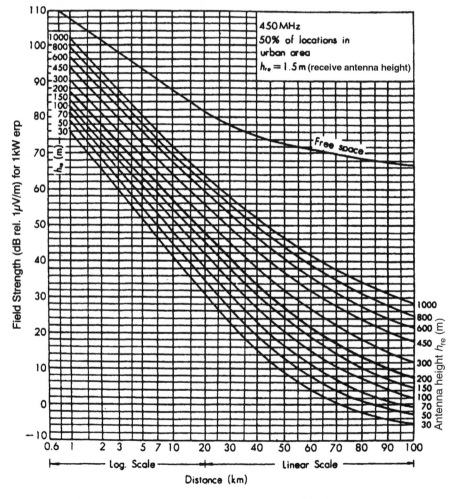

Figure 3.3 Okumura relationship between field strength and distance for a 450-MHz band.

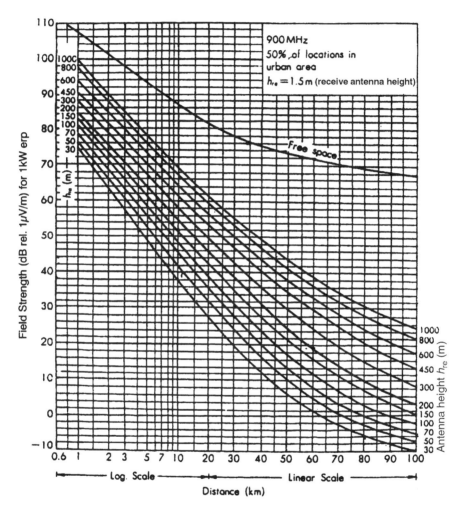

Figure 3.4 Okumura relationships between field strength and distance for a 900-MHz band.

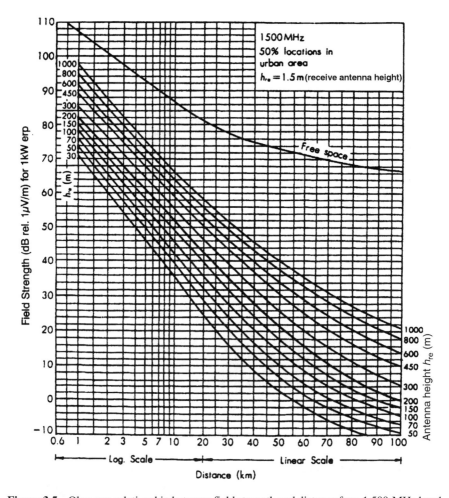

Figure 3.5 Okumura relationship between field strength and distance for a 1,500-MHz band.

CORRECTION FACTORS

The Okumura curves refer to the signal level in 50% of locations and therefore refer to the *median* signal level. The field strength at any range is almost log normally distributed, and thus it is virtually symmetrically distributed about the median. In that case the mean is equal to the median. It should be noted that the easiest thing to measure is the mean, but only because of the additional computations, and thus time, needed to calculate the median or other decile value.

It is more usual today to design for 90% coverage in 90% of locations, and this upper decile value will give a figure that more realistically reflects a consistent standard of coverage. The relationship between the mean and the upper decile value is given by

$$\text{Mean} = \text{Level } (90\%/90\%) + 1.28\ \rho$$

where ρ is standard deviation. This would be a fairly simple relationship were it not for the fact that the standard deviation of the level measurements is itself a function of the terrain. The values are given in Table 3.1.

To get a consistent upper decile value of field strength, this now requires that a range of median (mean) field strength values be used in place of a single value. Again this is something that is allowed for in very few computer models.

There are a number of other correction factors that need to be applied to the Okumura model, and fortunately most of these are fixed values.

Terrain Factor

This factor is a constant level in decibels which allows for the different attenuation of terrains encountered. It is best determined by measurement, and it ranges from around −20 dB to +20 dB.

ERP

The plots given are for a 1-kW transmitter, and the level sort needs to be increased by $30 - 10 \times \log$ (actual transmitter power in watts).

What the plots are good for includes

- Preliminary site studies
- Relative studies (e.g., the effect of increased antenna height, or additional ERP)

And what they are not good for is final designs.

MULTISITE SYSTEMS

Multisite systems require a good deal of iterative design. When central sites are chosen largely for handheld coverage, they will not figure significantly in the overall

TABLE 3.1 Field Strength Deviations for Various Terrains

Terrain	Standard Deviation (dB)
Urban	8–12
Suburban	6
Flat suburban	3–5
Rural	3
Water paths	1.5

coverage (in most instances). Primary sites are those that have been selected for good wide-area coverage. Often these will be determined by the availability of suitable towers or hilltops.

The design begins with identifying these primary sites, surveying their coverage, and thereby determining areas that need to be filled in by supplementary sites. Generally the objective is to fill these gaps in coverage with as little overlap as possible. This is not always easy because by the nature of the site selection (searching for sites with good wide-area coverage), it almost follows that some significant overlap will occur. Overlap is to be minimized, but because it cannot be totally avoided, it is comforting to know that it affords some redundancy in the event of a site failure.

It must be emphasized that predicted coverage (from a computerized "design tool") is not accurate enough to enable the selection of good supplementary sites. At best a prediction is a good first-order guess. If you attempt to predict the fringe coverage (and hence the "holes"), you will find it virtually impossible to get even a reasonable agreement with measurement. Because the accuracy of a computer design system is at best ±6 dB (and such accuracy would be rare) and an error of 6 dB is the equivalent of halving (or doubling) the coverage, it is *not* sufficient to rely only on the computer design if optimum coverage is to be achieved. Although some computerized design systems may be better than others, none are good enough to be used without field

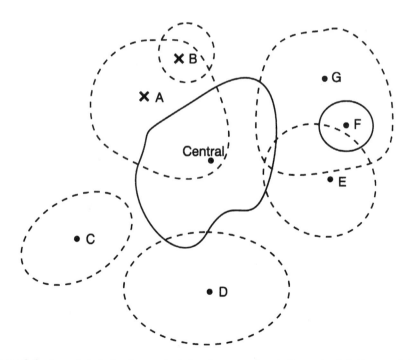

Figure 3.6 A central site has been selected and surveyed, and other potential sites are plotted (in dotted lines) from a computer or map study.

survey input. The fact that many designers are not aware of this accounts for the many poor designs seen today both in cellular and trunked radio systems.

Computerized systems, if used intelligently, will help identify likely sites and problems and, especially in very rugged, hilly terrain, (where shadowing is the main factor governing coverage) may even give a credible approximation of the coverage.

SURVEY PLOTS

Assume that a central site has been identified and surveyed. A number of sites are then found from map studies that are promising, and their predicted coverage is plotted as in Figure 3.6. Sites F and B can be eliminated from further study, because they are unlikely to be of any use. The more promising sites are then surveyed, and the results might appear as seen in Figure 3.7. The survey has now eliminated site C as being of little value, but the rest remain of interest. The survey has also revealed an area (shown shaded in Fig. 3.7) that is not adequately covered. The solution to this will probably be a new site to cover that area, and a repeater might be considered. However, a map study should be undertaken of the shaded area to see if the reason for the lack of coverage, in that area, can be identified.

The design should proceed in this way until the requisite coverage has been achieved, noting that redundant overlapping should also be minimized.

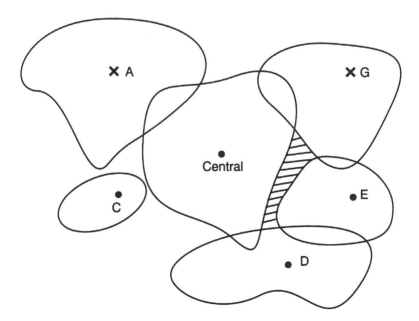

Figure 3.7 The same sites and their coverage after survey.

CHAPTER 4

RADIO SURVEY

It is essential that all trunked radio sites be surveyed in detail before being placed into service. All too often today, engineers choose to rely on the highly questionable computer designs, which virtually ensure a nonoptimum site layout.

Computer design tools *predict* coverage in the same way that the weather bureau predicts the weather. Both usually predict wrongly, and both do so for the same reasons: insufficient data and the chaotic nature of the problem they are set.

Computers have been used successfully for decades to predict microwave paths. These are obstruction-free and roughly follow an inverse square law, and the path losses in general are repeatable. However, even for microwave paths, reliability factors are calculated based on empirical formulae, which give a measure of how frequently the path can be expected to be radically worse than the original prediction. Land mobile paths, on the other hand, are subject to a vast number of obstructions and roughly follow an inverse fourth power relationship. The paths are highly terrain-dependent, and they can vary with the time of day and the time of year. Small variations in terrain data can induce big changes in the path losses.

The survey method must be one that can measure the field strength and its variations in real time. The far field is determined by the total path loss, near end obstructions, and multipath interference. Surveying is generally done from a moving vehicle, and the received signal may take the form indicated in Figure 4.1.

The rapid variations in level are largely caused by multipath interference. Because the signal can arrive at any given site from a number of different paths (hence the term multipath), each of which can be of slightly different lengths, the signals can arrive out of phase and thereby cancel each other out. This gives rise to rapid variations in level. If we assume that the measurement is taken from a vehicle moving at 70 kph and that the signal being measured is 500 MHz, then the speed of the vehicle is 70,000/3,600 m/s = 19.4 m/s. A wavelength at 500 MHz is 300/500 = 0.6 m, so from Figure 4.1 the vehicle could be passing a trough in the waveform 19.4/0.3 times per second, or 65 per second.

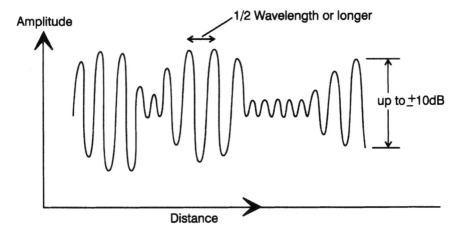

Figure 4.1   A typical field-strength pattern from a moving vehicle.

The Nyquist sampling rate for a random waveform is twice its highest frequency, and this would imply that a sample rate of at least 130 samples per second would be needed to completely characterize this waveform. Fortunately, the interference pattern is to some extent co-related, and this allows a lower sampling rate to be used. In practice, a rate of around 25 samples per second has been found to give good results, with errors (of an unpredictable nature) becoming obvious much below 15 samples per second. The point needs to be made that the sample rate is frequency-dependent, and at higher frequencies a proportionately higher sampling rate will be required.

DESIGN LEVELS

Mobile users have traditionally tolerated a rather poor standard of coverage. Essentially the boundary was traditionally defined as the point where intelligible speech could just be discerned. Today, with the widespread use of telephone interconnection and handhelds a much better standard is required. In-building usage is generally expected of handhelds, and an extra allowance must be made depending on the type of structures that the mobile will be expected to work in (this will be covered in more detail later).

There is no agreed design level for the acceptable field strength for a mobile radio. An upper limit can be set by looking at the analog cellular design level of –96 dBm. Because cellular radio is full duplex and also because "wireline" quality is expected, this level must be higher than what we are looking for. So how much lower will still produce an "acceptable standard of communications?

A number of subjective studies have been done, and there seems to be agreement that somewhere from –104 dBm to –112 dBm is an acceptable level for analog trunked radio mobile radios. For digital radios, a subjective equivalent must be determined. The actual figure will be, to some extent, dependent on the frequency chosen, the sensitivity of the receivers used, and maybe even the kind of service (and its criticality) being provided and customer expectations. A good starting point would be –108 dBm, but with the proviso that it is good practice to "fine tune" the design level in the field at the first opportunity. Digital systems will generally be less sensitive, particularly if they use TDMA; for these systems refer to the rated sensitivity of the mobiles. This can be estimated by taking the nominal sensitivity of an analog mobile to be –116 dBm and comparing it to the nominal sensitivity of the digital mobile. The difference between these two figures should be subtracted from the design levels quoted above (note that the nominal sensitivity of the analog mobile may be referenced to 12 dB SINAD, while the reference for the digital mobile may be referenced to a bit error rate).

Handheld radios, having a lower transmit power, a lower antenna gain, and generally a lower antenna height, need to operate at levels of around 8 dB higher than mobiles. When a design is undertaken, it must be clearly established if the coverage is to be for mobiles only (with incidental handheld coverage) or for handhelds, because the design requirements are quite different.

The above figures are in decibels and are useful for city coverage outdoors only. An additional 20 dB should be allowed for in-building penetration, a figure which often means that there will be dedicated "downtown" sites and/or repeaters.

WHAT TO MEASURE

When measuring field strength it is important to understand what it is that you are trying to measure. Let's assume that you set out to measure the average field strength. You may go and measure two sites and get a result similar to that shown in Figure 4.2. Both signals have an average field strength of –110 dBm. A quick glance will tell you that not only are the two signals not equal, but they would probably provide a very different *subjective* quality of signal, particularly because signal too frequently goes below the receiver threshold level (assumed here to be –120 dBm). Also, we need to question what was averaged. If we take the arithmetic average of the actual measurements in dBm, the result is actually the geometric mean of the measurement. This is because $(\log (a_1) + \log (a_2) + \cdots + \log (aN))/N = \log [(a^1 \times a^2 \cdots \times aN)/N]$; this is, of course ,very different from the log of the average value which is $\log ((a1 + a2 + \cdots + aN)/N)$. Most field-strength measuring sets measure the "log" average. Field strength is a log normally distributed variable, so the average of the log values will be the median value of field strength.

Median field strength, as seen in Figure 4.2, is not a definitive measure of the signal quality. It is also different from the true average field strength as we see next, by imagining that a signal that varies linearly like the one shown in Figure 4.3 is measured using both a true average level and then a log average level. The true average will simply be

$$\text{Average (true)} = \frac{b - a}{2}$$

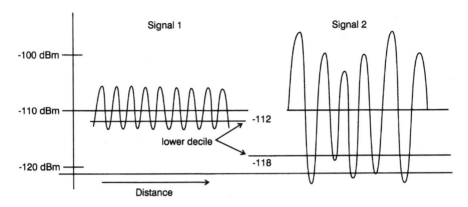

Figure 4.2 Measuring the lower decile.

A "log average" bears no simple relationship to the real average, and will differ from it by around ±20%, for the level variations encountered in radio survey.

Most commercial meters will be measuring the "log average."

The evidence then is that an "average" field strength does not tell us an awful lot about the signal, and even less if we are not sure what is being averaged.

A better way is to use a lower decile method. Looking back at Figure 4.2 and remembering that the two signals, although very different, had the same average value, we should look instead at the lower decile value. This is the value that is exceeded by 90% of the readings (or alternatively it is the highest value of the readings that are in the bottom 10%). Now the two signals give very different values (as it might have been expected that they should), with signal 2 reading –118 dBm while signal 1 has a value of –112 dBm. Unfortunately, not too many commercial meters give this value.

PRACTICAL SURVEYING

The survey can be conducted from any vehicle, provided that there is somewhere suitable to mount the survey antenna. Ideally, this is mounted centrally on the vehicle roof. Antennas mounted elsewhere, while they may better approximate a "real-life" situation, will give a distorted reading due to the body of the vehicle itself, interfering with the received signal.

For small systems a temporary survey vehicle can be improvised by using magnetic base antennas and a sedan. Larger systems can justify a purpose-designed vehicle fitted with receivers and PC hardware. A typical survey vehicle is seen in Figure 4.3.

The survey hardware may consist of something similar to that shown in Figure 4.4. The receivers are often simply mobiles, with the level being measured at the received signal strength indicator (RSSI). Generally, it is preferable to read the RSSI as an analog voltage and convert it with an A-D converter, unless the digital level is

Figure 4.3   The equipment in a survey vehicle.

refreshed at a rate greater than 100 samples/second. The advantage of using mobiles is that they are readily available and inexpensive. The disadvantages are that the RSSI voltage will probably have a rather narrow measurement range, and the level as a function of signal level may drift with time (this problem is less severe with modern radios because many of these have quite stable reference levels).

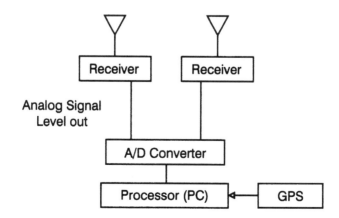

Figure 4.4   The schematic of a radio survey device.

RECEIVERS

If mobiles are used, they need to be calibrated regularly (at least weekly); if practical, two should be used in parallel, so that they can be used to cross check each other.

If the budget allows, commercial-quality survey receivers are available which are highly stable, accurate, and self-calibrating. For those on lesser budgets the ICOM 8500, a top-end amateur receiver priced at around $3,000, performs admirably. The ICOM has the analog level available from a socket in the back. If the equipment is also to be used for surveying cellular sites, be aware that the ICOM 8500 as sold in the United States has these bands blanked out, and special arrangements should be made to get one that has access to the full spectrum.

Multiple receivers enable sanity and calibration checks to be carried out easily, and they also permit the surveying of multiple sites (often very handy in mature systems).

ANTENNAS

The antennas used for survey should be relatively low gain (about 3 dB) to avoid errors due to a narrow beam width. High-gain antennas are more likely to have their patterns altered by near-end obstruction (by the vehicle), and the gain may even vary measurably when they are off-vertical.

Low-loss cables from the receiver to the antenna and top-quality connectors are a must. Cheap connectors can have unpredictable losses. Although the sensitivity of the survey receiver may be high, in practical surveys the received signals may be very low level due to a limited ERP from the survey transmitter.

Figure 4.5 A mobile antenna mounting platform.

Although the survey antenna is often just a regular antenna mounted as it would be for a mobile radio, more consistent and repeatable results can be obtained if a flat ground plane is established on top of the vehicle. This can be achieved by mounting a flat sheet of aluminum between ski racks as seen in Figure 4.5. Once the ground plane has been established, it is good practice to survey using two or more receivers, each with different antennas with each acting as a cross-check for the other.

GPS

It is common today that survey equipment comes with an on-board GPS, much like the one shown in Figure 4.4. The GPS will log the location of the data for future processing. It also enables a one-man operation of the survey vehicle.

A software package that reads the field-strength data and plots it on a map against the recorded position is used later.

SURVEY TRANSMITTERS

In mature systems with paging channels, the paging channel is radiating continuously, and thus it is ideal for survey. For systems without paging channels, it may be possible to designate a channel to continuously radiate (whether or not it is carrying traffic) or, failing that, to take one channel out of service and turn it on for survey purposes.

For the design of new systems, it will be necessary to set up a temporary transmitter to survey from. If possible, this should be a base station channel that has been designed for continuous radiation at high RF power levels.

Sometimes a mobile set to continuous transmit is used for survey. If this is done, it must be recognized that most mobiles (full duplex ones excluded) are not designed for continuous transmissions and that they will overheat if run at full power for more than a few minutes. Most modern mobiles will not self-destruct if this is done, but rather will automatically turn down (or switch off) the RF carrier. Some regulate their power in a way that enables continuous transmission but turns down the levels until the PA temperature reaches a safe level. If this happens, all survey results will be invalidated, because the transmitter reference level will not be known. Either way, this is not what is wanted for survey.

If a PTT mobile is used as survey transmitter, its RF output must be turned down to a level that is safe for continuous operations. The mobile vendor will generally be able to provide information on this mode, but if none is available, try a reduction to 20% power and monitor it over a few days to ensure stability. Any overheating will result in a reduced power level and thus can be easily detected.

CHAPTER 5

BASE STATIONS

The radio base station provides the link between the trunked radio system and the end user (often called the air interface or AI). Usually situated on a strategically high spot, it contains the transmitters, controllers, power supplies, and antennas. For most trunked radio systems, this will be where most of the cost of the system is. A typical system will cost around $10,000 to $20,000 per channel (inclusive of all common equipment) and have up to 20 or more channels per site. Physically, the base station will typically occupy one or more 600-mm (19-in.) racks. There may be three to eight channels on a rack.

A small base station site to accommodate only a few channels can be very simple to construct and may contain only a single rack of equipment. Larger systems may be the size of an average room. Whether the system is small or large, however, reliable operations must be a priority, and good construction practice should apply. The room must be absolutely waterproof, because water can totally destroy the equipment. The site should be clean and free of dust (which can contaminate switches, connectors, and equipment boards).

Although most trunked radio systems are rated by the manufacturers for operations without air conditioning, it is advisable to provide it, unless the operation is on a shoestring budget. Air-conditioning will significantly increase the service life of virtually all the components and will significantly increase reliability of the system. Heat cycling, caused by daily temperature changes, has been identified as a major stressor of electronic hardware (and also may reduce battery life to half).

Trunked radio bases can broadly be classified into three types. They may either have a local controller without local switching or have a local controller with associated local switching capabilities; or the intelligence may be on a per-channel basis, with a simple controller coordinating the channels.

SITE CONTROLLER

Where a site controller is fitted, it will be the interface between the main switch and the base site and will make decisions on whether the call can be processed locally or via the central controller, thereby checking caller validity and accepting site registrations.

Often the site intelligence is entirely with the controller, which totally controls the sites resources. If the controller fails, then the site will revert to a more basic "fall-back" mode of operation. The fall-back mode will generally allow the site to switch local calls only, and it may not have any billing or customer ID verification (if not, it will switch any mobiles that are on frequency).

LOCAL SWITCHING

A local switch at the base site gives the ultimate in flexibility. If the site controller has a local switch, then it will generally be able to switch directly to other sites (bypassing the main central switch) or, if required, to the PSTN or local PABXs. If the switch is powerful enough, it may permit direct access at the E1/T1 level, as well as interfacing as required at two- or four-wire E&M.

A typical base with a local switch is seen in Figure 5.1.

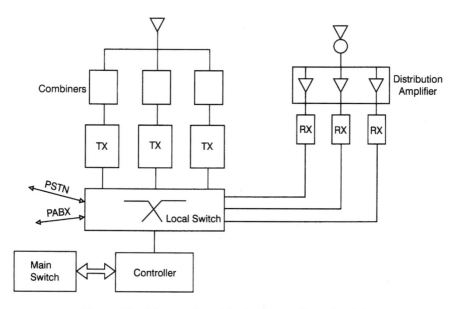

Figure 5.1   A base station with a local controller and switch.

INTELLIGENT CHANNELS

Some systems place the intelligence with each channel on a per-channel basis. These channels may have PSTN capability, which will generally be allocated on a needs basis. The PSTN capability is hard-wired into the line card, which includes things like ring tone generators, DTMF encoding/decoding, and line monitoring. While this configuration can be a low-cost approach for small systems, larger systems really need pooled resources, for economies of scale and flexibility of operations.

ANTENNAS

Trunked bases almost universally are fitted with omnidirectional antennas. Because wide-area operations are required, high-gain antennas are always preferred; as a general rule, the higher the gain, the better. The maximum gain is limited largely by the physical size of the antennas. The most common of these is the collinear dipole, which is an evolution of the basic dipole shown in Figure 5.2.

Antenna gains are measured as either dBd (gain relative to a dipole) or dBi (gain relative to an isotropic radiator). An isotropic radiator is one that radiates uniformly in all directions, and for practical purposes in mobile radio it can be regarded as a theoretical construct. A dipole has directivity and a gain relative to an isotropic radiator of 2.1 dB.

One way to obtain more gain using dipoles is simply to stack more of them together. Two dipoles can deliver twice the gain (or 3 dB gain) of a single unit. The dipoles are simply stacked on top of each other as seen in Figure 5.3. The antennas are combined in phase with a transformer that matches the input impedance to 50 Ω. The whole assembly is then mounted inside a fiberglass tube. A quarter-wave transformer like the one shown in Figure 5.3 is all that is needed.

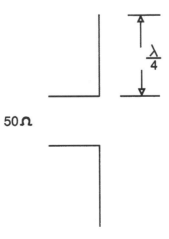

Figure 5.2 A simple dipole

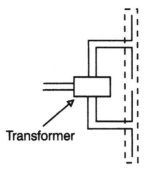

Transformer

Figure 5.3   Two stacked dipoles for 3 dB of gain

This process can be repeated for more gain; but to get an addition 3 dB, another pair of dipoles are needed (making four in all) for a 6-dB gain. To get the next 3 dB of gain, twice as many dipoles are needed (making eight in all) for 9 dB of gain. For frequencies of interest in trunked radio, 9 dB is somewhere near the upper limit of gain that is practical, because of the sheer physical size constraints (to get 12 dB, no less than 16 dipoles need to be stacked).

Other than physical size, there are other constraints on gain. High-gain antennas have very narrow beam widths, and so their orientation becomes critical. Unless they are mounted exactly vertically, they will be radiating either into space or into the ground, instead of into the service area. If you take the time to look at a few tower installations, you will see that even a casual inspection from the ground will reveal that a significant percentage of antennas are not mounted vertically. Some were probably installed that way, but most have shifted in the wind.

The least expensive antennas are usually fiberglass whips. Because they are base-mounted, large fiberglass antennas can bend alarmingly in high winds. This distorts the pattern and can interrupt fringe coverage in high winds. A solution to this is the use of bracing (fiberglass, plastic braces, or other nonmetallic material). Cellular radio fiberglass antennas are mounted inside very thick and substantial tubes, and thus they are not subjected to this problem. These cellular-style antennas are available in some trunked radio bands (particularly in the 800- to 900-MHz bands where wide-band cellular antennas can be used), but they are expensive (typically around $1000).

Another antenna widely used in trunked applications is the folded dipole. This antenna costs two to three times as much as the collinear whip; but it is very rigid and its elements are at DC ground level, making it highly resistant to lightning strikes. As seen in Figure 5.4, the folded dipole consists of a loop of hollow (usually) tubing grounded at the midpoint to the supporting pole and fed from the end points. Because the active element is grounded to the metal support pole, the whole antenna has a good path to ground in the case of a lightning strike. Like the dipole, the folded dipole can be stacked to give additional gain.

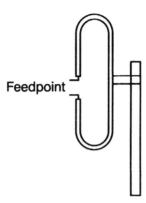

Figure 5.4   A folded dipole.

Fiberglass whips, on the other hand, are insulated, and they often melt or shatter when struck by lightning, even if the strike is an indirect one. The one disadvantage of the folded dipole is that the support pole itself will, to some extent, interfere with the radiation pattern, and a simple dipole or dipole array cannot produce a truly omnidirectional pattern. This pattern distortion is sometimes countered (to some extent) by mounting the dipoles at either 90 or 180 degrees apart around the support pole.

It should be noted that any antenna mounted on the side of a tower or other conductive mount will have its radiation pattern distorted by the mounting structure itself. Top mounting is the only way to avoid this problem, but it can be minimized by an offset mounting at other levels.

FEEDERS

Having taken care to select the highest-gain antenna that can reasonably be accommodated, it is now essential to minimize losses in the cables. It is false economy to use inexpensive high-loss cables, and ideally the cable loss should be kept below 1 dB. Table 5.1 shows some cables and their recommended lengths to achieve this loss.

TABLE 5.1   Maximum Cable Lengths to Achieve 1-dB Loss

Frequency (MHz)	Cable Length (meters)				
	RG213	3/8 Foam	1/2 Foam	7/8 Foam	1 5/8 Foam
150	13	23	35	60	110
400	8	15	22	40	65
900	5	9	14	25	38

PROTECTION

The cable shields are made of high-conductivity copper, and are ideal paths for lightning discharges that either (a) directly strike the antenna or (b) are induced into the cables by the fields from the lightning-rod grounding cable (which often runs close to, and parallel with, the antenna feeders). To minimize problems, the feeder cable should be firmly bonded to the tower, as well as being well-grounded at a point just outside the equipment hut. To protect the equipment, it is essential that each cable have a coaxial surge protector fitted. Ideally, this protector should be mounted as near as practical to entry point of the cable into the equipment room, but where this is not possible it is acceptable to place it anywhere along the cable, up to and including the first interface with the equipment.

TRANSCEIVERS

Most base stations are fitted with transceivers which were originally designed to act as conventional PMR repeaters. Some are in fact unmodified conventional repeaters with an add-on controller, but others, while still appearing to be relatively unmodified, have trunked boards fitted inside. Likewise, many trunked mobiles are conventional mobiles with a trunked radio controller board fitted.

The only real difference between a trunked radio channel (hardware wise) and a conventional one with processor control is that because of the high-speed signaling that often takes place, access to the audio line is required on the RF side of the mute-lift, and often before any frequency emphasis or deemphasis takes place. Some systems that use subaudible signaling may need access to the audio before any filtering takes place. Additionally, there may need to be access to a fast PTT input.

Because a large number of channels may be in use at any one site, it is usual that the RF channels are combined in such a way that a number of them can share a single antenna. Combining does introduce losses that increase as more channels are combined into a single antenna. The number that can be combined together is determined by the power-handling capacity of the antenna and by the total power loss that is tolerable. In turn the power loss will depend on the frequency spacing and will increase rapidly as the channel spacing decreases.

Unlike losses on the receive path, losses on the transmit path can be countered (to some extent) simply by increasing the output power of the final RF stage. The losses will result in heat generation; but because the total ERP permitted by the regulators is mostly well below the capabilities of modern power amplifiers, these losses can be accommodated without degradation of the transmission path gain.

Sometimes the combining is taken even further, and the transmitters and receivers are combined into a single antenna using a cavity device known as a *duplexer*. While this saves on antennas, it does so at the cost of receiver sensitivity, because the duplexer introduces additional losses in both paths.

Figure 5.5 A tower with top-mounted omnidirectional antennas.

IMPROVING BASE SENSITIVITY

Modern systems can be expected to have a significant number of handhelds in the network. Because of the relatively low ERP of a handheld, the range that can be achieved (or the signal quality from any point within range) will be determined by the handheld TX to base station RX path. Not a lot can be done on the handheld side, but good design on the base-station receive section can make a lot of difference.

On the base station RX path, a good-quality high-gain antenna, a low-loss feeder, and finally a low-loss preselector filter can make a lot of difference.

A preselector filter is connected between the antenna feeder and the receiver distribution amplifier, to filter out unwanted out-of-band RF energy that could cause

Figure 5.6 Mounting yagis on tower faces.

intermodulation or receiver blocking problems. The preselector is usually a bandpass filter, and any losses in this filter will directly degrade the noise performance of all the receivers it is connected to.

Tower top preamplifiers (often referred to as low-noise amplifiers (LNAs) can improve the sensitivity of the receivers by an amount equal to the feeder loss (typically a few decibels by the time connectors are accounted for). The downside of these is that they are a maintenance liability (particularly on high towers) and are subject to weather and lightning damage. A problem with intermodulation can also occur because the LNA is quite exposed to any transmitters that may be on the site.

Good base station design and construction are necessary for reliable operations; and because base stations are ordinarily sited in rather hostile environments, a good deal of care should be taken with them.

TOWERS AND MASTS

A tower is distinguished from a mast by being self-supporting. Masts will be partially (or completely) supported by guy wires. The mobile radio community often gives very little thought to towers and tower construction, and this results in some very poor installations which in turn will become long-term maintenance hazards.

The tower needs to be structurally sound and also needs to be rated to withstand the winds it is likely to encounter. The wind zoning can mostly be obtained from the local planning authorities.

Figure 5.7   An omnidirectional antenna offset mounting.

Omnidirectional antennas are best mounted at the top of the tower as seen in Figure 5.5. As is usually the case in trunked radio applications, a pair of antennas is used, one for transmit and one for receive. To improve the isolation between the transmitters and receivers, the transmit antenna can be mounted directly below the receive (see the mounting at the top of Fig. 5.5).

Microwave and single-channel links often do not require the full height of the antenna and should be mounted as low as is consistent with an acceptable path budget. Lower mounting levels mean less wind loading, as well as lower incident wind levels on the dishes (which can tend to move them around).

Yagis can be mounted at any level, and simple mountings can be devised to ensure that they are well clear of the tower (see Fig. 5.6). Omnidirectional antennas cannot always be top-mounted, and when this is the case it must be accepted that some field pattern distortion will occur and that the antenna will, to some extent, become directional. Some manufacturers produce the "patterns" that can be expected when the antenna is mounted at various distances from the tower. These are approximate only because there is no such thing as a standard tower or mounting.

Figure 5.7 shows a mounting arrangement that gets the antenna quite a few wavelengths away from the tower and thus minimizes the pattern distortion. The problem with this type of mounting is that large twisting moments are presented to the mounting bars, and the antenna is likely to move in both the vertical and horizontal planes. It is good practice to be aware of this twisting (whatever the mounting arrangement) and to routinely check that the antennas are vertical. This can easily be done from the ground by sighting the antenna against a vertical tower member.

Due attention should be given to the cable entry into the building. The tower, which is usually a tall, metallic, grounded structure, is likely to induce lightning strikes. Any voltage surges from these strikes can potentially damage the equipment, and precautions need to be taken. All cables should be grounded to the tower at the top and the bottom. Additionally, the cables should be grounded at the entry point into the equipment shelter (the cable window). Lightning surge arrestors should be placed on the inside of the window and as close as possible to it. Commercially available rubber boots provide weather protection at the cable window (see Fig. 5.8).

Figure 5.8 A base-station cable window, with protective boots and grounding.

BASE SITE SECURITY

It is rare that a base station has on-site personnel, and security can be a problem. Unattended sites are subject to vandalism and are sometimes subject to break-in attempts. It is best if the site can be equipped with a security fence. Deterrents widely used include the posting of signs on the gates and shelter doors with warnings such as "Danger: RF radiation," "Danger: high voltages within," and "Radiation precautions must be observed on this site." Once a catchy "warning" sign has been determined, the sign can be mass-produced and placed on all sites. Figure 5.9 shows a Vodaphone sign, posted in the United Kingdom.

Figure 5.9   A base-station warning sign.

TOWER SECURITY

The tower can be a serious security problem, particularly if unauthorized people attempt to climb it. The operator could be liable for excessive damages should someone fall while climbing, even if they were trespassing. Signs warning of "RF radiation" should be posted prominently, and, where possible, some device to prevent easy access to the tower should be installed.

CHAPTER 6

MAINTENANCE

All trunked radio systems will require regular maintenance. Most of the hardware used has an MTBF of around a few years; and because a lot of modules are involved in a complete base station, it should be expected that maintenance is an ongoing business.

Most systems will come with a central controller that will report major system faults. These will include sites out of service, channel failures, and high bit error rates. Some will give a reasonably sophisticated maintenance summary on a regular basis. As good as some of these diagnostics are, there is still a need for good basic maintenance, as some faults will always be missed.

ANTENNAS AND CABLES

Antennas are exposed to the elements; and because they are mounted high, they are prime targets for lightning strikes. The antennas used for trunked radio systems often are fiberglass and have relatively low wind ratings. It is not unusual that one or more antennas become damaged in a storm.

Generally, there will be a minimum of two antennas per site (combiners can be used to permit one antenna to serve both the TX and RX functions, but this is not a preferred configuration in most cases). Typically, each TX antenna can support eight channels, so there may be up to three TX antennas for a large site with 20 channels, or even more with systems like some MPT 1327 that can support 100+ channels at a single site.

A large nationwide or statewide system could, like the Victorian state government in Australia, have over 100 sites and thus have at least 200 antennas. If the MTBF of an antenna is taken to be 10 years, then out of a pool of 200 antennas, about 20 could fail in a year. Some of these failures will be catastrophic, while others will result in degraded performance that may be difficult to detect.

VSWR alarms are frequently fitted to the PA stages of a trunked radio transmitter, and while these may be useful, a faulty antenna can be missed particularly because the antenna cable and the combining equipment will provide a considerable degree of isolation from the fault.

Some care is needed when taking VSWR measurements to account for the cable loss. VSWR can be calculated from the forward and reflected powers as

$$\text{VSWR} = \frac{1 + (P_r/P_f)^{1/2}}{1 - (P_r/P_f)^{1/2}}$$

where P_r is reflected power and P_f is forward power. Any measurements made at the end of a cable will correctly account for the forward power, but will read the reflected power as a level that is lower than the actual reflected power, by an amount equal to twice the cable loss. So the VSWR recorded will always be lower than the true level. The effect is most pronounced when the VSWR at the fault is highest, and even small cable losses can disguise the true magnitude of the fault.

It is wise to initiate a regular routine check on the antennas, either by way of a field strength survey or by on-site VSWR checks. If field-strength testing is to be done, it may be necessary to swap a TX and RX antenna to ensure that a faulty RX antenna is not missed.

Cables are also very exposed and suffer a lot of the same problems as antennas. Connector problems and water ingress are relatively common. As a cost-saving measure, trunked radio systems mostly use foam-filled cables; and while these are adequate, they are not as waterproof as pressurized cables, which might be considered for the more critical sites. The routine testing of the antennas will mostly reveal any cable problems.

QUEUES

A system with excessive queuing may not necessarily be as overloaded as might be indicated. Systems with control channels can control queue length to some extent by limiting the queue length and the time spent in the queue. On systems that do not have control channels but rely on methods such as extended PTT to ensure access to the next free channel, sometimes, out of frustration, users develop the habit of attempting to grab a channel "just in case." Of course, on control channel systems users can stack the queues in a similar way. This artificially clogs the channel capacity.

Excessive queues can be an indication that the system is overloaded, but they may be a symptom of other problems. A base site that has a reduced ERP due to combiner or antenna problems will fail to carry its share of the traffic. This may force traffic onto another site and may overload it. Reduced receiver sensitivity can have the same effect.

Overloading can occur gradually; but as the overload increases to much beyond 10%, the queues will grow alarmingly (or shrink to no-queuing at all if the system is so programmed). The perceived quality of service will fall sharply, and the complaints department will be very busy. A more detailed look at queuing systems under stress will be covered in the chapter on traffic engineering concepts (Chapter 11).

Queues can rapidly overload and can seriously degrade the system performance. They need to be monitored regularly.

DATA PERFORMANCE AND AUDIO LEVELS

Increasingly, trunked radio systems are used for data transmission. It is vital for data that the transmission levels be correct (at the audio level), because overdeviation will result in data corruption and hence frequent retries. Over or under deviation on a channel used for signaling can cause call failure due to excessive BER.

Users will expect audio levels to be consistent whether a call is placed to another mobile, a PSTN line, or a PABX. This can only be done if the line-up levels are consistent.

Audio levels may drift for a number of reasons. A replacement transceiver may be set to levels that are different from those of the original one. Lines connecting the base site to the main switch can be rerouted in a way that changes the overall link gain. With leased lines, the trunked operator has no control and no knowledge of the routing configuration, which can be changed from time to time with alarming effects on the system audio levels. Sometimes the settings may just drift.

INTERFERENCE

Interference from third parties is a common experience. Sometimes these will be legitimate operators who are simply interfering from a distant site. Sometimes they will be illegal systems operating on the same band. It even happens that interference comes from other licensed operators who have strayed outside their allocated band.

Interference can sometimes be hard to track down. This is especially true if the interference is due to intermodulation. A good-quality communications receiver can be a handy asset for monitoring interferers. Couple this with a handheld Yagi antenna at the frequency of interest, and you have a basic transmitter tracker.

Simply monitoring the channel with the illegal users can sometimes be enough, because often they will refer to a phone number or company name or address, which is all that is needed to locate the offenders.

When all else fails, a few weeks of jamming, carried out by transmitting the control channel information on the offending frequency, alternatively a data stream or appropriate swept audio signal, will often clear the air. However, the legality of this should be checked with local authorities before being undertaken.

INTERMODULATION

Often hard to distinguish from interference, intermodulation can be a problem that grows as the system grows; because the more carriers there are, the greater the chance of intermods occurring. Intermodulation is the result of the mixing of two or more radio carriers in a nonlinear device (see Appendix 2 for more detail). In the real world, almost all devices are to some extent nonlinear, so intermodulation can happen almost anywhere.

As a rule, the most troublesome intermodulation products are the third-order ones. For frequencies A and B the third-order products are $2A + B$, $2B + A$, $2A - B$, and $2B - A$. To see why the third-order products are most troublesome, let's assume that A and B are two trunked radio transmit frequencies that are close together, so that we can write $B = A + \varepsilon$, where ε is small. Then $2B - A = A + 2\varepsilon$, so the product forms at a frequency very near to the transmitter frequencies. Where there is only a small separation between transmit and receive bands (e.g., 5 MHz, which is commonly encountered), this product may form in the receiver band.

An infinite number of other products exist, in the form $2A - 2B$ (a fourth-order product), $3A - 2B$ (a fifth-order product), and so on. As the orders get higher, it is fortunate that in most systems the levels get lower and it is usual in all but the most critical of applications to ignore any products after the seventh order.

Where there are more than two frequencies, other products can occur. For example, with three frequencies $A + B - C$ is another (often overlooked) third-order intermodulation possibility.

The calculation of intermodulation products is a job for a computer, once the number of products gets much above two, and the software included with this book can handle this problem for any sized system. Any transmitters that are combined in any way into a common antenna, a common feeder, or leaky coax cable should be subjected to an intermodulation study, and any products that fall into the receiver band must disqualify that combination of transmitters. Where there are shared sites, it is also advisable to run a study on all the co-sited frequencies to be forewarned of potential problems.

Tracking down intermodulation sources is a trying business, so it is worthwhile looking at some of the more probable sources.

Most phase-modulated transmitters have class C power amplifier (PA) stages, because of their efficiency. They are, however, also highly nonlinear and will generate intermodulation products. For this reason, most wide-band amplifiers, while they can accept a wide band of frequency inputs, cannot be used to amplify more than one frequency at a time.

Receiver RF stages, distribution amplifiers, and LNAs are generally Class A devices and are inherently linear. However, when subject to overloading (often by stray carriers) they will go nonlinear and can be a serious source of intermodulation. In this respect, tower-mounted LNAs that may be in a hostile transmitter dominated environment may be problematic. A very good preselector filter and good shielding are essential.

The intermodulation may not necessarily occur in the equipment but may occur externally in rusty fences, roofs, chains, towers, or almost any conductive material that is nearby the transmitter antennas. Look especially for things that may have corroded or bad joints, dissimilar metals in contact, or are not firmly bolted.

Occasionally the source of the intermodulation can be tracked easily with a wide-band receiver and a Yagi. Receiver-generated intermodulation can be distinguished from an external carrier generated problem by a simple test. Place a 3-dB pad before the receiver (or preamplifier) that is suspect and note the level of the intermod: If it is receiver-generated, then there will be a 6-dB drop in level; if it is external, then the drop will be only 3 dB.

If the problem is not on the receiver side, then the next place to look is the transmit side. A visual examination of connectors, antennas, coaxial surge protectors, and cables may reveal some corrosion or damage. A rusty tower, cable clamps, guy wires, or other structural features may also be suspect.

The combining cavities should be checked with a spectrum analyzer and tracking generator to ensure that they are properly tuned (especially in a growing system where new channels may have been added without proper attention to the cavity tuning).

More often than not, however, the solution is found by a tedious process of elimination.

SITE NOISE

The receiver sensitivity is vital to the overall performance of the base station. Induced noise from co-sited radio systems and from other electrical apparatus can seriously degrade the receiver performance. At lower VHF frequencies, cosmic noise can be the dominant source.

A test of the receiver sensitivity using a SINAD (*signal noise and distortion*) meter will reveal the receiver performance as a stand-alone item. This is done by using a signal generator and a SINAD meter as shown in Figure 6.1. The signal generator level is adjusted until the SINAD meter reads 12 dB, and the noted level of the generator is then the receiver sensitivity. Modern receivers can be expected to have a sensitivity between –116 and –123 dBm. When the higher figure applies (> –116 dBm), the receiver is probably deliberately of low sensitivity and is meant to be preceded by an LNA.

This test, however, will not detect noise injected into the receiver from the combiners and antenna system. To measure the external noise component, it will be necessary to measure the SINAD with the antenna connected. To do this, a device to

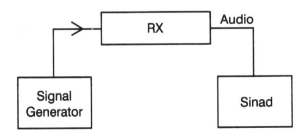

Figure 6.1 Measuring the receiver SINAD.

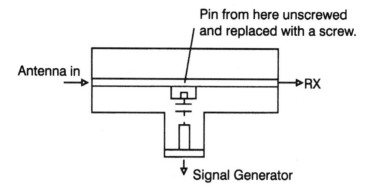

Figure 6.2   A lossy-T connector.

allow the connection of the antenna and the signal generator will be needed so that the test can be performed with the antenna (plus all combining equipment) connected. One way of doing this is with a lossy-T connector. This connector is simply a standard T-type connector with the pin at the T junction removed and replaced with a screw and given the equivalent circuit shown in Figure 6.2. Effectively, this allows a straight-through connection between the antenna and the receiver and a high-impedance connection for the signal generator.

If the SINAD is measured with the lossy-T, the result of this measurement might be −80 dBm (see figure 6.3); we now note that the lossy-T itself will influence the SINAD reading, and so it must be accounted for. To achieve this, simply disconnect the antenna system and replace it with a 50-ohm dummy load. Now measure the SINAD again. The result may be −85 dBm. The difference (5 dB) between these two readings will be the noise contributed by the antenna/combiner.

From the measurement of the receiver alone of −116 dBm, it can be concluded that the antenna system is additionally desensing the receiver by an amount of 5 dB. This would be a serious level of desensing, and its source should be investigated.

For digital systems, BER will be measured in place of SINAD.

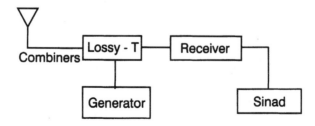

Figure 6.3   Measuring the antenna system desensing.

TEST SETS

The most valuable piece of test equipment that a trunked radio operator can have is a communications test set. A wide variety of sets are available on the market today, and some are remarkably powerful. A typical test set will feature the following:

- A signal generator to about 2 GHz
- A signal monitor
- A spectrum analyzer
- An FM deviation monitor
- A VSWR meter
- DTMF and CTCSS decoder
- Oscilloscope
- Level meters
- Frequency counter
- For digital systems, a BER tester

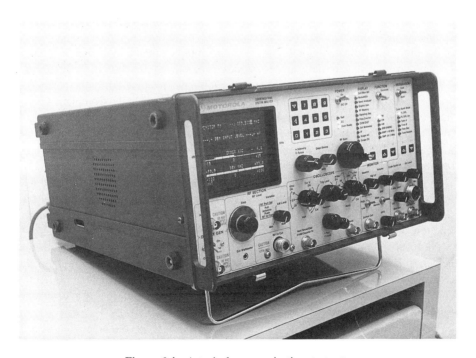

Figure 6.4 A typical communications test set.

Additionally, the more upmarket units will have optional modules that allow them to become powerful network analyzers for the major trunked radio protocols. They can read and display the overhead messages as they are sent. Typical of these monitors is the Motorola test set shown in Figure 6.4.

Unless the communications test set has a spectrum analyzer with a tracking generator (the better ones will have it), a separate spectrum analyzer will be needed to tune the cavities at the base station.

MOBILES AS MONITORS

On most trunk systems it is possible to set up a mobile at remote sites to auto-answer and thereby confirm operations. This mobile can be controlled by appropriate software to generate calls and monitor their progress in a systematic way. Often the mobile can be made to perform loop-back testing. Where this is an available option it should be considered for use in all base sites, this being particularly true for the more remote sites.

In some systems a mobile can be forced to home onto a particular site, so that a number of sites can be tested from a single mobile.

MOBILE FAULTS

Even the best-designed trunked radios can be rather complex when compared to a simple single-channel press-to-talk operation. From an operational standpoint, some of the more sophisticated mobiles are far more complex than a mobile phone; and it must be expected that, initially at least, some customers will find this all a bit daunting.

Experience shows that a lot of the early problems encountered by the customers will be due to failure to understand fully how to operate the mobile. Complaints range from "unable to call a particular mobile" (because the called mobile did not know how to answer) to "unable to call anybody" (the mobile was not turned on). Expect the unexpected and be prepared to retrain the users within weeks of the first training session.

A good indication of how well the lessons were learned is to follow the customer traffic for the first few weeks. Often unused or underutilized mobiles tell of poorly trained customers. A moderate increase in traffic should be expected weekly over the first few months as the customer becomes more competent with the unit.

Mobiles themselves can be a source of network problems. A mobile that is defective in sensitivity or frequency stability may cause multiple resends on the control channel each time it is called. If there are a number of such faulty mobiles, the control channel may become overloaded with these retries. The only way to detect this is with a control channel monitor and subsequent data analysis.

Mobiles that are out of specification can also cause problems on the speech channels, particularly if they overdeviate when signaling. Most of these problems will go unnoticed unless there is some routine off-air monitoring.

Some operators use low call timer limits to prevent long calls. This will work for most customers, but some are simply frustrated by it. If their call is interrupted, they simply redial immediately. This redial will cause a heavy loading on the control channel. Where such customers are detected, it is often better to extend their access time rather than increase their frustration by restricting call times even more.

It is not unknown that some customers will deliberately set up multiple calls in order to congest the network. Surprisingly, this behavior may be hard to detect. Someone who generates frequent calls (e.g., the dispatcher for a taxi company) cannot be distinguished from someone just making multiple nuisance calls—that is, by looking at the paging channel alone. The call has to be followed to the speech channel in order to see if there is any activity there. A billing system that charges for excess airtime is one solution to this problem.

Data can be sent over timed channels if the subscriber is determined enough. Software that packetizes the message into segments small enough to get through in the time allowed for one call does exist. The software then automatically reestablishes the call for each packet until all the data are sent. Once again this can cause serious network congestion if enough customers do it.

Although overusage by a subscriber may be seen by the operator as errant behavior on the part of the user, it may also be a sign that the operator is not providing the kind of service that the customer really wants, and it may be time for a complete review.

ACCEPTANCE TESTING

A vital part of maintenance is proper acceptance testing. By signing the acceptance-test sheets the operator has accepted the quality of the installation and the maintenance staff also accept responsibility for the ongoing system performance. It is important that the system is not accepted into service unless the installation has been efficiently done. It is also a vital part of the quality control that any faults found during acceptance testing be fed back to the installers to prevent future recurrences of the problems.

An acceptance test can *never* be too detailed, and this is the one time that nit-picking is a virtue. Any faults or shortcomings that are not detected during the acceptance test can become a future maintenance liability.

The operator should be particularly careful not to fully rely on testing procedures of the supplier, which usually are not very exacting. Remember that the supplier's objective is to get paid as soon as possible, and this may not be consistent with spending extra time remedying installation shortcomings. Additionally, it has been my experience that the supplier's "acceptance tests" are generally no more than the test procedures that are used in the factory before delivery; they rarely take much account of the real world, as the installed environment. Often they fail to check building suitability, grounding, power supply integrity, rack security, cabling terminations, lightning and surge protection.

The acceptance-test procedures that follow will form a good basis for the operator to ensure the quality of the installation, and they are designed to be photocopied for use in the field.

ACCEPTANCE-TEST PROCEDURES

Acceptance-Test Sheets   The following acceptance-test checklists can be used by acceptance-test personnel for trunked radio base sites.

Site (Name) _____

Location _____

Switch (Location) _____Tel. _____

Installed by _____

Installation supervisor _____Tel. _____

Inspected by _____

☐ on completion   ☐ work still in progress

Acceptance date _____

Signed _____

Conditional acceptance date _____

Signed _____

(Subject to rectification of items _____
indicated on attached sheets)

Date in service _____

Power Rectifiers

	OK/NOT OK	COMMENTS
1. Mounting and layout	_____	_____
2. Cabling and terminations	_____	_____
3. Alarm extension	_____	_____
4. Designations	_____	_____
5. Commissioning test results	_____	_____
6. Handbooks	_____	_____
7. Load sharing	_____	_____
8. Safety signs	_____	_____
9. DC distribution	_____	_____

Batteries

	OK/NOT OK	COMMENTS
1. Fusing		
2. Battery spacing		
3. Electrolyte level		
4. Battery vents (also applies to sealed batteries)		
5. Battery leadburning/connections		
6. Battery cabling		
7. Hydrometer and thermometer		
8. Cell voltmeter and millivoltmeter		
9. Destinations		
10. Drip trays		
11. Safety equipment		
12. Safety signs		
13. Water supply		
14. Fuse alarm extension		
15. Accessibility for testing		
16. Battery function and continuity (test)		
17. Floor loading of batteries within limits		

External Plant

	OK/NOT OK	COMMENTS
1. Tower, mast, or pole		

2. Gantry _____ _____

3. Guys and anchors _____ _____

4. Lightning protection _____ _____

5. Tower, mast, or pole grounding _____ _____

6. Mains surge protection _____ _____

7. Tower lighting _____ _____

8. Corrosion protection _____ _____

9. Safety signs _____ _____

10. Site RF radiation records _____ _____

11. Equipment shelter _____ _____

12. Base security, fences, locks, gates, etc. _____ _____

13. Cable window and seals _____ _____

14. Grounding of equipment rooms _____ _____

15. Access to tower restricted _____ _____

16. Antennas correctly mounted _____ _____

Internal Plant

	OK/NOT OK	COMMENTS
1. Mounting and layout	_____	_____
2. Cabling and terminations	_____	_____
3. Alarms and telecontrol	_____	_____
4. IDF cabling	_____	_____
5. IDF labeling	_____	_____
6. System performance	_____	_____
7. Commissioning test results	_____	_____
8. Station logs	_____	_____

9. Designation   _____   _____

10. Handbooks   _____   _____

11. Drawings   _____   _____

12. Test cables and extender   _____   _____
    cards

13. Spares

14. TRX plug crimps

15. RF N-connectors   _____   _____

16. Independent link to switch   _____   _____
    (control center) functional
    (telephone line or radio
    link)

17. Lightning arresters on all   _____   _____
    incoming cables

18. Door alarms functional   _____   _____

19. Suitable fire extinguishers   _____   _____

20. Redundant control channel   _____   _____
    functional

Item To Be Checked

Power Rectifiers

Mounting and Layout

- ☐ Correct positioning of cabinets.
- ☐ All mains terminals covered.
- ☐ All cabinet components supplied, including tops and coverplates for unused positions.
- ☐ No cracked or nonworking meters.

Cabling and Terminations

- ☐ Cabling runs are satisfactory, cable ties used where necessary.
- ☐ Correct size of cable used for current to be carried for maximum size of installation.
- ☐ Cable crimps not loose.

☐ No undue mechanical stress by heavy cables on circuit breakers.

Alarm Extension

☐ Correct settings and operation of all alarms provided on power cabinets (mains fail, float low, high volts, etc.)

Designations

☐ All circuit breakers labeled.

☐ Cabinets (if more than one) are numbered.

☐ Switch plates clearly marked.

☐ Modules are all numbered, if modular-type rectifiers.

Commissioning Test Results

☐ Should be provided by the installers and be on site.

Handbooks

☐ Should be left on site by the installers (and usually contain the test results).

Load Sharing

☐ If power supply is modular, all modules should supply approximately the same current (but not necessarily equal). Turn the rectifiers on one at a time and note reconfigured load sharing and current limiting functionality.

Safety Signs

☐ Mains hazard.

☐ –Ve 24 V ground (where applicable).

☐ Any other that are required by local regulations.

DC Distribution

☐ Cabling from power cabinets to radio cabinets for correct current rating.

☐ Trays used where necessarily to support the cable correctly between cabinets.

☐ Bus bars and battery feeders insulated and suitably protected against accidental short circuits.

Batteries and Distributions

Fusing

☐ Current rating of fuses supplied for battery capacity and current drains.

☐ If indicators supplied, these should be clearly visible.

☐ Spare fuses available.

Battery Spacing

☐ Clearance over and around cells is sufficient for maintenance work (SG readings, etc).

Electrolyte Level

☐ Correct in all cells.

Battery Vents

☐ These can be frail and can be easily broken in installation. Check that none are broken.

Battery Lead Burning/Connections

☐ Each cell is correctly connected to the next via the lead V-connection. There should be no cracks or breaks capable of producing a high-resistance across the connection.

Battery Cabling

☐ Cables correctly tied and supported on cable trays.

☐ Correct size cable used.

☐ No loose crimps.

Hydrometer and Thermometer

☐ Correct size hydrometer is on site.

☐ Thermometer is present in each pilot cell.

Cell Voltmeter and Millivoltmeter (Wet Cells)

☐ Both on site.

☐ Spiked voltmeter 0–3 V to measure call voltage.

☐ Millivoltmeter to measure the volt drop across the intercel connections.

Designations

☐ Each cell and battery is correctly labeled.
☐ Pilot cell is indicated.
☐ Fuses are labeled.

Drip Trays

☐ Size and capacity are adequate to catch and contain one battery cell full of acid if a cell container cracks.

Safety Equipment

☐ Check presence of equipment as required by company regulations (can include rubber gloves, rubber apron, face shield, first-aid kit, etc.)

Safety Sign (Wet Cells)

☐ Acid precautionary warning signs posted.

Water Supply (Wet Cells)

☐ Fresh, clean water and a small washbasin available.

Fuse Alarm Extension

☐ Alarm given when fuse OC or CB operated.

Accessibility for Testing

☐ Batteries should be placed in such a position to allow cell replacement without impediment.

Battery Function and Continuity

☐ Gradually reduce the rectifier output voltage and note changeover to battery-powered operation. A voltage drop of not more than 8% should occur and current should remain the same. This checks battery and battery feeder continuity as well as the changeover mechanism. In systems not yet commissioned, the rectifier power should be turned off for this test. The battery voltage will initially drop rapidly and will then rise stabilizing about one volt above the minimum.

Floor Loading of Batteries within Limits

☐ Battery loads (kg/m^2) should be confirmed as being within floor structural limits.

External Plant

Tower, Mast, or Pole

☐ State which, and give height. Check that all iron-work is galvanized, no evidence of early rust, and all nuts and bolts are in place.

Gantry

☐ Feeders between structure and building are adequately supported by a gantry or tray.

Guys and Anchors

☐ Check masts having multiple guys and concrete anchors. Guys should be examined for correct tightness and rusting, and concrete anchors for flaking and cracking.

Lightning Protection

☐ Ensure lightning rods will not interfere with antenna pattern.

Tower, Mast, or Pole Grounding

☐ Structure to be strapped to ground at ground level. Feeders to be grounded top and bottom of tower and at cable entry. Tower grounding connected to building ground.

Mains Surge Protection

☐ Usually provided on mains powerline into building.

Tower Lighting

☐ Provided on structures near airfields, in accordance with local aviation regulations.

Corrosion Protection

☐ All ground strap connections are sealed with anti-corrosion kit.

Safety Signs

☐ Relevant safety signs are prominently displayed at foot of structure and on site fence.

Site RF Radiation Records

☐ When required, this record contains the maximum radiation from each antenna, along with the safe working level or radiation on the tower top.

Equipment Shelter

☐ The equipment shelter is properly and completely finished.

Base Security, Fences, Locks, and Gates, etc.

☐ The site fencing and security are ensured.

Cable Window and Seals

☐ Cable window and seals are properly fitted to prevent water seepage.

Grounding of Equipment Rooms

☐ The equipment room is adequately grounded and the ground resistance less that $10 \, \Omega$. The test results should be available.

Access to Tower Restricted

☐ The tower is separately fenced and locked.

Antennas Correctly Mounted

☐ Antennas are correctly spaced and either vertical or at the correct level of downtilt.

Internal Plant

Mounting and Layout

☐ Positioning of cabinets and supporting framework is as per design drawing.
☐ Blank panels, covers, etc., provided as required.
☐ Feeder supports provided above cabinets and up to the cable window.
☐ Frame racks firm and secure.

Cabling and Terminations

☐ Neat distribution and positioning of all intercabinet cables; tied down at regular intervals.

☐ Cable plugs to be complete, no missing components.

☐ Plug labels correctly supplied and marked.

☐ Combiner-module and combiner-star connector tails are free and unstrained; N-connectors to be tight.

Alarms and Telecontrol

☐ All alarms as specified are correctly returned to the mobile switch or monitoring center.

IDF Cabling

☐ Neatness and tying of cables.

☐ Terminations correct for the type of termination applicable at IDF.

IDF Labeling

☐ All circuit correctly labeled appropriate to the type of IDF system.

☐ Record book correctly made out.

System Performance

☐ All channel modules within specification.

☐ All combiners within specification.

☐ Feeder-antenna return loss within specification.

☐ Base station controller, redundancy functional.

☐ PCM or link system.

☐ System line-up levels on each channel or port correct.

☐ All alarms functional.

☐ Final call-through test on each channel module prior to cutover.

Commissioning Test Results

☐ To be left on site by installers for subsequent use by maintenance staff.

Station Logs

☐ To be provided at cutover by installation staff for batteries, attendance, or other as locally required.

Designations

☐ Cabinet numbering.

☐ Module numbering.

☐ –Ve ground signs.

Handbooks

☐ To be left on site by installers for maintenance use.

Drawings

☐ Copy of floor layout and cabling records to be left on site with handbooks.

Test Cables and Extender Cards

☐ Available where needed.

Spares

☐ Check any spare parts ordered are available before cutover.

TRX Plug Crimps

☐ Check crimping of wire to plug connection.

RF N-connectors

☐ If available, use N-connector gauge on all feeder connectors and tails. Some center pins may be out of specification and can cause damage to the socket.

Independent Link to Switch (Control Center) Functional (Telephone Line or Radio Link)

☐ Check for functioning of voice link to switch or control room.

Lightning Arresters on All Incoming Cables

☐ All cables entering and leaving the building should have suitable lightning arresters.

Door Alarms Functional

☐ Test all door alarms and confirm the proper working.

Suitable Fire Extinguishers

☐ Suitable nonconductive fire extinguishers are in place. Check currency of extinguisher testing.

Redundant Control Channel Functional (Where Fitted)

☐ Turn off the operation control channel and confirm the proper functioning of the changeover.

CHAPTER 7

FILTERS, COMBINERS, AND PRESELECTORS

Trunked radio systems will, by their nature, ordinarily have a large number of co-sited channels. These channels will be relatively close in frequency and must be able to operate concurrently. Additionally, there is generally a need to minimize the number of antennas in use, and this will be done by combining several transmitters and the receivers into one or more antennas. It is the filters, combiners, and preselectors that permit all this to happen.

Modern trunked radio systems will have a significant proportion of handheld radios among the terminal units. Unlike fixed vehicle-mounted mobiles, handheld radios need to be economical with battery power. This means that the transmitted radio-frequency (RF) power level (the most dominant factor in power consumption) will need to be limited. Typically, radios are rated at power levels from 1 to 5 W, with some having switchable power levels.

The limiting factor in the range that can be obtained from a single site is the talk-back range of the handheld. In most trunk systems a fixed mobile will have a power output of 10–25 W; and because it is vehicle-mounted, it will probably be using a gain antenna. This gives the vehicle-mounted unit a net advantage of 6–10 dB. In order that handhelds work as well as possible, it is vital that the base station receiver be optimized for sensitivity. This can be achieved by the use of high-gain base receive antennas and by minimizing all losses in the base receive path. While it is common in conventional public mobile radio (PMR) to use a combiner to allow the transmitter and receiver to share a common antenna, the losses in the combiner, particularly in the receiver path, make this option less than desirable. Combiners are only used where antenna space is at a premium.

Generally then, a trunked base site will have separate antennas for the receiver and transmitter; and in the case of sites with a large number of channels, it may have multiple transmitter antennas (again because losses accumulate with the number of transmitters combined).

How the combining and filtering is done can be of some significance to the operator who is seeking to maximize coverage and minimize cost. Except for some digital systems the channel combining is done on a per-channel basis, and so costs go up in direct proportion to the number of channels in service (for most digital systems to some extent the same applies, because the combining is done on a channel group basis).

Very few trunk radio manufacturers are also manufacturers of the filtering hardware. When a turnkey system is purchased, the RF plumbing (as the filters and combiners are known) is second-sourced by the supplier and probably marked up by 50% or more on the original equipment manufacturer (OEM) cost. Buying the plumbing directly from the original manufacturer makes good economic sense.

Shared sites can be particularly hazardous for a trunked radio system because not only is it necessary to contend with the RF generated by the system, but also it is necessary to operate with the interference from other operators as well as possibly deal with complaints from them regarding interference to their equipment from the trunked system. Good filtering and combining techniques will cost effectively minimize these problems while maximizing the transmitted RF levels.

TRANSMITTER COMBINERS

At many rented sites today there is a "per antenna" rental charge, which might well be as high as $10,000 per year. This can be a very significant cost to the operation of a large system and can be the driving force to keep the number of antennas down. Even if the operation is at a site owned by the operator, competition for the best antenna locations, wind loading, and aesthetics may all be factors mitigating against too many antennas.

It is not possible to simply directly connect several transmitters into a single antenna as shown in Figure 7.1. One reason that this cannot be done is that both the antenna and the transmitters are designed to work into a particular load impedance (usually 50 Ω) and connecting the transmitters in parallel would play havoc with the impedances. Furthermore, in this configuration each transmitter would effectively be delivering power to every other transmitter, resulting in wasted power and incredible intermodulation problems.

What is needed is a way of delivering the power directly to the antenna so that only a minimal amount of power is fed back to the other transmitters, and that the impedance at the point of combining remains at 50 Ω.

If the transmitters were sufficiently far apart in frequency, a simple series tuned circuit might do the job of preventing the transmitters from interacting, and the circuits could be coupled by an impedance transformer. Such an impedance transformer can be a quarter-wave transformer, which is made up of a small matching quarter wavelength of coaxial cable as seen in Figure 7.2. The matching section has a characteristic impedance equal to the square root of the product of the input

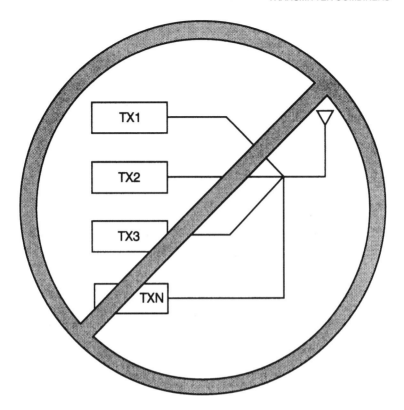

Figure 7.1   You can't do this.

impedance (50 Ω) and the output impedance (25 Ω), or 35.5 Ω. When connected together, this transmitter combiner would be as shown in Figure 7.3.

It is easy to see that this method could be extended to any number of transmitters. Simply add another quarter-wave transformer to the antenna lead as shown in Figure 7.4 to connect four transmitters. This process can then be repeated as necessary for additional transmitters.

Although the configuration in Figure 7.3 would rarely be used in practice to combine two transmitters, in principle it could be; and because in essence (as we shall see later) it is the same as the more common cavity filters, it will be worth exploring the basics of this circuit just a little more.

As stated before, the LC combiner requires some considerable frequency separation, but how much is enough? To begin we have to consider the sharpness of the LC circuit involved. A measure of this is its Q (or quality factor). The Q is a measure of the efficiency of the tuned circuit and is defined as

$$Q = 2\pi \times (\text{maximum energy stored/energy dissipated per cycle})$$

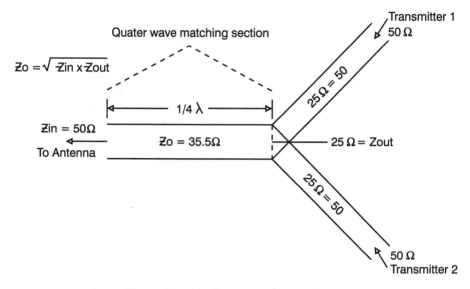

Figure 7.2   A basic quarter-wave transformer.

For the frequencies we are concerned with here, nearly all the losses are due to the coil resistance, and the Q can be calculated as

$$Q = 2\pi L/R.$$

where R is the coil resistance and L is the coil inductance. Typically, for such a coil the Q will be about 200. We can now use another parameter, the bandwidth of the tuned circuit, to see just how broad the tuning of the circuit is. For a tuned circuit we

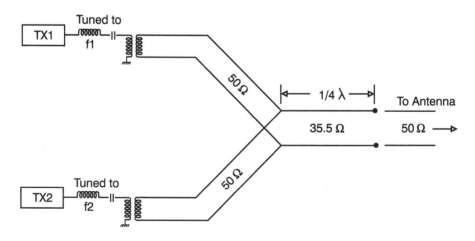

Figure 7.3   A method of connecting two transmitters into a common antenna.

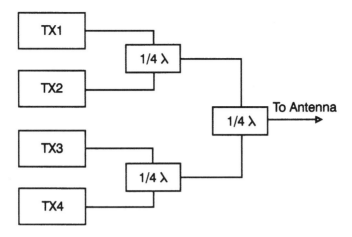

Figure 7.4. Four transmitters combined.

can define the bandwidth to be the point where the power through the circuit is 50% (3 dB) down (see Figure 7.5). The frequencies F_l and F_h indicate the lower and higher frequencies for which the power throughput will be within 3 dB of the power at the resonant frequency. A very convenient relationship is that

$$F_h - F_l = F_0/Q$$

In other words the bandwidth is inversely proportional to the Q. The concept of Q can be extended to other more complex tuned circuits and is a good indicator of the efficiency of the device.

Thus, if we assume a transmitter that operates at 400 MHz, then the bandwidth for a Q of 200 is 400/200 or 2 MHz. This tells us that the second channel must be at least

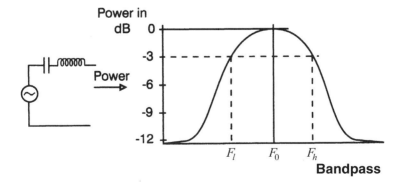

Figure 7.5 An LC response curve.

2 MHz away. In practice it would be a lot further because 3-dB isolation is hardly enough. So is it practical to use such a filter? If we consider a transmitter at 2 MHz, then we similarly find that the bandwidth is only 2/200 MHz or 10 kHz wide. So the problem comes down to the fact that at higher frequencies, higher Q values are required to achieve a reasonably sharp filter.

RESONANT CAVITIES

Resonant cavities are routinely made with Q values of the order of 5,000 to 10,000. With this sort of Q it is possible to do some serious filtering even at high UHF frequencies. A simple resonant cavity is, for practical purposes, a series tuned circuit and works in the same way as the circuit in Figure 7.3. It consists of a metal container, usually of simple geometric construction (drums or box-shaped are the most common). The resonance occurs because the internal dimensions are *electrically* some odd multiple of a quarter wavelength at the frequency concerned. All that is needed additionally to the cavity is a means of getting power into the cavity, along with a means of extracting the signal. At UHF frequencies a single half-turn loop is generally sufficient to do this. Figure 7.6 shows the construction of a simple cavity using a 44 gallon drum.

There are a number of reasons why a 44-gallon drum would not make an ideal filter. The most immediately obvious one is that it is too bulky. Because currents are flowing in the skin of the drum, causing energy losses, it might have been better if it were made of copper or some better conductor than its steel construction. And speaking of steel and the magnetic fields that will be produced by those currents, steel is a high-loss material for magnetic fields at RF frequencies. Temperature variations will cause changes in the dimensions of the drum, and we have no compensation mechanism.

Commercial cavities are mostly made of copper, a nonmagnetic material with good conductivity. To further enhance the conductivity, the inside surface is often coated with silver. Temperature compensation is added, along with a tuning mechanism. The

Figure 7.6 Shows the construction of a simple cavity.

tuning consists of a plunger on a screw thread that can be screwed to change the electrical length of the cavity.

To see how closely a resonant cavity is related to a tuned circuit, consider the evolution of the tuned circuit in Figure 7.7.

Starting with a simple tuned circuit in Figure 7.7, we can increase the frequency by making the coil have a few less turns (Fig. 7.7b). This process can be continued until the coil has been minimized to half a turn (Fig. 7.7c). In each of these iterations the resonant frequency will rise. To bring down the frequency we could increase the inductance by using two half coils in parallel (Fig. 7.7d) and continue this process with a number of half coils around the capacitor plates. Taking this process to its limit, we have a drum (Fig. 7.7f). In fact, currents do flow in the walls of the cavity just as they would in a coil, and an electric field exists inside the cavity much as it would in the capacitor (of course, magnetic fields exist as well, because a changing electric field always is accompanied by a changing magnetic field).

By suitably arranging the physical characteristics of the cavities, a variety of different filters can be devised. The most common are the bandpass, notch, and pass reject filters. The bandpass filter has a characteristic curve as seen in Figure 7.8a. A notch filter (Fig. 7.8b) is sometimes added in series to stop a powerful nearby transmitter from intermodulating with the transmitters. It is commonly seen as a harmonic filter in trunked systems where it is used to attenuate the second and higher harmonics of the transmitters. Where a notch is required to be close in frequency to the transmitters (again generally because of some co-sited interferer), the best option might be a pass reject filter, which is essentially a series bandpass and notch filter (Fig. 7.8c). The pass reject filter will most commonly be used in a mobile radio duplexer where the notch is used in the receiver path to take out the transmitter RF.

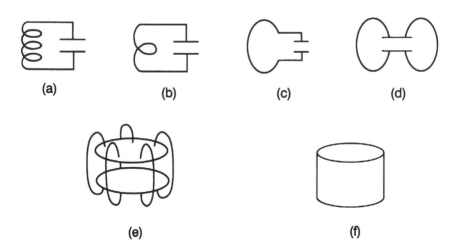

Figure 7.7   The evolution of a cavity from a tuned circuit.

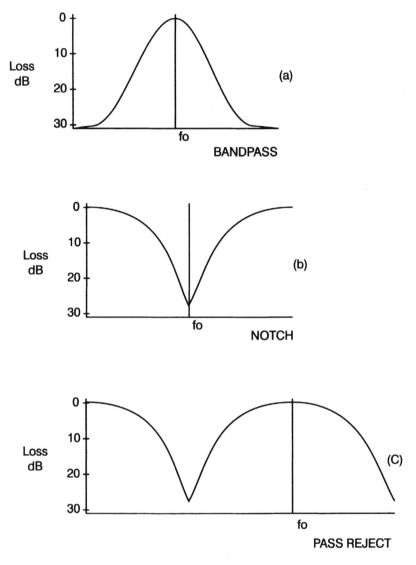

Figure 7.8 The characteristics of various filters.

PRACTICAL COMBINERS

The simplest practical combiner is electrically equal to the LC combiner discussed previously. It consists of one or more bandpass cavity filters connected in series and joined together via a star harness (which is a multiport electrical equivalent to the quarter-wave transformer discussed before). This simple combiner is inexpensive, has low insertion loss, and is readily reconfigured. Its disadvantages are that it is physically bulky and is suitable only when the frequency spacing between channels is large. A typical cavity combiner is shown in Figure 7.9; notice that unused terminations on the star combiner are connected to dummy 50-Ω loads. This simple combiner is good for channel spacings down to about 0.5% of the center frequency. Figure 7.10 shows an actual combiner from Deltec.

Where closer frequency spacing is required (and that's almost always in trunked radio situations), the addition of an isolator (producing a cavity-ferrite combiner) may be necessary. An isolator is generally a circulator with a dummy load connected to the reflected power port.

Circulators are devices, with the geometry arranged so that the transmitted and reflected power appear at two different points. Figure 7.11 shows the operation of a circulator. Typically, the isolation between the ports is about 20 dB, and the loss through the circulator is around 0.6 dB. To turn a circulator into an isolator (see Fig. 7.12), all that is necessary is to connect a dummy load to the reflected power port. It should be noted here that while in general the reflected power is (hopefully) low, under fault conditions such as antenna open circuit, the full transmitter power may appear at the reflected power port. For this reason it is best to use a dummy load rated at the full RF power of the transmitter. Manufacturers frequently offer one- and two-stage circulators (which of course can be used as one- or two-stage isolators). When two stages are used, the isolation is increased to about 40 dB, but the loss will rise to 1 dB or higher.

A cavity-ferrite combiner is seen in Figure 7.13. The diagram shows the addition of a harmonic filter. This filters out the second harmonic of the transmitter and significantly reduces intermodulation. This may not always be included in the

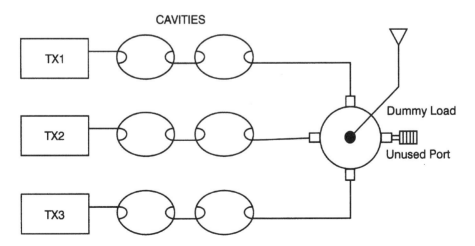

Figure 7.9   A typical cavity combined system.

Figure 7.10 A combiner from Deltec.

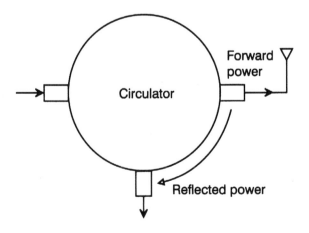

Figure 7.11 A circulator.

Figure 7.12   An isolator. (Photograph courtesy of Deltec.)

combiner. This combiner can be used for channel spacing down to about 0.04% of the center frequency, with the actual figure being manufacturer-dependent. Because they are versatile and easy to reconfigure, these combiners are widely used in cellular radio systems and in trunk radio when the channel spacing is sufficiently wide.

When channel spacing is really close (and this happens particularly with small trunked radio systems, or with those where a small number of contiguous channels have been allocated), the only option is a hybrid ferrite combiner. Hybrid ferrite combiners can combine and isolate transmitters of any frequency spacing. They can even be used to combine two transmitters on the same frequency. These devices dispense with the cavities altogether, and they rely on hybrid couplers to do the isolation. A hybrid coupler is a two-way power splitter that combines and isolates two input signals. A hybrid coupler is shown in Figure 7.14. The downside is that they are costly and are high-loss devices. A typical hybrid coupler, seen in Figure 7.14, may have a loss of around 4 dB, but when they are interconnected to combine more transmitters they are connected in series and so the losses grow rapidly.

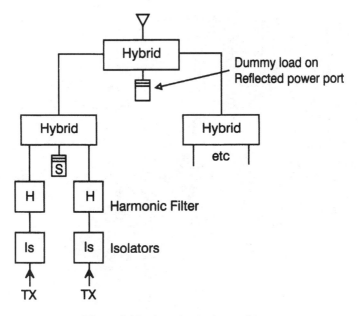

Figure 7.13   A cavity-ferrite combiner.

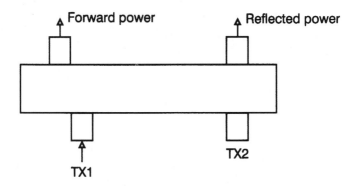

Figure 7.14   A ferrite coupler.

HOW MUCH POWER LOSS IS ACCEPTABLE?

All transmitter combiners will have some power loss, and some (in particular the ferrite hybrid) will have substantial loss. Although not as critical as receiver losses, transmitter losses do have a cost. The first cost may be the transmitter itself. If high-loss combiners are used, it may be necessary to use a higher-power (and hence higher-cost) transmitter power amplifier just to ensure that the ERP is sufficient for the system balance between transmit and receive paths. Typically, the power amplifier stage of a transmitter is around 30% efficient, and it will be responsible for the bulk of the power consumed by the transceiver. Thus a 100-W transmitter will probably require 300 W of main power to drive it. When there are a large number of transmitters on site, the cumulative power can be substantial. For example, for a 20-channel site operating at 300 W per channel the power consumption would be 6 kW (at the going rate of 10 cents per kilowatt hour, this is 60 cents an hour for power, which, depending on usage patterns, could amount to $1,000 to $2,000 per year). Another cost associated with high transmitter power is the cost of battery backup. Ordinarily, the size of the batteries is calculated on the basis of a predetermined outage time, based on peak hour usage. The size and cost of the battery backup will increase in almost direct proportion to the power consumption.

When the loss in the combining is high, that loss is dissipated as heat. The heat itself can be a problem: In air-conditioned rooms it will add to the cooling costs, while in un-air-conditioned rooms it may lead to premature aging of the base station hardware with consequent maintenance bills.

Thus, there is no clear-cut answer to the question of how much loss is acceptable; but as a general rule, 3 dB is a good target, unless other factors (such as the need to combine very closely spaced channels) dominate.

RECEIVER COMBINING

This is the most critical part of a modern trunked base station. Unlike transmitter losses—which, while inconvenient, need not be performance-affecting—any receiver losses or impairment can be translated directly to reduced handheld coverage. It is usual practice to combine all receivers onto a single antenna; for reasons that will become clear later, this can be done without any significant losses at the channel level.

A typical receiver combiner is seen in Figure 7.15. The essential elements are the preselector filter (which is a bandpass filter, passing only the receive channels), a low-noise amplifier (LNA), and the distribution amplifier.

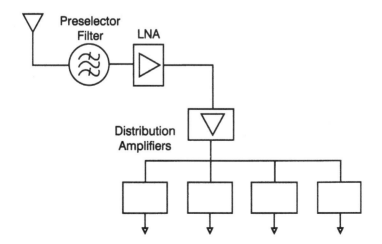

Figure 7.15   A typical receiver combiner.

In practice a receiver combiner ordinarily has a number of channels in a single rack-mounted frame. Figure 7.16 shows a Deltec eight-way receiver multicoupler usable at 25–520 MHz.

The preselector filter typically passes the whole of the trunked radio system receiver bandwidth in a single window as per Figure 7.17a and will have a loss of less than 3 dB. Sometimes the receiver channels will not all be contiguous and they may be split into a number of different groups. This can be accommodated by multiwindow preselectors (Fig. 7.17b) that will come at a somewhat higher cost and with additional losses. In most systems, following the preselector will be a low-noise amplifier. It pays to use the best LNA available. For practical purposes the noise figure (and thus performance) of everything that follows is determined by this device. The contribution to the overall noise figure of the system by devices after the LNA is reduced by a factor that is (nearly) inversely proportional to the gain of the LNA, and so the gain of the LNA is also important. Offsetting the positive contribution of the LNA gain is the intermodulation distortion produced by it. An LNA that is subjected to high levels of spurious signals or high levels of wanted signals will produce intermodulation signals (the most troublesome being third-order products) that will degrade the receiver performance. Ideal devices do not produce intermodulation distortion, but virtually all

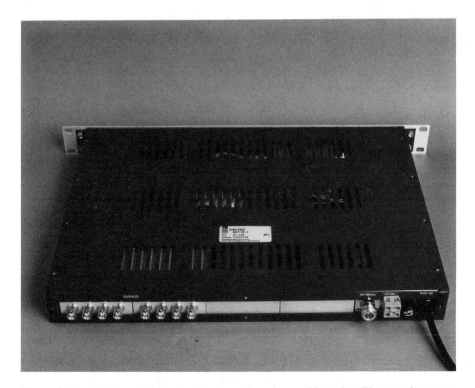

Figure 7.16 A Deltec wide-band, eight-channel receiver multicoupler. (Photograph courtesy of Deltec.)

real devices do. An ideal device is one where the output level is directly proportional to the input. For most real devices the relative amount of distortion produced increases as the input power level is increased. This occurs because of inherent nonlinearities. For a typical RF amplifier the relationship between the power in and power out follows a simple linear relationship, while the third-order intermodulation level follows a log-linear relationship. Figure 7.18 shows that for low level of input the distortion products are relatively low; but as the level rises, the distortion levels rise more rapidly, until at 27 dBm input the distortion power is equal to the fundamental (or wanted signal). Obviously, nobody is going to operate an LNA at such a level; but this level, called the third-order intercept point, is a good measure of the power-handling capabilities of the amplifier. Its value is significant when determining the preselector filter characteristics. Fortunately for system operators, most LNAs have a third-order intercept around 35 dBm, and so this is one less variable to worry about.

It is not only active components like amplifiers that contribute to intermodulation distortion, because to some extent all real-world devices are nonlinear. Even

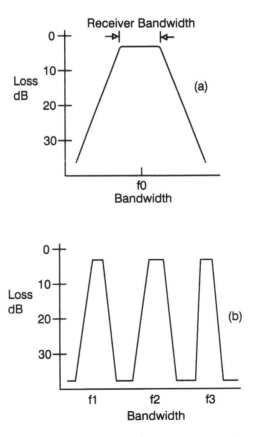

Figure 7.17   (a) A single preselector window and (b) a multiple window.

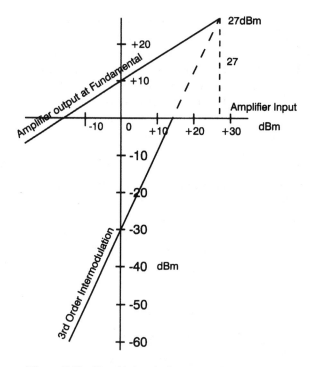

Figure 7.18 The third-order intercept of an amplifier.

components such as cavity filters and connectors will produce some third-order products, and in most cases they will have a measurable third-order intercept.

NOISE FIGURE

The other parameter that measures the performance of an LNA is its noise figure (NF). This is usually expressed in decibels, and it is the additional noise that is injected into the system by the amplifier itself. A high-quality LNA will have a noise figure of around 0.5 dB, while an average one may be more like 3–4 dB. A typical LNA that is mounted inside a receiver multicoupler is shown in Figure 7.19.

Passive components such as cavities and cables introduce additional noise that is equivalent to their loss. So a 3-dB attenuator (or a length of cable with a 3-dB loss) connected in series with the antenna will add 3 dB to the total noise figure of the receiver. So if this were followed by an LNA with a noise figure of 3 dB, the overall noise figure would be 6 dB. However, the importance of the LNA can be seen by comparing the loss if the same 3-dB attenuator were connected after the LNA. Assuming that the LNA had a gain of 10 dB (equals 10×), the total noise figure is only 0.962 dB. For a more detailed explanation of this, see Appendix 2. The important thing to realize is that anything that follows the LNA contributes only marginally to the overall noise factor (and hence the sensitivity) of the receiver.

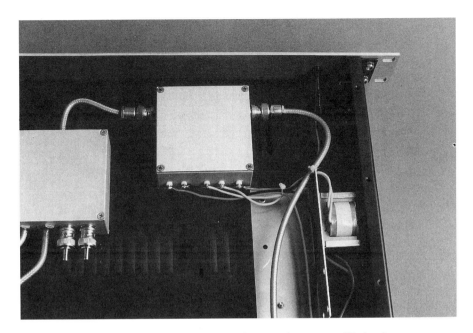

Figure 7.19   A typical LNA. (Photograph courtesy of Deltec.)

RECEIVER MULTIPLEXER

Figure 7.15 shows a series of distribution amplifiers (also known as receiver multiplexers) immediately after the LNA. Because, in the way that has already been explained, these amplifiers contribute little to the overall noise performance, a good number of them can be cascaded in series. It is usual that they branch in multiples of 4, 5, or 8, with the number being determined by the rack layout expected. In total there may be 60 or more receivers fed from a single antenna, but in trunked radio systems a maximum of 20 is most common. The gain of these amplifiers is generally low, but sufficient to overcome the distribution losses.

For example, if the amplifiers are branched in groups of 4, a minimum gain of 10 × log 4 (6 dB) will be necessary to maintain the LNA levels. Too much gain here can be a problem because it could overload the first RF amplifier of the receiver, thereby causing intermodulation distortion.

CUSTOMIZED RF

If you need to customize your RF, there are a few things that need to be quantified before contacting a supplier:

1. The actual frequencies to be used need to be known; and if these are not available, then at least the band needs to be identified.

2. The maximum transmitter power level and the maximum acceptable combining loss.

3. Where space is at a premium, the limitations on size need to be specified.

4. The number of channels to be combined as well as the long-term expansion plan (the maximum number of channels on any one site).

5. The frequency separation between transmit and receive.

6. Any channels that need to be notched out (co-sited potential interferes).

7. The maximum preselector filter loss.

8. If an LNA is to be used, whether it will be tower- or rack-mounted.

9. Preferred connector types (N type are the most common).

10. Do you want the supplier to do an intermodulation study on the chosen frequency plan (there may be extra charges for this)?

FREQUENCY SELECTION

For some small operators, this will not be much of a problem because they will have only a few channels to choose from. For large nationwide operators, the selection of frequencies and the resultant groupings can be very important. As explained in Appendix 2, intermodulation is inevitable. It is also unpredictable because the signal can intermodulate at any of the devices it passes through, or even at some external location like the tower legs, between layers of a metal roof, at the joints of a wire fence, and countless other places. Because you have to live with this problem, it is important to be sure that the intermodulation products from the transmitter do not fall inside the receiver bandwidth. The most troublesome intermodulation products are usually the third-order ones, although this is not always necessarily the case.

If there are two transmitters on frequencies A and B, then the third-order products will be $A \pm 2B$ and $B \pm 2A$. Mostly it will be the $2A-B$ or $2B-A$ products that will be likely to cause a problem.

Assuming that a base site that used -5-MHz spacing from transmitter to receiver had one channel on 300 MHz and another at 302.5 MHz, then the third-order product $(2 \times 302.5) - 300$ equals 305 MHz. This is exactly inband for the receiver of the 300-MHz transmitter and is likely to cause problems. Higher-order products may also be a problem; for example, a fifth-order product $3A-2B$ is likely to cause a problem when the two transmitters are closer in frequency.

ADVANCED CAVITY TECHNOLOGY

There are two areas of advanced cavity technology that may be of interest in some trunked radio applications. Both result in high Q cavities, which gives a sharp cutoff characteristic with low loss.

Cavity design techniques now allow for elaborate cross-coupling between cavities which results in higher Q. A low-tech way to produce a high Q is simply to make the cavity bigger. High Q preselector filters will permit closer spacing of channels with minimal loss.

Using superconductor technology, a number of new companies have been able to produce some spectacular filters. In principle, all that is needed to improve the sharpness of the filter is to cascade more cavities. The downside of this is that it also introduces more loss. By using superconductors, the losses can be minimized even when a large number of cavities are used.

High-temperature superconductors will work at 100 degrees Kelvin, a relatively easy temperature to attain. It does mean, however, that along with a cavity it will be necessary to have a refrigeration plant. Adding to the bulk of the arrangement is the need for the cavity to be able to continue to work even in the event of a power failure; this means building in thermal inertia and very good insulation. Cavities with 6 hours of power-off capability have been demonstrated (naturally this means that normal operations are at temperatures somewhat lower than the minimum required for superconductivity).

It is interesting to note that while superconductors have zero resistance to DC currents, they offer some resistance to AC (and so they offer resistance to the RF currents that are circulating in a cavity). Of course the resistance to the current will result in heat dissipation, and this heat increases the cooling load. Another disadvantage of superconductors in cavities is that they produce intermodulation distortion in the superconducting material itself.

Superconductor filters have been demonstrated in practice in cellular and PCS preselector filter applications; and in the laboratory, transmitter combiner cavities with Q values of 30,000+ have been built. Such Q values are, however, not much higher than the best conventional filters, some of which have Q values of the order of 11,000.

CHAPTER 8

REPEATERS

Repeaters (also known as *bidirectional amplifiers*) can be a cost-effective way to achieve more range, or better in-building coverage, without adding a new base-station site. In trunked radio applications, every site added in a given service area means additional costs for group calling. Less directly, there are also additional costs for mobile-to-mobile calls, because an increase in the number of base sites means that the probability of calls passing through two base sites increases. It is a difficult juggling act to provide "just the right number" of base stations, to give acceptable coverage, while at the same time trying to minimize the total number of bases, provide an adequate signal quality, and provide sufficient traffic channels.

Repeaters are significantly less expensive than base stations, and some can even provide a similar ERP. While repeaters are the technology of choice for tunnels, valleys, and in-building coverage, it is worthwhile considering them, whenever a site is to be added for the purpose of extending the coverage area.

Repeaters can be very simple bidirectional amplifiers similar to the one in Figure 8.1. This repeater simply picks "off-air" the incoming signal, amplifies it and sends it on. While such a simple repeater can be very cost effective, it will only be suitable for very local area operations. It can potentially oscillate because the input at one end is the output of the other. The antennas therefore need to be arranged for the maximum possible isolation between input and output, and there needs to be at least a 10-dB margin between the isolation and the net gain.

Another problem with this type of repeater is that the repeater signal and the base station signal will ordinarily have some regions where the signal levels are within ±3 dB. In these areas, destructive interference may occur, causing patchy coverage.

Perhaps the most effective use of this type of repeater is to improve coverage inside a building or tunnel. In this case a repeater may be connected to a leaky cable within the building/tunnel, and it is relatively easy to ensure good isolation between the two antennas.

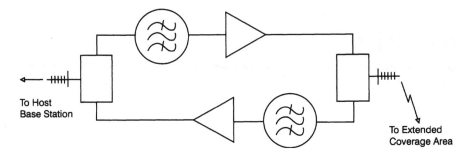

Figure 8.1   A simple active repeater.

The effectiveness of the simple repeater is highest when there is a large separation between the transmit and receive frequencies. At lower frequencies the TX/RX separation may be a problem in isolating the input from the output. One solution would be to use very sharp, high-performance filters. These, however, will be both bulky and expensive.

A more effective way to deal with close RX and TX frequencies is to use a repeater where the gain is in an intermediate-frequency (IF) stage. The IF will operate at a frequency comparable to the bandwidth of the TX/RX band, so that filtering will not be a problem. I once had Aerial Facilities (United Kingdom) custom make one such repeater for a system with only 5 MHz between TX and RX and operating at 500 MHz, but where the TX (and RX) bands were 2 MHz wide (there were 200 channel pairs allocated to the system). The repeater worked well, but it is unlikely that any other type of repeater could have been made to work at all. Figure 8.2 shows such a configuration.

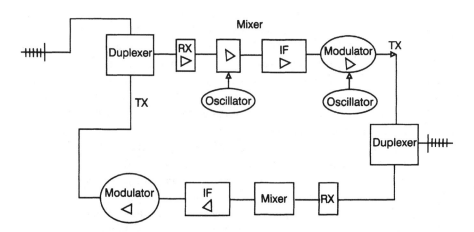

Figure 8.2   A repeater with an IF stage for narrow RX/TX frequency separation.

FREQUENCY TRANSLATORS

The principle of a frequency translator repeater is that the incoming signal is received at the repeater site and is then retransmitted on a different frequency. Figure 8.3 shows this principle. This method of repeating does not have the potential to oscillate, which the others do, but for most trunked radio systems there are some additional complications.

The repeater will be "copying" a distant base station site, and it may be sending an instruction to a mobile telling it to switch to channel N. The problem here is that because of the frequency translation the mobile working off the repeater will go to the wrong channel if it simply follows this instruction. This means that the repeater must "examine" all incoming instructions and must modify them, if necessary, before passing them on.

In its simplest form, the repeater would have one channel for each channel at the donor site and simply repeat all channels. Because there is intelligence in the base station, it is not necessary to repeat all of the channels from the donor base site—only those that are needed for traffic capacity. To accomplish this, it will be necessary that the repeater have frequency agile channels and that it be capable of distinguishing the repeated traffic from the donor site calls.

In principle, this can be done for any trunked radio system, and translator repeaters of varying degrees of sophistication exist for most protocols. A particularly elegant translator repeater for MPT 1327, which uses full duplex mobiles, or a repeater channel, as the transceiver is manufactured by Radio Systems Technologies (see Fig. 8.4). This repeater is completely transparent to the mobiles using it, and all the usual facilities including PSTN and PABX functionality are available. One of the channels is designated as a control channel, and this can scan for secondary control channels (or even alternate donor cells) if necessary. As in the base sites, the repeater control channel can revert to a traffic channel in the case of all channels busy (if it is so programmed). The number of equipped channels at the repeater can be determined by the traffic served, with the repeater even having the ability to send its own "all

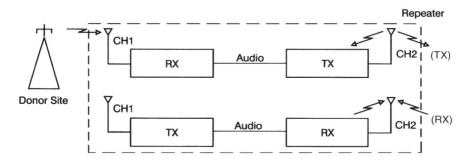

Figure 8.3 A basic frequency translator repeater.

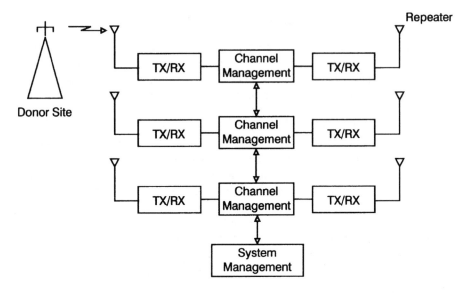

Figure 8.4 A trunked radio-frequency translation repeater.

channels busy" tone in the event that the repeater becomes fully occupied before the donor site does.

The economics of the translation repeater is of interest. While the repeater will have two transceivers per channel—and the transceivers are, for most systems, the major cost of a base station—it might appear on the surface that there is little economic advantage over a base site (because it will cost about the same). Furthermore, the coverage of a translator repeater can be expected to be similar to an actual base site, so the repeater could actually be considered to be a replacement for a base, were it not for the fact that while it provides additional coverage, it does not provide additional traffic handling capacity.

While the decision to use a simple repeater for local coverage extension in tunnels and buildings might be straightforward, the selection of a more expensive (and capable) translator repeater may in some cases not be so straightforward. In this case, some other factors do have to be considered. First, a base station will require links, and for a small system these links themselves are probably transceivers and will have a significant cost, particularly if the links need to be carried over a number of hops. Next, while it is true of many systems that the major cost of a base station is in the RF part, for some systems the base station controller can be very expensive, and this can tilt the economics in favor of a repeater. As a general rule, even a translator repeater will cost about half as much as a fully trunked base station that it may replace (assuming that each has the same number of channels).

Most trunked radio systems have evolved with very unpretentious switching capabilities, and the systems usually come with rather small switching modules. It is not unusual that the main switch can only accommodate six to ten sites, and that for further sites a new system is needed and/or an expensive central switch upgrade is required. In this case the marginal cost of adding a new site can be very high, and a repeater will prove economic.

It is important to remember that the traffic capacity of the repeater is provided in the first instance by the donor site. If a lot a traffic is being carried by the repeater, the marginal cost of the extra channels at the donor base may become significant.

CHAPTER 9

LINKING SITES

All but the simplest of trunked radio systems will require some links. Once there is more than one site, it will be necessary to provide intersite links. These may be wireline (most wireline network providers offer permanent links for this purpose), radio, or fiber-optic.

Ordinarily the system of choice for a trunked radio system will be radio. This is because it is generally the least expensive and because it is familiar technology. To embrace other forms of links not only involves having to broaden the staff skills, but often requires the purchase of new test equipment and a whole new range of spares holding.

SINGLE-CHANNEL LINKS

Single-channel links are usually the first thing that a trunked radio operator will consider. They are familiar technology, inexpensive, and quick to install.

In the simplest form a single-channel link can consist of a pair of back-to-back conventional mobile radios (one transmitting and one receiving), operated at something less than 20% of their maximum power rating. The reduced power level is necessary because most mobiles are not rated for continuous service and will overheat if held in the transmit mode at full power for prolonged periods. In regular press-to-talk usage the mobile will have somewhat less than a 50% transmitter-on duty cycle and will rarely be transmitting for more than a few minutes continuously.

To complete the system, these mobiles are then connected to a duplexer (so that only one antenna is needed), and the duplexer feeds a Yagi pointing at the remote site, and so you have an instant link.

A slightly up-market version of this is to use a base-site repeater as a link. This has the advantage that higher power levels can be used (because the transmitters are rated for continuous transmission thus longer range is obtained), and that the base-station hardware is likely to be a little more reliable. A lower cost alternative is to use full duplex mobiles in place of the base-station transceivers.

Things become a little more complex when more than one channel is needed. Because of the need to avoid using multiple antennas, multichannel single-site links will generally have a combiner to join all the links into a single antenna. These combiners will probably be of the same type that is used on the base site for the trunked radio channels.

In this way, single-channel links will use technology that is very familiar to the trunked radio operator. Depending on the frequency, single-channel systems should have little problem linking up sites 50 km away if the frequency is not above 500 MHz. Under good line-of-sight conditions and at low frequencies (VHF), a single-channel system may work as far as 100+ km.

Single-channel systems are rightly the link of first choice for small systems; but for larger systems, a thin-line microwave system may prove to be a more economical answer. The reason is that a thin-line microwave system costs about the same as four to six single-channel links, yet it provides 30 channels (or 24 if T1's are used). At the small link end of the range there are few suppliers of links with capacities other than one or two channels, and for some reason a two-channel system will often cost more than a pair of single channels. A typical thin-line microwave system (a system without redundancy and built to a budget) can be purchased for $25,000 (or less) complete.

Minilinks can be considered. They have an integrated transmitter/receiver/ antenna in a single case. This means that the whole system has to be tower-mounted, and so it is not a great idea for high towers (from a maintenance view). The advantages are that this will be a very cost-effective way to go. There are no RF cables (which can be expensive and lossy at the higher frequencies), and there are virtually no transmission losses as the transmitter is coupled directly into the antenna. The cabling to the link consists of power leads and audio (ordinarily at the T1/E1 level). The downside is that the whole unit is sitting out in the weather and vulnerable to rain and lightning (more so than it would be if all the electronics was safely mounted indoors). However, this technology has matured a lot lately, and MTBFs of several years are not uncommon.

Here is a word of caution when costing links: Suppliers often quote link costs "per end." This is the price of only one-half of the link. The reason for this is that the links need not be symmetrical. For example, you may choose to use a large dish at one end of the link and a smaller one at the other. Because both the transmitter and receiver pairs use the same dish, the system gain will still be balanced (that is, it will be the same each way). Reasons for minimizing the dish size at one end would include the need to reduce wind loading, because of space restrictions or building owner objections.

Frequency allocation may be a problem, and particularly so if only a small fraction of the microwave capacity is being used. Some manufacturers get around this by having as an option partial modulators that deliver a smaller system using less bandwidth. An example of this is the MAS, a 10-channel link that is otherwise identical to their 30-channel, 2-Mbit link except for the modulator.

However, don't overlook the other costs associated with microwave. These include staff training, spares holding, and maintenance contracts. Taken together, these could easily total another $20,000. This cost will need to be added (as virtually a fixed cost) regardless of the number of microwave links in service. This simply pushes the break-even point a bit further out; but because many trunked systems are quite small (and likely to remain that way), this cost should not be overlooked.

The other negative for microwave is that it will invariably permit significantly less range than single-channel systems, and you may find that a two-hop microwave system is needed to replace a single-hop, single-channel link.

DESIGNING A LINK

The philosophy that needs to be applied to link design is somewhat different from that used for mobile coverage. By definition, one part of a mobile link is mobile; and if the user happens to be in the region of a local fade, then simply by moving it is often possible to reestablish connection. Fixed links also suffer from local fading (most likely caused by atmospheric conditions), and thus they must be operating with a margin to accommodate these fades without dropping the link.

Generally, trunk systems will have a requirement for a data link between base sites and the central controller. This link will be continuously polling, and a drop in the link will cause the base site to be temporarily unavailable to the system (although for most systems the site will continue to function as a local stand-alone site; and in some systems, existing calls using that link may be held up). This can be a source of considerable annoyance to the network users and can be an embarrassment to the network operator. The link integrity is of considerable importance.

A major determinant of the link reliability is the fade margin. This is usually expressed as the margin in decibels that the link operates in average conditions above that needed to sustain a specified signal-to-noise ratio (or for digital systems a specified bit error rate). The fade margin bears a rather complex relationship to the average field strength seen by the receiver, and mostly is expressed as an empirical formula derived from experimentation. This means that one of the factors affecting fading is the climate. Needless to say the larger the field strength at the receiver the less likely it will be that a fade will cause the loss of the link. However regardless of how good the link is, there will always be some probability of outage due to fading.

A fade margin that gives a 99.99 reliability (equivalent to 53 minutes outage per year or outage probability of 0.01) is generally regarded as acceptable for trunk radio applications, but some digital systems may require a higher level of reliability. When links are connected in series, the reliability will decrease in accordance with the laws of probability. So if link A has an outage probability of 0.01 and link B a corresponding figure of 0.005, then if these links are connected in series the probability of outage will be $P = 0.01 + 0.005 - (0.005 \times 0.01)$ or 0.01495. The rule generally is that the probabilities of outage for the two links are added and then the joint probability (the product of the two probabilities) is then subtracted, to give the

1111111

outage probability of the two links in series. If more than two links are connected in series, any two for which the series reliability has been calculated as above can then be considered a single link for the purposes of reliability calculations. So, for example, if the two links above are then connected to another link with an outage probability of 0.008, then the outage probability of the three links is 0.008 + 0.01495 – (0.008 × 0.01495), or 0.02283.

There is not a one-to-one relationship between fade margin and reliability. Because fading is largely due to atmospheric effects (principally refraction and rain) and reflections, then the longer the path, the larger the fade margin needs to be to maintain any given reliability standard. Higher frequencies cause more interaction with the environment and thereby increase the expected level of fade over any given link.

The other major cause of fading is multipath. Although it is tempting to simplify a microwave beam to a single ray of light (much like a laser), it is not an accurate description. The microwave beam spreads significantly as it propagates; and if there are suitable reflectors along its path, there is a significant probability that a significant level of signal will arrive at the receiver from more than one path. Because these paths are of different lengths, they arrive "out of phase." As such, they may add constructively (or otherwise) at the receiver.

Reflections most likely will occur from flat, regular ground somewhere in the path of the rays. For this reason, propagation over water (a good reflector) or over flat plains is most likely to cause multipath interference and hence fading. In farming districts, this fade mode may be highly seasonal, with freshly planted crops being seen as effective reflectors, while the larger, less regular mature crops may disperse the RF energy and nullify multipath fading.

It should be clear by now that coping with multipath is as much an art as a science. It can be as variable as the weather itself. Consequently, the formulas used to derive the "estimated" multipath are really just mathematical guesses. Furthermore, most of the mathematics of fading were condensed from a field study of strictly limited number of paths, but over long periods of time.

For a link without diversity, the percentage of outage time can be approximated by the following:

$$P = 3 \times 10^{-7} \times C \times Q \times F^{1.5} \times D^3 \times 10^{-M/10}$$

where

$P =$ percentage of time of outage
$C =$ climatic factor
$Q =$ terrain factor
$F =$ frequency (GHz)
$M =$ fade margin (dB)
$D =$ distance in km

and where

$Q =$ 0.25   for hilly or mountain paths
$Q =$ 1   for rolling terrain
$Q =$ 4   for very smooth terrain and over water

and

$C =$ 1/8 for a dry climate
$C =$ 1/4 for a temperate climate
$C =$ 1/2 for humid or coastal areas
$C =$ 1   for coastal regions

The above equations indicate a very strong dependence on distance (increasing as it does with the third power of the distance) and a lesser but still strong dependence on frequency.

DIVERSITY

Link diversity is one way to significantly improve the outage rate. It comes in several guises. Path diversity is the most common, and this relies on having two separate RF links separated in space (vertically). The signal can then be selected from the link with the highest RF level (or lowest bit error rate); this is called *switched diversity*. Another arrangement is to combine both signals from each path in a way to minimize the bit error rate; this is known as *combiner diversity*.

The space diversity improvement factor I_d can be given as

$$I_d = 0.0012 \times S^2 \times F/D \times 10^{M/10}$$

where S is the vertical space separation in meters. With diversity the probability of outage is decreased by a factor of I_d.

Yet another diversity arrangement uses two receivers at each end (but only one transmitter) with vertical space diversity. This is know as *hybrid diversity*.

Most major routes have various forms of redundancy, and one form is often to have one or more bearers on standby in case of failure. This kind of redundancy is called $N + 1, N + 2$, and so on; here N stands for the number of active bearers, while the number 1, 2, and so on, signifies the number of standby bearers. Because the standby bearers are operating on a different frequency from the active ones, in addition to providing equipment redundancy, they also offer frequency diversity (fading is frequency-dependent).

Frequency diversity is even more vaguely accounted for than space diversity. In the G. T. E. Lenkurt book *Engineering Considerations for Microwave Communications* (GTE Lenkurt, San Carlos, 1970) an improvement factor for frequency diversity, put forward as simply an educated guess, is

$$I_f = \frac{1}{8} \times \frac{F_d}{F} \times 10^{M/10}$$

where I_f is the improvement factor for outage reliability and F_d is the difference in the frequencies of the main and protection bearer.

Polarization diversity can provide increased reliability because fading is also very much dependent on polarization.

Having elaborated on all the hazards of links, maybe a rule of thumb would be appropriate: Aim for at least a 30-dB fade margin, and relax if you can get 40 dB. Additionally, for longer hops (>10 km), use the lowest frequency available. Higher frequencies mean not only a greater path loss, but also a greater tendency toward multipath fading and attenuation in rain. In high rainfall areas (such as the tropics), rain limits the frequencies that can be used reliably over paths any greater than a few kilometers to those below 11 GHz.

SITE SELECTION

In general, radio links should be line-of-sight, so it is relatively straightforward to evaluate the potential for a link. The starting point is to look at a topographical map and confirm that there are no obvious terrain obstructions. Assuming that the path looks feasible on the map, the next move is to visit the site to be connected (most likely a base station), and with binoculars try to establish if line of sight exists to the desired point of connection. While on site, it is also wise to carefully consider the probability that the development in the area might lead to future loss of line of sight (including the growth of near-end foliage. The rapid expansion of many major cities has meant that many a link has had to be relocated.

Problems can arise if the base station site was selected without consideration of the links. While it is most probable that a good trunked radio site will also be a good link site, it does not necessarily follow that this is so. While the trunked radio site may give the desired local coverage, its path back to the switch/controller may be obscured. It is good practice to involve the link designer at the site selection stage.

If line of sight cannot be established visually, then there are two ways to proceed. Either an attempt can be made to establish the link in two hops via an appropriate third site or it may be that the ocular inspection (as the binocular check is more formally known) may be supplemented by a link survey. Where analog links are being used, it must be remembered that for each link added to the chain, there is an additive noise component from each, and so the number of links in the chain needs to be kept to a minimum.

Often line of sight (or something approaching line of sight) will exist even though this was not established visually. This is particularly true in many of the world's large smog-filled cities.

A link survey can be done simply by placing a transmitter at one site and measuring the received field strength at the distant end. This level can then be compared to the free-space path loss to determine if there is any significant obstructions. Quite often it will not be practical to do the path survey at the same frequency as the final link; but by using the formula below, the loss can be corrected for frequency.

The free-space path loss L is such that

$$L = 32.5 + [20 \times \log(d)] + [20 \times \log(f)]$$

where L is the path loss in decibels, d is the link distance in kilometers, and f is the frequency in gigahertz.

So if a link survey is done at frequency F_s, F_a is the actual link frequency, and L_s is the loss at the survey frequency, then the actual link loss (L_a) at frequency F_a will be

$$L_a = 20 \times \log(F_a/F_s)$$

The survey transmitter is often simply a mobile connected to a Yagi antenna. The equipment used for these surveys must be accurately calibrated. This means that the transmitter power must be measured at the beginning and end of each survey (any discrepancy here necessitates a resurvey) and that the loss of the cables and connectors is known. It is good practice to have a selection of calibrated cables, with their losses clearly marked. Connectors contribute to the losses and must be accounted for. Finally, the gain of the antenna must be taken into account.

At the receive end, much the same applies. Antenna gains and cable losses must be accounted for. The receiver must be a device that has a current calibration, and it should be checked regularly (in the case where the receiver is a mobile and the levels are read from the limiter voltage, the receiver calibration should be done before each measurement, because considerable drift can be expected).

RANGE GAIN

The range gain of a link system is the total gain of the microwave system. This gain is made up of the RF power level or the transmitter, plus antenna gains and minus any losses in the cables and connectors. The gain is expressed as the RF level at the receiver above the threshold level for a route over which the path loss is zero.

WHEN LINE OF SIGHT IS NOT POSSIBLE

It is not actually necessary to achieve a line-of-sight path. This is because the RF energy disperses as it travels and because energy traveling just a little skyward of the line of sight encounters thinner air, and hence a lower refractive index. The effect of this is to cause this energy to refract (or curve) downwards. This ducting allows the RF to propagate further than line of sight. To approximate this effect, a traditional way of accounting for it has been to imagine the radius of the Earth as being larger than it actually is (hence yielding a larger value for the line of sight distance). Commonly, a

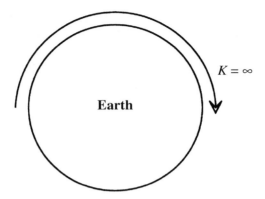

Figure 9.1 The effect of $K = \infty$.

curvature of 4/3 (meaning line of sight based on an Earth radius 4/3× bigger than it actually is) is used.

However, if the path is further than the line of sight, a penalty will be paid in terms of reliability. In abnormal atmospheric conditions (such as temperate inversion) the propagation may be such that the effective Earth radius decreases even below the line of sight, and so the path will fail. Traditionally, designers of links have taken into account various possible values of effective curvature (designated K factor, where K is the ratio by which the effective Earth radius is increased) from $K = \infty$, $K = 4/3$, $K = 2/3$. $K = \infty$ corresponds to perfect ducting where the radio path follows the Earth curvature, as in Figure 9.1.

FREQUENCIES

Link frequencies vary from megahertz to 38 GHz (and rising). From the path loss equation it can be seen that the path loss will increase with frequency and that doubling the frequency results in a 6-dB additional path loss. The choice of frequencies is more limited in tropical areas where the high rainfalls make anything much above 11 GHz too susceptible to dropouts in heavy downpours, to be usable for anything but the very shortest of links.

The most widely used microwave frequencies are given in Table 9.1.

Frequencies below 1.0 GHz are sometimes available for links, but there is a current tendency by the regulators to push link designers above 3 GHz, which is currently considered to be the upper limit for mobile operations. Some years ago, 900 MHz was a widely used link frequency, but mostly these systems have been recovered to make way for cellular and trunked radio applications.

Although high frequencies are associated with high link losses, it is possible at these frequencies to obtain antenna gains that are unheard of lower in the band. Microwave dishes can offer 35 dB and more of gain; and with this gain comes high directivity, which means that the link frequency can be reused with only a minimum

TABLE 9.1   The Most Widely Used Microwave Frequencies

Frequency (GHz)	Usage
2	General purpose, long-haul, small capacity
6–7	General purpose, long-haul, medium capacity
10	Small capacity
11	Medium capacity, medium haul
18, 23, and above	Short haul, various capacity

of spatial separation. The use of polarization will enhance the reuse capabilities. In fact, it is generally possible to reuse the link frequency on every second hop.

DESIGN SOFTWARE

A wide variety of commercial microwave design software tools are available that will automatically look at all these parameters and more. If a number of links are to be designed or if there is any reason to consider a link marginal, then these must be used. The software comes in two basic forms. The first is one that requires the manual entry of a selected number of reference points and their elevations along the proposed path. These are then used to calculate the path loss and the existence of any obstructions. Notice here that "line of sight" does not ensure that there will not be obstructions in adverse atmospheric conditions, and the software will look at the worst-case scenario. Alternatively, fully automated versions exist that contain digitized regional maps, and all that is necessary to input are the coordinates and the height of the link ends.

INTERCONNECTING TO THE LINK

Generally, in a trunked radio environment the connections will be four-wire E&M. This expression derives from its original telephony usage and refers to two wires for the transmission circuit, two wires for the receive circuit, and "E" (or ear) lead plus an "M" (or mouth). The basic functionality, as well as the terminology, has survived to this day. Radio paths, of course, only have a send and receive channel, but the functionality of the E&M leads can be derived in a number of ways, including via subaudible or superaudible tones sent over the voice channels. Another way is to send inband tones and to notch them out at the receive end as part of the voice processing. On modern digital microwave links the E&M functions will be sent over separate signaling channels.

To illustrate how this configuration might be useful in a radio environment, consider the controlling of a base-station voice channel. In order to conserve power, it will be desirable to switch the base-station voice channel off when it is not in use. The channel activity is readily monitored by looking at the state of the mute lift, which is ordinarily presented at the transceiver as a contact that opens or closes in response to a carrier being detected. Some systems, however, do not permit switching and require a continuous duplex path to be provided.

From the trunked radio controller, it may be desired to activate the voice channel, transmit and receive audio, and then drop the call if there is no carrier present from the called mobile within 20 ms of its instruction to transmit an acknowledgment on this channel.

This can be done with wires as shown in Figure 9.2.

The four-wire E&M concept has been around for a long time and was a standard means of signaling over copper wires. In telephone local circuits, often the E&M wires were a pair of physical wires. For cable routes, because copper is expensive the E&M part of it would mostly be derived, using the voice channel wires. A simple derived circuit is shown in Figure 9.3.

Many variations on this type of derived signaling circuit existed. Unfortunately, there were also many different ways of signaling, and there are no firm standards for the states that the E&M wires can assume. For example, the M end of the circuit in Figure 9.2 has two states: open circuit and ground. Variations of this concept include open circuit and battery, battery and closed circuit, and open circuit and loop back, and of course each of these can have its meaning reversed (open circuit might mean "on" and ground might mean "off," or vice versa). To further confuse matters, some E&Ms use three states such as ground, battery, and open circuit.

You would be excused for thinking that with these limited number of alternatives, modern E&M hardware would be programmable (or at least strapable) for each of the alternative configurations. But no. Generally, the E&M equipment of today will have one or a few of these options. This means that for a successful E&M connection it is necessary to match the signaling type (if possible). Sometimes matching is not possible and it will be necessary to build simple relay circuits to match the conflicting signals.

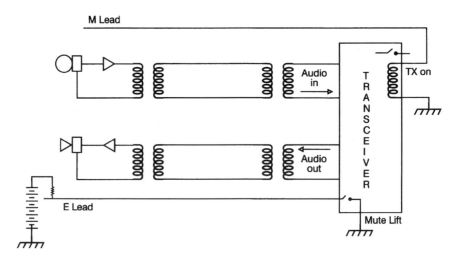

Figure 9.2   A four-wire E&M configuration.

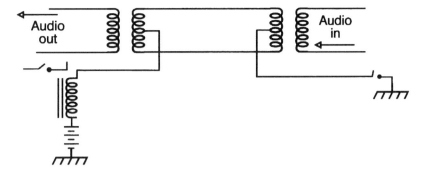

Figure 9.3   A derived DC signaling circuit.

Some, but very few, trunk radio systems have T1 or E1 interfacing capability built in. Most have two and four-wire E&M. If it is necessary to interface to a E1/T1, then it will be necessary to use a multiplexer or MUX. The MUX is a piece of hardware that interfaces between the E1/T1 data stream and the two- or four-wire analog world. The MUX is not usually an integral part of the microwave equipment and is often provided by a different supplier. A wide variety of these are available from various suppliers, and their purpose is to take an E1 or T1 data stream and convert it to four-wire E&M. Again it is very important to be sure that the E&M of the MUX matches that of the equipment. Signal levels could be a problem, but most MUX equipment can accommodate a wide range of input and output levels.

Some MUX equipment can also convert to two-wire E&M, which can be useful in some circumstances (primarily where the equipment to be interfaced is expecting to see a conventional telephone line).

As with four-wire E&M, be aware that there are several versions of two-wire E&M. The most common is one that simply gives a telephone line with loop dialing, dial tone, and ring tone. Other two-wire systems are designed to connect telephone switches to each other at the two-wire level. They will have signaling capability, but it will be very different from a telephone loop.

PSTN INTERCONNECTION

Trunk systems differ markedly in their PSTN interfacing capabilities. Some have facilities that are little more than telephone patching into the network, while others have PSTN-style switch capabilities with all the signaling enhancements of such a switch. Some trunked radio switches are limited to only two- and four-wire E&M, while others have E1/T1 capabilities.

Those with more sophisticated switches have group level access, which enables them to connect to the PSTN in the same way that a cellular switch would. These are usually capable of routing to multiple PSTN destinations, which can be used to maximize the network security and, in some cases, to reduce trunk call costs.

For small systems, interconnection is simple because it is just connected at the subscribers level, in exactly the same way as a landline subscriber. There are, however, variations on this type of connection which can offer advantages.

To obtain a better call quality, it is usually possible to interconnect at the four-wire E&M level. At this level there is a reduced probability of hung lines (lines that fail to clear down after a call); and because no call hybrid is involved, the call voice quality will usually be higher. This level is effectively the subscriber's level.

The main disadvantage of connecting at the subscriber's level is that an incoming call from the PSTN will need to overdial (send extra digits) after connection to reach the desired mobile. This can ordinarily only be done when DTMF (tone dial) facilities are available.

The next level up is to connect at a PABX level, so that each mobile number has a unique calling code. A typical PABX number may be 9234-4XXX, where the last digits (XXX) are individually allocated to the respective mobiles. This is called in-dial or sometimes inward direct dialing (IDD) or direct in-dialing (DID). If this option is chosen, it is best to connect at the four-wire E&M level, or better still for bigger systems at the E1 or T1 level.

E1 and T1 levels are the digital levels with 30 and 24 time slots, respectively. When connecting at this level, it is best to use an appropriate signaling system such as R2 or SS7, although it is possible to retain E&M signaling. E1 and T1 are suitable both for PABX level and the next highest to a local exchange (class 5 in the United States) or even to trunk level. The difference is that at local exchange level a code (usually four digits) will be assigned. For example, the switch may be 9234-XXXX, where XXXX represents the four digits of the trunked radio switch. At the trunked level a full access code is assigned. For example, the access code may be 0114-XXXX-XXXX. Here the termed "trunked level" refers not to the trunked radio switch but to the level in the PSTN hierarchy. Naturally, this level would only be applicable to the very largest of the trunked radio networks (but such systems do exist).

Finally, in very large trunked radio systems (such as nationwide) there may be a need for multiple interconnections to the PSTN—for example, in each major city. The trunked radio internal switching network may even have the capability of routing calls via its own links to and from these cities. When this stage is reached, the PSTN connection is virtually the same as a cellular interconnection, and again would only apply to large nationwide or statewide systems.

PABX INTERCONNECTION

Because a large proportion of trunked radio customers are corporate, it follows that they are likely to also have a PABX. It is often a requirement to link these PABXs to the trunked switch independently of the PSTN (mainly to avoid PSTN charges). This can be done in two ways. The first is to connect the trunked radio switch directly (via land lines or radio) to the PABX, and this would be the way it is usually done. The other way is to use an "off-air" device to link a port on the PABX to the trunked network via a trunked radio.

CHARGING

Where the interconnection is via a PSTN subscriber line, the charges to the trunked radio operator will be standard PSTN charges, and the only "variable" to consider is what to charge the mobile subscriber for incoming calls. At higher levels of interconnection, it may well be that the PSTN operator will view the trunked radio system as a cellular-like interconnection and may want to structure the call charging accordingly. If this is done, the trunked radio operator must be fully aware of the capabilities of the billing system, because mutual access to billing records is often required for reconciliation.

It may be that an enhanced billing package may be needed just to satisfy the rules of the interconnection agreement.

DERIVED CIRCUITS

Leased lines can be very expensive, particularly when they are long haul lines. One way to improve the economies of this is to lease 64-kbit/s lines and, using a transcoder, derive some extra channels. By multiplexing down the voice channels, it is possible to get two, four, or even eight voice channels onto a standard 64-kbit/s link. There is, however, a price to pay. The more the compression, the poorer will be the voice quality. Also, cascading compressed circuits (using two or more in series) further degrades the voice quality, and it is not recommended.

It is worth noting, however, that while compression of the 64-kbit circuits is often an economically viable way of getting better use of leased circuits, it generally is not the best way to go when you are putting in your own links (except for very long hops). In microwave systems, added capacity is relatively inexpensive, and it would be very much cheaper, for example, to use an 8-mbit/s microwave link than to use a 2-Mbit/s link with 4× capacity gain—except on very long haul routes.

DATA

Data cannot be transmitted directly over an analog radio link, and what is needed in this case is a modem. A modem effectively converts the "1" and "0" logic states into audible tones that are then transmitted over a standard radio link. Radio modems are ordinarily different from wireline modems because the harsher environment of radio requires a more robust error correction technique. Also, the long fades that can be anticipated in the radio environment mean that long blocks of code may need to be re-sent, and the ability to do this is built in. Additionally, there are often timing constraints in data communications that can be violated by the radio propagation and internal signaling delays. The modem must be able to handle these interfacing problems.

DATA ON THE CONTROL CHANNEL

Most systems that have control channels allow for short messages and status messages to be sent over the control channel, in a packetized mode. This is usually the fastest data mode, but it has limited capacity due to the control channel availability (when it is sending data, it is not doing its control functions).

Longer data messages will be sent on the voice channels; and depending on the modulation technique used, the data rate can range from 300 baud to >9,600 baud for analog systems, and much higher for the latest digital networks.

INFRA RED LINKS

A recent development has been the availability of infra-red links. Like microwaves, these links are transmitted through the air, but have the advantage that they require no licensing and so can be deployed very rapidly. This makes them especially suitable for emergency links, unplanned installations, and short hops where cable is not an option.

The light is in the 860–920 nm band and is generated by lasers. The whole unit is about the size of a shoe-box and can handle data rates up to 622 Mbps. The cost is a function of the capacity, but broadly the range is $7000 for a low power low capacity link to $50,000 for the highest capacity high power links. The links can serve as a direct replacement for fiber optics. Naturally the link requires a clear line of sight to operate at all.

Reliability figures in the range of 99.1% to 99.9% are quoted by one supplier, ICS. These figures are relatively low compared to most telecommunications links, but would certainly be adequate for temporary and emergency links.

A specification for a link supplied by Infrared Communications Systems Inc., for a $4 \times T1$ link (E1 links are also available) is:

Range	1000 meters
TX power	30 mW
RX sensitivity	−45 dBm
Capacity	$4 \times T1$
Power consumption	50 watts

The link is transparent to the rest of the system, and "looks" like a part of the transmission cable, and no special interfaces to it are required. The links come in a number of models with different outputs which give different ranges. These models include those for 200, 500, 1200, 2000, 4000, and 6000 meters.

CHAPTER 10

POWER AND PROTECTION

The power requirements of a base-station site can, to a first approximation, be taken to be the power consumed by the transmitter. The total load is about three times the total RF power (measured at the final stage of the power amplifier). For small stations, this will only be a few hundred watts, but it rises to kilowatts for larger stations.

Most of the base-station equipment will be 12-V DC-powered, but there may be a need for some 24-V or 48-V power for switching, modem, and MUX equipment for the microwave. If these supplies are needed, it is best to derive them from the 12-V DC rail so that only one set of backup batteries is required. Most base stations will be powered from the AC mains and have rectifiers to convert the voltage as necessary.

The usual way to power the base site is to float it off a battery bank, which is charged by mains AC. The AC supply can be single-phase if the load is in the low kilowatt range, while two- or three-phase supplies will ordinarily be required for higher loads.

The batteries should have a reserve capacity of at least 4 h on full load (more in places where the power supply is unreliable). Lead-acid batteries are most commonly used because of price. These batteries have a life that is a function of how often and how deeply they have been discharged. Deep-cycling a lead-acid battery will greatly reduce its life; and so when calculating the capacity required, account should be taken of this by assuming that it will be discharged to only 80% of its rated capacity. To prevent complete battery discharge (which does serious damage to the battery), devices can be fitted that disconnect the batteries before serious damage is done.

For most trunked radio systems the manufacturer can supply a power rack as part of the hardware. These power supplies are mostly second-sourced and are marked up extensively. It will be well worthwhile investigating a direct-sourced alternative.

Although car and truck batteries are commonly seen in conventional mobile repeater installations, they are definitely not recommended for trunked applications. Telecommunications-quality sealed batteries are to be preferred. Similarly, the practice of using computer-grade UPSs as a power backup source is only suitable for

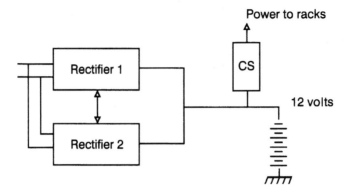

Figure 10.1 A typical base station power supply.

low-priority systems. Telecommunications-grade inverters cost only a little more and are far more reliable, especially if they are brought into service for prolonged periods.

A simple power supply is shown in Figure 10.1. The rectifiers in the diagram are configured in the 1 + 1 redundant mode, so that the failure of one will not interrupt the supply. Generally, for trunked applications an N + 1 redundancy (one more than is needed for the maximum load) will be sufficient.

The power supply should additionally have some surge protection against power line surges and lightning strikes. Lightning strikes occur on the Earth at the rate of 100 per second, or 40,000 per day. Some areas are far more prone than others to strikes, but no area is totally free of the hazard. Strikes can occur directly onto the powerlines, or indirectly through induction. Power-line surges occur during switching and line faults.

The simplest form of protection is a shunt surge diverter. These are connected in parallel to the powerline and dissipate surges as they occur. One such device by CRITEC has five internal metal oxide varistors (MOVs), each separately fused and

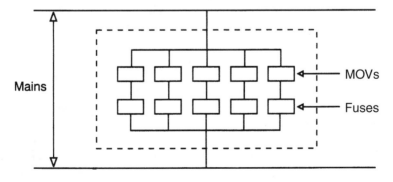

Figure 10.2 A shunt power protector.

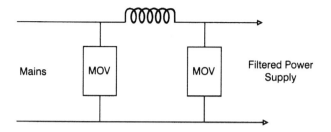

Figure 10.3   A surge reduction filter.

each with an LED to indicate correct functioning (see Fig. 10.2). The LED indicators are most important because the effectiveness of a surge diverter degrades with usage.

More elaborate protection is available through filtered surge protectors. These are essentially a π filter as seen in Figure 10.3. Although MOVs will take the brunt of a surge, they have a relatively slow response time; and because lightning surges can have rates of current rise as high as 10 kA/μs, there will be a period of time between the strike and the clamping of the MOV when the equipment can be subjected to high voltages, and a filter can prove invaluable. The extra protection provided by these filters is probably justified for switching centers and major base station sites.

It must be stressed that all surge protection devices require a good ground connection to work efficiently. It is also essential that all grounds are connected together to avoid damaging ground loops.

MAINS SUPPLY

For small systems consisting of a few channels, the mains supply will invariably be single-phase. For larger systems the power authorities may insist that two- or three-phase connections be used. Multiphase power distribution systems use less copper and smaller transformers than single-phase ones.

Two-phase supplies are ordinarily treated as two separate single-phase sources and require no special treatment. A three-phase supply can be treated as three separate single-phase supplies, but the operator can gain some economy of scale by deriving the DC supplies for the radio network from three-phase rectifiers, which will be cheaper on a per-ampere basis than the single-phase alternative.

It should be noted also that it is possible to derive a three-phase supply from two phases. A special transformer to derive the third phase will be needed.

SOLAR POWER

Remote hilltop sites are often tempting for good coverage, but it is not unusual that power is not available at such places. In some instances, solar power may be the answer.

The angle of tilt for a solar cell is optimized if it is equal to the latitude of the site tilting toward the equator. In the case where the latitude is less than 15 degrees, a

15-degree tilt should be used to ensure the free run-off of water and dust particles. In the case of low-latitude sites the tilt need not necessarily be toward the equator, but rather optimized for local conditions. For example, if the site has a rainy season characterized by afternoon rains, the tilt might well be chosen to face the morning sun, thereby optimizing the solar energy collected in the rainy season.

The solar radiation measured above the ionosphere (the region 80–180 km above the Earth) is 1,370 W/m^2, so this represents an upper limit on the energy that can be collected. The actual power that can be extracted by a solar cell is highly dependent on its geographical location and local weather conditions. Meteorological records are kept worldwide of the amount of surface solar radiation. Known as *insolation* (*incoming solar radiation*), it is measured in various ways. The value of interest in solar panel design is the peak-hours per day, which is measured in kilowatt-hours/square meter. The values of insolation vary from about 7.0 at the equator to 3.0 in Northern Europe. Very large local variations may also occur due to local cloud cover and rainfall. Naturally, the values vary also with the seasons, and for reliability the worst month is the one that should be designed for.

Assuming a 12-V installation, the solar panel will be good for charging at the following daily ampere-hour level.

Ampere-hours/square meter of panel = insolation × panel efficiency/12; or for panel with a known peak ampere rate, it will be that rate multiplied by the insolation.

Solar panels are available in various sizes up to around 80 W (peak power) per panel. The cost of the energy supplied by solar means is about twice the cost of that provided by the regular power grid. However, solar costs are slowly falling, while the main grid supply costs are rising.

Batteries and their maintenance are a major part of the overall cost of a solar installation. The battery not only has to have sufficient reserves to keep the equipment going when the sun is down, but also needs reserves for overcast days. It is usual to design the solar battery to have reserves of five days.

The life of the battery will be seriously compromised if it is allowed to fully discharge, so the five-day capacity should be calculated after allowing for the discharge depth. The permissible depth will depend to some extent on the actual characteristics of the batteries in use, which for lead-acid batteries is about 80%. It is best if some protective device is used to ensure that the batteries are disengaged before they are fully discharged.

Even though the solar panels are tilted and washed by the rains, some dust and dirt will stick to the surface. To ensure peak efficiency, solar cells require regular cleaning, and it is advisable to do this once every three months. The safest way to clean them is to use water and a soft cloth only. Soap and other detergents should not be used, unless specifically recommended by the manufacturer.

BATTERIES

The batteries need to be able to provide power at nighttime and at times when, due to weather, the incident solar energy is well below the design levels. The amount of reserves can vary from about two days in dry arid regions to typically five days in coastal regions. This can result in quite substantial battery requirements.

EXTENDING BATTERY LIFE

It is sometimes worth taking a little more time with a solar installation to get the most out of it. For instance, a base site will have channel capacity adequate for the peak hour, and thus it will have excess capacity most of the time. If remote control facilities are provided that can switch off excess channels in the quieter periods, a good deal of battery power can be saved. Also, during periods of prolonged cloud cover, it may be practical to take some channels out of service for the duration, to conserve power. A combination of automatic control and manual operation will achieve the best results.

Battery life is decreased by around 50% for every 10°C above 20°.

WIND POWER

There are not very many places where wind power is suitable for trunked radio applications. Some small installations have been wind powered, but usually only in isolated windy places. Particularly on small islands, there can be a problem with bird droppings on solar panels, which renders them unreliable; wind power may be used as an alternate or supplementary power source.

CABLES

The cables used for power must be selected for their current-carrying capacity (otherwise they will overheat) and, in some cases, to minimize the voltage drop (this generally only applies when the cable runs are long). Cables are available in standard metric or American wire gauge (AWG). The ratings of the more popular cables are listed in Table 10.1. The voltage drop can be calculated from the resistance/meter:

$$\text{voltage drop} = \text{current carried} \times \text{ohms/meter} \times \text{length in meters}$$

TABLE 10.1 Ratings of the More Popular Cables

Metric (mm²)	AWG	Current Rating (A)	Ohms/Meter
1.0	18	10	0.0194
1.5	16	15	0.0131
2.5	14	20	0.0078
4.0	12	30	0.0050
6.0	10	35	0.0033
10.0	8	50	0.0020
16.0	6	70	0.0012
25.0	4	90	0.0007

There may be a specific voltage target; but if not, the voltage drop should not exceed 10% for most applications.

AIR CONDITIONING

Air conditioning is not always fitted to trunked radio sites, but it will significantly increase equipment lifetimes and is recommended.

For practical purposes the air-conditioning load will be roughly equal to the total power consumed by the site.

It is not usual in trunked radio system to allow for battery backup for the air conditioning (mainly because most of the equipment is rated for continuous operation without air conditioning), and thus the cooling will be down during power outages. In stations with emergency power generators, the air conditioning should be connected; and as a precaution where possible during extended outages of mains power without air conditioning, fans can be used in hot climates to cool the station. This applies particularly to the larger stations, which will be the ones to generate the most heat.

LIGHTNING

A fully developed thundercloud will generally be positively charged at the top and negatively in the middle. Thunderclouds or cells generally occur in clusters, arising independently of each other. Cloud-to-cloud discharge is common, but it is the cloud-to-Earth discharges that are of most interest to us. A local region of positive charge at the bottom of the cloud is often noted, but it is not particularly important for this consideration of the subject.

The bottom of the cloud will typically be at a potential of around 20–100 MeV relative to the ground. A strike happens when a small initial discharge makes its way along a "leader" which conducts some negative charge to the ground at around one-sixth of the speed of light, in a stepwise fashion moving about 50 m on each step. As the leader leaves behind it an ionized high-conductivity path, it prepares the way for the main strike which starts when the leader initially contacts the ground and then travels upwards from the ground toward the cloud.

Once the first stroke has occurred, others may follow at intervals of around a few hundredths of a second.

The strike will typically carry around 10,000 A and deliver to the ground a total charge of 20 coulombs. About 90% of strikes deliver a net negative charge. The cloud, having discharged, can build up for a new strike in as little as 5 seconds.

THE HAZARDS

Because trunked radio systems work best on high and prominent sites, they will be subjected to significant risks of lightning strikes. Although there are devices on the market which claim to be lightning diverters, there is in reality little one can do to

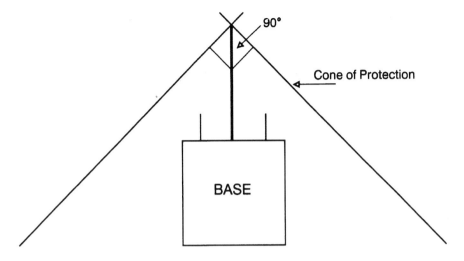

Figure 10.4 The zone of protection provided by a lightning rod.

avoid a strike, if one is imminent. The best that can be hoped for is to be prepared for the strike and to minimize its impact on the equipment.

The first line of defense against lightning is the lightning rod. This needs to be mounted higher than any of the equipment it is to protect, and traditionally it is regarded to have a cone of protection defined by a right-angled cone, as seen in Figure 10.4.

A lightning strike can consist of a current pulse of more than 200 kA, with a pulse width from 10 to 350 μs. In order to sink the current from the strike, it is essential that a good ground be provided for the lightning rod and that the connecting cable be adequate to carry the surge current. The waveform of the pulse is such that even the inductance of a straight cable will be significant, and any bends or kinks may present themselves as high impedances.

Because the lightning rod conductor will run more or less in parallel with the antenna feeders, the high current of the strike can induce considerable currents into the feeders. A shielded conductor, which minimizes induced fields, can reduce this problem significantly. However, coaxial surge protectors should be fitted to all feeder cables to minimize the chance of damage.

The most vulnerable parts of the systems are the antennas and cables (including power cables).

GROUNDING

Effective grounding is essential for good performance of any protection system; and while the basic rules are quite simple, they are frequently misinterpreted.

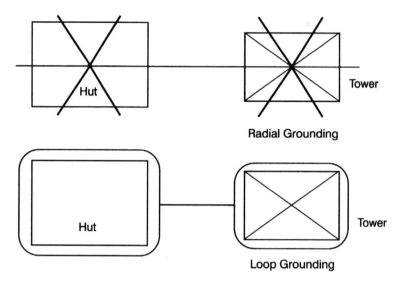

Figure 10.5   The cone of protection of a lightning rod.

The point of connection to ground must ensure a low-resistance path to the Earth. This can sometimes be hard to verify, because most ground conductivity meters measure the DC resistance only, while the waveform of a lightning pulse is such that it is really the impulse reactance that is of significance. Nevertheless, a low DC resistance is generally (but not always) an indication that the connection is a good one. Impulse ground resistance meters are available from a few suppliers.

An ideal grounding system will either loop or radiate around the area of the base site and tower, as seen in Figure 10.5. Where buried conductors are not practical, a number of grounding rods can be placed around the building and tower and linked together with a cable.

What is most essential is that all grounding systems are linked together, and it is preferable if the number of connection points can be kept to an absolute minimum.

With lightning currents as high as 200 kA, even the small resistance that might exist between two isolated grounding systems can result in high voltages developing between the two systems.

The internal building grounding system must also be bonded to the outside one, and any subsystems must also be bonded.

STATIC AIR CHARGES

The air has a nearly constant field potential of around 100 V/m (measured vertically). This potential gradient is continuous up to about 50 km, where a region of high conductivity prevents a further buildup of voltage. At this altitude the voltage is around 400,000 V with respect to the Earth. The natural conductivity of the air which is brought about largely by particles ionized by cosmic rays ensures that this voltage difference causes a current of the order of 1,800 A to be flowing continuously between the air and the ground. This is the equivalent of a 700-MW generator!

The source of power for this field-strength generator is the lightning discharges delivered continuously over the surface of the Earth. The field peaks (by about 15%) at 7:00 P.M. universal time (UT) and is at its minimum at 4:00 A.M. UT. Notice that the field follows the intensity of the thunderstorm activity worldwide and that the peak occurs everywhere at the same time (local time is not a factor).

When the air is dry the conductivity goes down and the potentials build up. These gradients can destroy high-impedance, modern integrated circuits in seconds. All printed circuit boards should be treated as static-sensitive, and the proper handling procedures should be followed. Damage to integrated circuits (ICs) can be immediate; but sometimes internal damage can occur, which may take months or even years to lead to ultimate failure.

Precautions that must be observed are as follows:

- Transport all PC boards in conductive plastic bags.
- When handling PC boards, ensure that a wrist strap is used and that conductive mats are used on the benches.
- Ensure that wrist straps and connection points are fitted to all equipment racks.
- Keep all spare ICs on conductive foam rubber sheets. Aluminum foil is not as effective because some leads may not be contacting the metal and the foil will transfer charges very efficiently during wrapping and unwrapping.

Dry conditions also allow local charges to build up on insulated objects such as carpets and clothing. Discharges from these sources can induce very high instantaneous currents, which can permanently damage ICs by puncturing semiconductor surfaces.

CHAPTER 11

TRAFFIC ENGINEERING ON TRUNKED RADIO SYSTEMS

Although the traffic engineering aspects of trunked radio are often dismissed as trivial, they are far from it. To take a simplistic approach like a rule that says there will be 70 users per channel, or even to use the Erlang B table (located at the end of this book) will cause the system to be designed nonoptimally. Does this matter? Considering that a channel costs from \$10,000 to \$15,000, a network with only 10 channels too many will cost upwards of \$100,000 more than one optimally designed. On the other hand, one that is 10 channels short of the optimum will provide poor performance, leading to low customer satisfaction, which in turn will lead to fewer subscribers and therefore less revenue.

The basic unit of traffic is the Erlang. An Erlang can be defined as one circuit (or channel) occupied for a given time. So if at any one instance, seven channels are occupied, then the *instantaneous* traffic is 7 Erlangs. If, however, over the next hour the number of channels occupied *on average* was 4.7, then the average traffic for that system would be 4.7 Erlangs.

WHICH TABLE?

There are a number of different relationships that are used to calculate circuit requirements for a given traffic. By far the most commonly used is the Erlang B; unfortunately, it is also widely misused. Erlang B tables can be properly applied to circuit dimensioning for the full availability trunk circuits to and from the switch, provided that the following assumptions are met.

Assumption 1. That the switch is full availability.

This means that any call which appears on one side of the switch will always have a path through the switch. This condition is true of most modern switches, including those used in trunked radio (but is not true of modern wire line subscribers switches at the subscribers end). Full availability switches are also referred to as *nonblocking*.

Assumption 2. That subscribers generate calls individually and collectively at random.

This assumption is true for most large populations of callers. It may not necessarily be true if the system has less than 100 users.

Assumption 3. Lost calls are cleared with zero holding time.

This assumes that the subscriber, on encountering congestion (no available circuits), will immediately hang up *and will not attempt to redial*.

The last part of Assumption 3 does not hold up well in the real world. It is common experience that when a caller encounters any problem with a call, the immediate reaction is almost invariably to redial. Unless all the assumptions agree with those of Erlang *B* table, it is not the table to use. For trunked radio systems with queuing, Assumption 3 definitely does not apply.

A. K. Erlang was an engineer with the Copenhagen Telephone Co. who developed a series of relationships for the call-carrying capacity of a group of circuits under a number of different assumptions about the nature of the traffic. The Erlang *B* model (which is also sometimes referred to as Erlang *A*—it's best not to ask why) was simply the first of these models. It is simple to calculate and produces good results with the step-by-step switches that were common at the time. It does begin to fall down in the modern switching environment with relatively fast tone dialing and telephones equipped with a redial button, making the assumption that a failed call attempt will not be repeated somewhat dubious.

The Erlang *B* table, which can be found in the tables section at the end of the book, is derived from the formula

$$B = \frac{P(N)}{P(0) + P(1) + P(2) + \cdots + P(N)}$$

where *B* is the blocking (or congestion) probability and *P* is the probability that the *n*th circuit will be blocked when offered a traffic equal to *A* Erlangs:

$$P(n) = \frac{A^{n/n!} \times e^{-A}}{N!}$$

And so *B* simplifies to

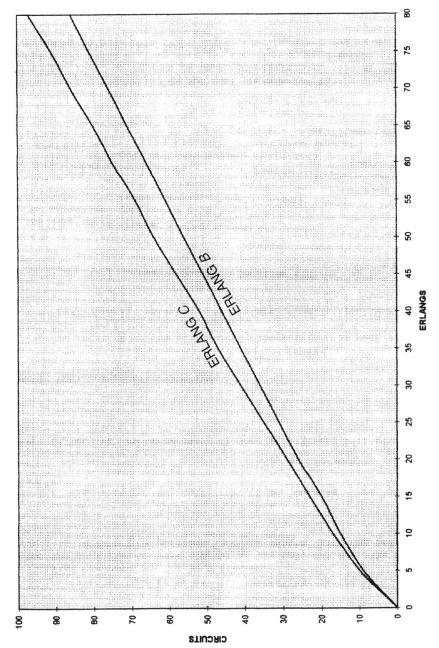

Figure 11.1 The Erlang *B* and Erlang *C* circuit requirements for a GOS of 0.05.

129

$$B = \frac{A^{N/N!}}{1 + A + A^{2/2!} + A^{3/3!} + \cdots + A^{N/N!}}$$

The expression for B can be compactly put into computer code, as seen as the last subroutine in the code that follows later in this section.

Erlang produced a number of equations for various traffic behaviors; and one, the Erlang C, has the assumptions of the B table, except that it is assumed that the caller will retry if unsuccessful. Except for the case when the number of circuits is very small, the Erlang C table, which assumes that unsuccessful traffic is immediately re-presented, will require more circuits for the same traffic. This means that use of the Erlang B table for base-station dimensioning will always result in a worse grade of service than was intended. To see what difference there is between the two tables, consider Figure 11.1.

The Erlang C function is closely related to the Erlang B equation by the following relationship:

$$C(A, N) = \frac{B(A, N)}{1 - (A/N)(1 - B(A, N))}$$

where $B(A, N)$ is the probability of call loss for N circuits offered A Erlangs, as calculated using the Erlang B relationship and $C(A, N)$ is the corresponding Erlang C function loss. An important thing to grasp with the Erlang C relationship is that it assumes that no calls are actually *lost*, but rather that they will be subject to retries until they are carried. Also, the Erlang C function has no physical meaning for the case when the traffic is greater than the number of circuits, because this would result in the build up of an infinite queue.

Because of the ease with which modern phones can redial, it does appear that the Erlang C table should be used and not the common Erlang B table, because the Erlang B table seriously underestimates the circuits required. To a good approximation, a queued system can be dimensioned using Erlang C tables. In general, however, Erlang C is the most appropriate table for connections to and from the PSTN, while a queuing calculation should be done for the RF channels.

DIMENSIONING

Dimensioning is the process of allocating circuits and their quantities to particular routes. There are a number of things to consider when dimensioning a trunked radio system. We begin the discussion here with a single-site system, and then later it will be expanded to include multisite systems.

The first thing that needs to be considered is the number of RF channels needed. To do this we first need an estimate (or measurement) of traffic. Traffic, measured in Erlangs, can be defined in a number of ways.

The instantaneous traffic is the number of calls carried at the time of interest. So if 15 circuits are active, then the instantaneous traffic is 15 Erlangs. More generally, an average peak traffic is of interest. There are numerous definitions of average traffic. One common definition divides the day up into half-hour slots and finds the busiest half-hour. The total number of call minutes in that half-hour is then divided by 30 to give the "average" calling rate for this period.

The traffic measurements may be available as part of the software package that comes with the operating system. In some cases it is provided as an "optional extra," and from some vendors it is part of the billing package. For systems that have autonomous base sites (ones that can switch local traffic locally without reference to a central controller), it may be necessary to download the traffic data from each base site individually on a regular basis. In any case, it is most likely that some traffic information will be available.

Once the traffic is known, it is then possible to calculate the number of circuits (or channels) needed to carry it, after the exact nature of that traffic is determined.

TO QUEUE OR NOT TO QUEUE?

Most systems that have a control channel will also have fully automatic queuing. Queued traffic behaves in a very special way, and in particular it *cannot* be handled as grade of service traffic using the conventional Erlang *B* tables. This is particularly so if we look at one of the major assumptions of the Erlang *B* table, which is that any unsuccessful calls *never* retry.

In a queued system, calls that cannot be allocated a channel immediately are placed in a queue, and they are connected when a resource becomes available. It should be noted that while this ensures maximum convenience for the subscriber, it is possible in theory for such a system to develop an infinite queue. Imagine for a moment a system that has 10 RF channels, but which is steadily offered just a little more than 10 Erlangs. Assume that it is offered 12 Erlangs. This theoretical system will carry 10 Erlangs (all channels busy) constantly and queue up another 2 Erlangs each hour. Now, assuming that this traffic is consistent for a full 24-h period, it can be seen that while 10 Erlangs is constantly carried, the queue builds up at a rate of 48 Erlangs per day, and in time it will grow to be infinitely large.

In real systems this does not happen; but because of the potential for *very large* queues to build up, most systems have a limit on the number of users that can be queued and the total time each is permitted in the queue. In the worst case, badly overloaded systems will effectively cease to queue most users most of the time.

HANDLING QUEUES

There are a number of excellent texts on queuing theory, and these should be referred to by those seeking a deeper understanding of the background to this subject. Unfortunately, however, it is not possible to dismiss the theory altogether so simply.

The reason for this is that while conventional traffic tables are two-dimensional (that is, the number of circuits depends only on the traffic and the GOS), queuing tables, are three-dimensional. To determine the number of circuits for a given queued system, it is necessary to consider not only the traffic, but also the desired queue time (the average time that a call is in the queue) and the call holding time. Note that at no point has GOS been mentioned; this is because it has no relevance in a queued system, at least not in the same sense that the term is used in wireline systems. A GOS of 0.05 implies that 0.05 (or 5%) of all calls will be lost. In a properly functioning queued system, no calls are lost; instead, some have delayed access (that is, they are queued). An alternative definition of GOS for queued systems that is sometimes used is that it is the percentage of calls that have to wait in the queue longer than a specified minimum time. It is best, however, not to use the term GOS at all for queued systems because it is liable to cause confusion.

Because of the three-dimensional nature of queued systems, there is really no good alternative to computerized circuit dimensioning. What follows is a routine written in BASIC that will enable the calculation (note that if an interpretive BASIC is used, it will be necessary to add line numbers).

```
' Here is a simple test program for the subroutines
CLS
INPUT "traffic";t
INPUT "queue length";ql
INPUT "callholding time";ch

CALL QCCTS(t,ql,ch,ccts%)
print "required number of circuits=";ccts%
END

' These are the subroutines needed to run the program

SUB pwrthing(A,K%,V#)

' evaluates A^K/K!

V#=1
FOR N%=1 to K%
    V#=V#*A/N%

NEXT N%
END SUB

SUB QUEUER(traffic,circuits%,callholding,Q,waiting)
' returns call holding time and number in queue
IF circuits%<1 OR traffic=0 THEN RETURN
' traffic=the traffic offered in erlangs
' circuits=the circuits provided
' Q is the calculated queue length as the number in the queue
```

```
' waiting is the calculated average time in the queue

IF circuits%>1 THEN
     Vxtot=0

     Vx#=1
     FOR N%=0 to circuits%-1
         IF N%=0 OR N%=1 THEN Vx#=1 ELSE
Vx#=Vx#*traffic/(N%-1)
     Vxtot# =Vxtot#+Vx#
     NEXT N%

     CALL pwrthing(traffic,circuits%,Vx#)
     Po#=(Vxtot#+Vx#/(1-traffic/circuits%))^-1
     Q=Po#*Vx#*(traffic/circuits%)/(1-traffic/circuits%)^2
END IF

IF circuits%=1 THEN Q=traffic^2/(1-traffic)
waiting=Q*callholding/traffic

END SUB

' this routine returns circuits for a given traffic and queue
time

SUB QCCTS(traffic,waiting,callholding,circuits%)
' Neil J Boucher Jan, 1994

' waiting=queue time
IF traffic=0 THEN
                 circuits%=0
                 RETURN
END IF
' does first approximation of circuits

IF waiting<15 THEN G=waiting*.02 ELSE G=.4

CALL ERLANG(traffic,G,Nguess)
circuits%=Nguess
IF circuits%<=traffic then circuits%=traffic+1

CALL QUEUER(traffic,circuits%,callholding,Q,waitx)
' calculates circuits for given traffic and queue
'callholding =call holding time
' loops until new estimate is on the other side of the
correct value
' to the first guess.

IF Traffic=0 THEN
```

```
                ccts%=0
                RETURN
END IF
Ngx=circuits%
IF waitx>waiting THEN K%=1 ELSE K%=-1
Alp%=K%

WHILE Alp%=K%

IF waitx >waiting THEN k%=1 ELSE K%=-1
IF Alp%=K% THEN
Nguess=Ngx+k%
Nguess%=INT(Nguess+0.5)
Ngx=Nguess
END IF
' this prevents convergence to traffic= circuits which makes
the queue infinite

If Nguess%=<traffic THEN
                Nguess%=Nguess%+1
                Alp%=6
END IF

LOCATE 20,10:PRINT"Current estimate of circuits";Nguess%

waitNg=waitx

CALL QUEUER(traffic,Nguess%,callholding,Q,waitx)

WEND
circuits%=Nguess%
LOCATE 20,10:PRINT"                          "
END SUB

SUB Erlang(A,G,N)
' Does Erlang B table for fixed circuits and GOS
' and used here as a first approximation for circuits

' A=traffic in Erlangs
' G= grade of service(or blocking, B)
' N= Number of circuits
IF A=0 THEN
            N=0
            RETURN
END IF
C=1
N=0

B=2
```

```
WHILE B>G
N=N+1
C=1+N*C/A
B=1/C
LOOP

END SUB
```

Note that the routines used to calculate the number of circuits are far from straightforward and that the calculation proceeds by iteration. Also, it can be seen that the Erlang equation is used only as a first approximation, in the process of iteration.

To be a little more precise, on a trunked radio system there are other things that should be considered as part of the call holding time. Once the mobile has received the call on the paging (or control channel) and acknowledged it, is very soon directed to a voice channel. This channel is effectively removed from the channel pool as soon as it is assigned, so that any time from the assignment that is spent waiting for the mobile to answer (whether this be system delays or end-user response time) needs to be considered. This will be small if only network times are involved. So for systems that do not switch to the voice channel until the called party has answered, it will be virtually negligible. However, for those systems that send ring tone on the voice channel, the average response time of the user should be included as call holding time, and this may be quite a few seconds.

While queuing theory is the correct approach for the base-site channels, the situation is less clear for the PSTN/PABX lines. Outgoing calls to the PSTN may be queued and, if so, should be dimensioned using queuing theory; incoming calls are not ordinarily queued and should be treated differently.

Incoming calls can be dimensioned using the Erlang B table, and to a first approximation this is a reasonable approach. However, modern phones are often equipped with a redial button, which would make the Erlang C relationship more appropriate.

Very few trunked radio systems have separate incoming and outgoing PSTN/PABX lines, and the bothway traffic has to be treated as a single entity. In this case the Erlang C relationship will provide a reasonable compromise.

PSEUDO-QUEUED SYSTEMS

Pseudo-queued systems are generally those that do not have control channels but rather have some means of allowing users to grab channels once they become free. This usually involves some active participation on the part of the users (as opposed to fully automatic queuing), and so it can be assumed that not all users will activate it. This is an extended Erlang B situation; and unless a large percentage of users invoke the pseudo-queuing, it would be appropriate to use the Erlang B relationship and to speak of a grade of service for these systems.

TRUNKED RADIO TRAFFIC EFFICIENCY

Traffic (or spectrum efficiency) is often quoted by regulatory authorities as the main reason for the support of trunked radio. To understand this efficiency, it is necessary to look at a real scenario. Take the example of a 50-channel system in an area where frequency reuse is not possible. Assume that the users have an average calling rate of 0.012 Erlangs each and have a call holding time of 15 s. To follow the calculations through, you might like to run the software provided for GOS calculations using Erlang C or use the tables at the end of the book.

Case 1.   A conventional radio system using these channels would effectively consist of 50 individual channels. If we require a GOS of 0.2 (a reasonably congested system), then the Erlang C formula (here it is assumed that there is no queuing, but the users keep trying until they get through) tells us we can have 0.2 Erlangs of traffic per circuit, or 0.2/0.012 or 16.6 users per channel. So in this case, 50 channels would service 50×16.6 or 833 users.

Case 2.   In a single-site trunked radio system, if we assume a GOS of 0.2, then the traffic carried is 42.69 (or 50 circuits) Erlangs (Erlang C), which is equivalent to $42.69/0.012 = 3557$ users. This is an increase in capacity of around 4.3 times.

Commercial services generally require that most calls get through almost immediately. This means that a better grade of service is required, and it can be shown that as the required grade of service improves (that is, the value of the GOS goes down), the relative advantage of trunk over conventional radio increases. The converse is also true. For very heavily congested systems (and some conventional systems are *very* heavily congested), a trunked radio system with the same number of channels may offer only a minimal improvement.

Case 3.   Assume that the trunked radio system is covering the same area, but has been designed for good handheld coverage and has a large number of sites (let's say 10). Each site then can be assumed to have five channels. Now there are two new factors at play. First, the advantage of a large trunked group is gone and the five channels can carry only 2.85 Erlangs each, or a total of 28.5 Erlangs (compared to the 42.7 for the single site). Here you see the pooling efficiency of large traffic groups. Just as a two-lane highway can carry more than twice as much traffic as a single lane, so splitting channels into smaller pools reduces their traffic carrying ability.

The second factor to consider is that as the sites are widely distributed, the callers will most likely be using different sites to originate and terminate the calls. In fact, 90% of calls will be intersite. Because intersite calls use two channels instead of one per call, this has the effect of doubling the traffic for 90% of the calls. So the average calling rate is now $0.1 \times 0.012 + 0.9 \times 2 \times 0.012 = 0.0228$. The capacity of this system is then only $28.5/0.0228 = 1250$ users. This figure represents only an improvement of 41% over the conventional single-channel system.

Case 4. Now look at the situation described in Case 3, but with the added complication that a significant number of the users have regular wide-area calls. Assume that 30% of all calls are wide area, involving all base sites. The average effective calling rate is now $0.3 \times 10 \times 0.012 + 0.7 \times 0.9 \times 2 \times 0.012 + 0.7 \times 0.1 \times 0.012 = 0.052$, which gives a capacity of $28.5/0.052 = 548$. Note that this system now has *less* capacity than the conventional network with the same total number of channels.

As trunk systems get bigger, there can be a point of diminishing returns. More sites mean more traffic, but by its nature it also means more intersite calls. Particular care needs to be taken with wide-area calls, because these can rapidly consume network resources. The important thing here to realize is that unless the configuration of the trunked system is carefully monitored, the traffic efficiency can get very low. Some systems have in-built ways to minimize these efficiency losses such as forcing mobiles onto a designated "home" site where possible.

At this point, it cannot be stressed too strongly how important it is to optimize the RF design of a trunked radio system, so that the traffic capacity is optimized. A poorly planned RF will result in *extra* sites, which as seen above will reduce the traffic efficiency. It can be very difficult, particularly with extensive handheld use, to get a truly optimum RF design. Repeaters can be an effective way of improving coverage without adding base sites, and the use of them is to be encouraged (refer to Chapter 8 for details).

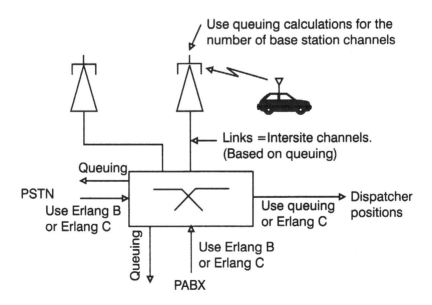

Figure 11.2 The dimensioning of a network.

DIMENSIONING A NETWORK

You may like to write your own software to dimension your systems, and for small systems maybe a good way to do this is to use a spreadsheet and write a macro to do the calculations. Alternatively, the software package provided with this text can do virtually any dimensioning that you might need. It is suggested that you start with the software provided, and see how well it suits your needs. Figure 11.2 shows a typical trunked network and the dimensioning techniques that are appropriate.

THE EFFECT OF OVERLAPPING CELLS ON GOS

In the next chapter you will see that in any large network, significant overlapping of coverage is inevitable. It is intuitively obvious that a mobile that has access to more than one base station will have a better grade of service than one that has access to only one. Consider a network where $N\%$ of the mobiles have access to two base sites. Assuming that the GOS that the base-station channels have been designed from is G, what is the average GOS that a mobile on the network will experience?

A mobile that has access to two base sites has an equal probability of failing to get served on either site of G. Therefore, its probability of failure on both sites is simply $G \times G$.

So the average GOS is for G for $N\%$ of the users and G^2 for $100 - N\%$ of the remainder. This gives an average GOS of

$$\text{Average GOS} = \frac{N}{100} \times G + \frac{100 - N}{100} \times G^2$$

Working the other way, if we know that $N\%$ overlap exists and we desire an average GOS of G_n, what should the individual base station GOS be? We have the relationship

$$G_n = [(1 - n) \times G] + (n \times G^2)$$

where G is the base-station designed GOS, G_n is the net GOS considering the overlap, and n is the fractional overlap ($n = N/100$). This equation can be restated

$$-G_n + G (1 - n) + (n \times G^2) = 0$$

which is a quadratic equation in G and which, in its general form $ax^2 + bx + c = 0$, has a solution

$$x = \frac{-b \pm \sqrt{(b^2 - 4ac)}}{2a}$$

Thus,

$$G = (n - 1) \pm [(1 - n)^2 + (4 \times G_n \times n)]^{1/2}/(2 \times n) \tag{11.1}$$

As an example, consider a system with 30% overlap and a desired average GOS of 0.05. Then from equation (11.1) the base-station design GOS is

$$G = (0.7 \pm [(0.7)^2 + 4 \times 0.3 \times 0.05)]^{1/2}/(2 \times 0.3)$$

$$G = 0.069$$

Sometimes there will be overlapping by more than one cell. Looking at the general case where the fractional overlapping of two cells is n, that of three cells is n_1, and that of four cells is n_2, we have the general expression

$$G_n = (1 - n - n_1 - n_2 - \cdots - n_k) G + (n \times G^2) + (n_1 \times G^3)$$

$$+ (n_2 \times G^4) + \cdots + (nk \times G^k)$$

Noting that G is a small number so that the higher powers of G will tend to zero, and replacing $(1 - n - n_1 - n_2 - \cdots - n_k)$ with n_0 (the fractional part of the cell with no overlap) we now have the general equation

$$G_n = (n_0 \times G) + (n \times G^2) + \{\text{terms tending to zero}\}$$

or

$$-G_n + (n_0 \times G) + (n \times G^2) = 0$$

for a general solution

$$G = \frac{(-n_0 \pm [n_0^2 + (4 \times G_n \times n)]^{1/2})}{2 \times n}$$

So, for example, if a system has

50% of the coverage with no overlap
20% of the coverage with two cells overlapping
20% of the coverage with three cells overlapping
10% of the coverage with four cells overlapping

and we want an average network GOS or 0.05, then the site design GOS is

$$G = (-0.5 \pm [(0.5^2 - 4) \times (0.05 \times 0.2)]^{1/2} / (2 \times 0.2)$$

$$G = 0.09625$$

It is generally the rule that most system designers ignore the effect of overlap. However, as can be seen from the above analysis, the effect of overlap can be significant. A station with 10% blocking would generally be regarded as heavily congested; but if it is operating under conditions similar to the one in the example above, it can be seen that in fact the customers are still experiencing an effective GOS of 0.05. This information is as valuable to the network operator as it is to the system designer.

Of course, if queueing is provided the GOS means retries, not lost calls.

CHAPTER 12

SITE PLANNING AND TRAFFIC EFFICIENCY

The traffic capacity of a multisite trunked radio system will depend very significantly on the number of sites, and to a much lesser extent on whether or how much they overlap. The real problem is not so much mobile-to-mobile calls, because they have a high probability of being intersite regardless, in any large system; rather it is the swamping effect of wide-area calls on a large network.

OVERLAPPING SITES

In trunked radio except for small systems with only one site, the sites will overlap to some extent; otherwise there will be gaps in the coverage. To examine the effect of this, consider first a system of omnidirectional sites with no overlap. A coverage grid based on four sites is assumed; other patterns could be used, but they would not materially affect the outcome of this study. The shaded area in Figure 12.1 shows that no overlap also means significant holes. It is easy to calculate the area of poor coverage as a percentage, by assuming that the square grid has sides of length 2 units, and hence an area of 4 square units. There are within each square 4 quarter circles, and the area of each circle is $\pi \times (1)^2$ or π square units. So the area covered as a percentage is $\pi/4 \times 100\%$ or 79%.

In cellular radio, the reality that real sites will either have some overlap or some holes is conveniently side-stepped by assuming hexagonal coverage from each site. As unrealistic as this is in cellular, it would be even more so in trunked radio, where the omnidirectional antennas can, on average, be expected to give something approximating circular coverage. Clearly, however, 79% coverage means 21% not covered and is unacceptable.

Suppose we propose a model with circular coverage, but this time with overlap and no "holes." Figure 12.2, where the coverage from sites on opposite corners of the grid

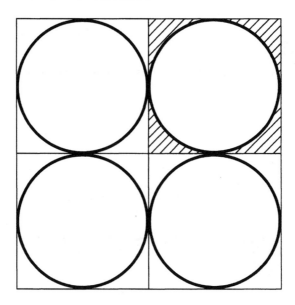

Figure 12.1 Circular coverage with no overlap.

Figure 12.2 A circular pattern with no holes.

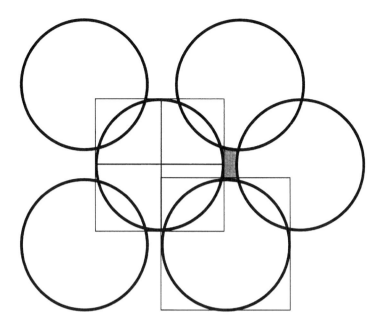

Figure 12.3 Gaps occur if the sites are just a little further apart.

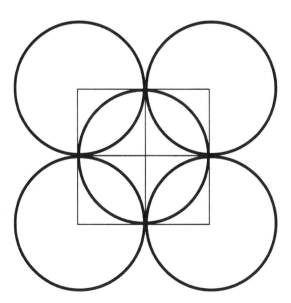

Figure 12.4 The detail of a plan with no holes.

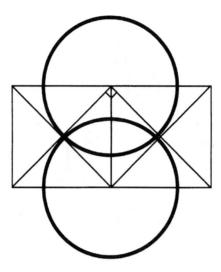

Figure 12.5   The construction of two sites with no holes in coverage.

just touch, is the most efficient model that leaves no uncovered areas. Figure 12.3, shows the gaps that occur if the sites are just a little further apart. Now we would like to know how much of the coverage is overlapping when we design for no holes as in Figure 12.2. Figure 12.4 shows the detail of one of these sites.

From Figure 12.5, it can be seen that the overlapping area is confined to a 90-degree sector of each adjacent site. The area of this overlap can be found by noting that the arc of overlapping area is a composite of a sector and a triangle. The area of this sector is then $\frac{1}{4} \times \pi \times (2^{1/2})^2 = \pi/2$ square units.

The area of the triangular portion is 1 square unit. The total overlapping area is then $2 \times [(\pi/2) - 1]$ square units. Expressed as a percentage, the overlap is $2 \times [(\pi/2) - 1]/(\pi/2) \times 100\% = 72.7\%$. This rather alarming result means that the aims of minimizing overlap and of ensuring seamless coverage are in serious conflict. While real-world coverage patterns are rarely circular and terrain can be used to provide some screening, nevertheless there will be significant overlap in coverage. It will be shown shortly that this inevitably leads to an inefficient use of channel resources.

TRAFFIC EFFICIENCY OF A LARGE TRUNKED NETWORK

Initially we consider only mobile-to-mobile calls. Assume that all of the calls are simplex (PTT) and that they can share a common channel only if both the called and calling party are within and using the same site. If an area, say a city, is divided into N sites, which have no overlap, and assuming that the mobiles are homogeneously spread across the city, the chance that any mobile calling another and both will be in the same site is $1/N$. Conversely, the chance that the called party will be in another site (and therefore require two base-site channels to complete the call) is $[1 - (1/N)]$. So the probable number of channels required per call is $\{1 \times [(1/N) + 2]\} \times [1 - (1/N)]$. This is illustrated in Figure 12.6.

It can be seen that the traffic efficiency of a large network rapidly approaches two channels per call, particularly if it is bigger than six sites.

However, as previously pointed out, in the real world the sites overlap, so let's now examine the case where there is sufficient overlap to ensure no holes.

In this case the probability of an intersite call is higher, because even if the called and calling parties are on the same site, there is a reasonable probability that regardless of their locations, they will still make an intersite call, because they are logged onto different bases. We have already found that there is a 72.7% overlap of coverage in a large network.

For a finite-sized network the relative area of overlap can be expressed as a fraction of the infinite network percentage. Referring to Figure 12.5, we find each site has one overlap, compared to four each for an infinite network, and so for two sites the overlap fraction is 1/4. For three sites, a seen in Figure 12.7, the overlap is four zones compared to 3×4 for the infinite network, for a fractional amount of $4/(3 \times 4)$. For four (see Fig. 12.8), the amount is $6/(4 \times 4)$, and for five it is $8/(5 \times 4)$. Once the sixth

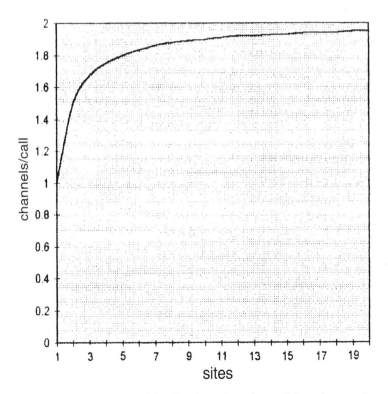

Figure 12.6   The average number of simplex channels used per call for a given number of sites in the network, assuming zero overlap, and mobile-to-mobile calls only.

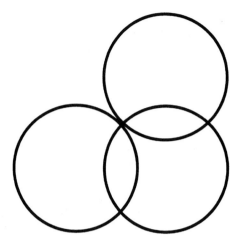

Figure 12.7   A three-site system.

site (Fig. 12.9) has been added, some symmetry emerges and the fractional amount is $(8 + 4)/(6 \times 4)$. This can be continued up to the Nth site, noting that each new site adds four new overlapping areas, which will have the overall fractional amount of $[8 + (N - 5) \times 4]/(N \times 4)$ of the overlap of an infinite site system.

If we let the overlap probability be P (here $P = 0.727 \times$ the fractional amount as determined above by the number of sites), then the probability that a caller in a given site will be logged onto that site is the probability that the mobile is in the

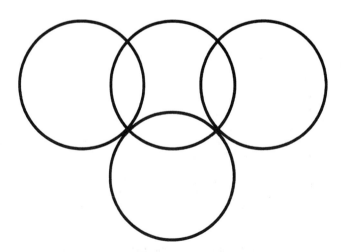

Figure 12.8   A four-site system.

Figure 12.9   A six-site system.

nonoverlapping area $(1-P)$ plus the probability that the mobile is in a overlapping area but still is logged onto the site $(P/2)$, or a total probability of $1 - (P/2)$.

The probability that this mobile then calls a mobile from a homogeneous population that is also in the same site is $1/N$. So the net probability is $(1/N) \times [1 - [(P/2)]$. The probability that the called mobile is also logged onto the local station is also $1 - (P/2)$. Therefore, the *net* probability that a call is local is $(1/N) \times [1 - (P/2)]^2$. So the probable number of channels per call is $(1 \times \{1/N[1 - (P/2)]^2\}) + [2 \times (1 - 1/N[1 - (P/2)]^2\})]$. Figure 12.10 shows a plot of this function, which is very similar to the nonoverlapping case except that as would be expected it converges more rapidly to two channels per call.

What is clear from all of this is that once there are more than about four sites, most calls will be intersite and overlap or additional sites will not make a lot of difference to the traffic efficiency, which tends toward 50% regardless of the configuration. Conversely, when the sites are few, significant gains can be obtained by keeping the number of sites small, but there is not a lot to be gained in minimizing overlap.

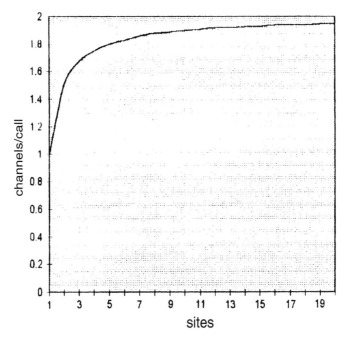

Figure 12.10 The number of channels used per call with overlapping and no holes, and zero wide-area calling.

WIDE-AREA CALLS

The next factor to look at is the one that consumes a lot of channel resources in large systems—namely, wide-area calls. For simplicity, assume that all wide-area calls are over the whole service area and that the percentage of wide-area calls is known. The overlapping model will be used. The average number of channels per call will then be proportionately the same for the mobile-to-mobile calls plus the additional channels used for wide-area calls. Shown graphically in Figure 12.11, it can be seen that as might have been expected the number of channels per call is dominated by the wide-area traffic. For low site numbers (less than 3), wide-area traffic does not represent a serious load; but for any system larger than that, this traffic is dominant and expensive.

The study so far suggests that large systems become increasingly inefficient as they get larger and carry more wide-area traffic. To see the extent of this effect, let's look at a system where the maximum size of a base station (or site) is 20 channels and see how the traffic capacity of such a system grows as the network grows, while carrying wide-area traffic.

The information is already contained in Figure 12.11 because the number of simultaneous calls is equal to (20 × the number of sites/the number of channels per site). Figure 12.12 depicts this. It can be readily seen that even small amounts of

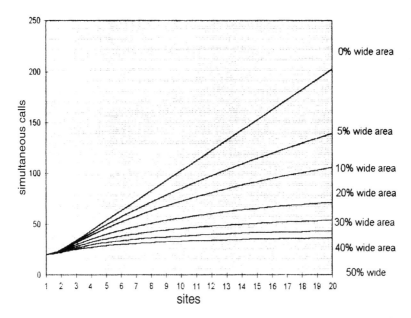

Figure 12.11 The maximum simultaneous calls on wide-area network with group calling (20 channels per site).

wide-area traffic significantly reduce the network capacity, while for wide-area traffic greater than 20%, additional sites add almost no additional traffic.

Other than telling us that big systems are not a good idea, this study also suggests shortcomings in any system that has a limited path budget (e.g., TDMA systems such as TETRA), because they will necessarily require more sites than their analog counterparts. It also suggests that any system that tries to be both cellular and trunk-like (e.g., iDEN) will have problems, unless the trunk part is a completely separate overlay on the cellular part, in which case you really have two systems anyway. In this context, the meaning of cellular-like can mean either high-density mobile-to-mobile calling or high-density mobile-to-PSTN calling. Of the two, mobile-to-PSTN calling is the less taxing of resources because it only ties up one channel.

Conversely, it shows that any system that can *extend* base-station coverage will be very advantageous, particularly when used for wide-area calling. The current trend toward compressing spectrum at the expense of path budget, as TDMA systems do, probably produces no gain in spectral efficiency and may even decrease it.

You might be tempted to suggest that the overlap in this study is excessive (tending to 72.7% for a large system), and that might be the reason for these conclusions. However, as already pointed out, overlap is really only a serious factor for mobile-to-mobile calls, and even then only on very small systems. What we are seeing here is the total dominance of wide-area calling, and that overlap would only be significant if it could be reduced enough so that the total number of sites could be

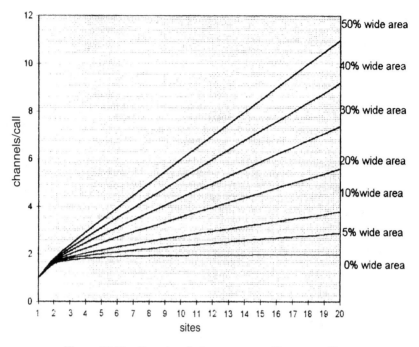

Figure 12.12 Capacity of a large system with group calling.

reduced (in which case the gain for large systems is nearly proportional to the relative reduction in the number of sites).

So far we have seen that large systems and wide-area traffic just don't mix—so what can be done about it?

MAKING WIDE-AREA TRAFFIC WORK IN LARGE SYSTEMS

It would be possible to ban wide-area traffic, and some operators have done just that, by making wide-area calls possible only on the site of origin. This solves the traffic problem for the operator, but it means a rather poor service for the user.

Another solution is not to let the system get too big. Let's assume that a city needs five sites for adequate coverage. When traffic demands more sites rather than simply make the system bigger, we might note that a five-site system can carry 40 calls, with 20% of them being of the wide-area type, while a 10-site system only carries 56 calls. Two five-site systems will carry 80 calls, and hence they are significantly more efficient. Here the economies of scale are working against the large carrier.

What is really needed is a systemic solution. I don't think that this has been done as yet by any system manufacturer (except perhaps iDEN). Because large wide-area systems simply will not work, there is a case for an *overlay* structure of high-powered sites to cover wide-area calls only. Unless the user is satisfied with broadcast calls only, an additional structural change is needed to ensure that wide-area mobile-originated calls can be successful. Because there will be a big imbalance in the path budgets talk-out to talk-in, calling mobiles will need access to the high-powered cells directly and also indirectly via non-wide-area cells, in the instances where direct access is not possible. There would also be a need for a signaling protocol to direct all mobiles in a group to the nearest wide-area cell when called, regardless of which cell the mobile was currently homed on.

CDMA offers the promise of better path budgets without loss of spectral efficiency, and it may hold some promise for better systems.

CHAPTER 13

BILLING

Traditionally, conventional mobile community systems have charged a fixed monthly fee, typically of around $15.00. This was fine until a few network enhancements came along, including interconnection and data. Telephone interconnect calls always tend to use a lot more airtime than the dispatch mode. Even with conventional repeaters, many operators charge an additional fee for telephone interconnect, and mostly this fee will be on a per-minute basis.

A very "easy" way to allow PSTN access, but still have full call accounting, is to make the users have a dedicated PSTN line, which is accessed by a PIN number. Each line will then be billed separately by the PSTN operator. It is more efficient, however, to have a common route into the PSTN, along with an in-house method for determining who has access.

Telephone interconnection calls have raised the average income per interconnected trunked radio system significantly, and net monthly revenues from these subscribers are similar to those of an average cellular subscriber, at around $50.00 per month. While some operators still bar or severely restrict PSTN access in order to conserve airtime, others are actively promoting its usage and also profiting from it.

Restricting high airtime utilization is lack of frequency allocations (often imposed by the regulators) along with hardware restrictions. Most trunked radio hardware still poses Draconian restrictions on PSTN access, although there are a few notable exception to this.

Often, because of limitations in the billing package, the PSTN access is limited to local-area calls only, so that a flat rate can be charged per minute of airtime. Some operators allow a relatively wide-area PSTN access (e.g., statewide or nationwide), and they cover the costs of this by charging a high per-minute fixed fee, which may be as high as $0.50 per minute.

Most trunked radio systems will come with a basic billing package that will permit airtime charging. Often the billing is an integral part of the controller, but sometimes it may be an optional extra. The basic billing will usually distinguish between

TABLE 13.1   A Typical Basic Trunked Radio Bill[a]

Mobile 234567	Time	Weight	Total
Mobile-to-mobile calls	87.50	1	87.5
Mobile-to-PSTN calls	15.10	3.5	52.85
PSTN-to-mobile calls	0.55	3.5	1.92
Group calls	2.7	1.3	3.51
Total equivalent minutes			145.78

[a]Total charge based on $0.09 per minute = $13.12.

mobile-to-mobile calls, mobile-to-PSTN calls (and vice versa), and group calls. In order to distinguish the charging, weighting factors are applied to each type of call. Thus a mobile-to-mobile call could have a weight of 1×, a mobile-to-PSTN call could have a weight of 3.5×, and a group call could have weight of 1.3×. The bill may then appear as in Table 13.1.

Alternatively, and perhaps more commonly, the operator might charge a fixed fee for all mobile-to-mobile calls and group traffic calls and, while permitting only local PSTN calls, charge an additional fixed rate per minute for PSTN calls.

Where wider-area calling is required (e.g., statewide), it may be possible to restrict access to just the codes that are needed for statewide dialing and to make the PSTN per minute charge high enough to cover any permitted destination. This can, however, make local calls very expensive.

WHY HAVE SOPHISTICATED BILLING?

Provided that each and every user is a "typical" dispatch-type user, it may make economic sense to just charge a fixed monthly fee. More recently, with the advent of data via trunked radio, some users have strayed far from the traditional operations. It is not unusual for very large files to be sent to air. A simplistic way to "limit" this traffic is to put timers on each call. This only limits the unwary, because the data can be packetized and sent out via a number of separate calls. In fact, installing timers can make things even worse as system resources are used to set up, clear down, and then reset calls.

Provided that the system has an adequate billing capability, these data calls can become a source of revenue rather than annoyance. Also, long-distance calling can be permitted, to add value to the service and add revenues.

In some countries, legislation rather than technology has limited the possibilities. In particular in the United States, where a maximum of 20 channels is permitted per operator, maximizing returns may have a lot to do with minimizing usage. India appears to have adopted a similarly restrictive approach to trunk radio (where as little as five channels are permitted). In most countries, there are few such restrictions, and it is merely up to the operator to prove that the frequency resources are being used effectively in order to obtain more spectrum.

A powerful billing system will not only permit the operator to offer more services to the user, but with large systems it can be an important tool for keeping track of cash flow, outstanding debts, fraud, and even traffic statistics.

A full billing system for a trunked radio operator will need to be similar in capability to a cellular system. The only real difference is the size of the subscriber's base (which may be a magnitude smaller than cellular). In hardware the difference is more pronounced because a trunked radio billing system can run on a high-end PC, a Pentium processor, while a cellular billing system will invariably need a minicomputer.

Larger wide-area systems may have linked base sites thousands of kilometers from each other. Where this happens, it may be necessary to charge users on a per-kilometer basis for the long-distance traffic in the same way that the PSTN operators do. This will probably mandate a full billing system.

A number of companies exist that specialize in providing billing systems. Most of these were originally devised for cellular systems, and they come at a cellular price. Many have capabilities well beyond basic billing, and they can interact directly with the switch to terminate nonpaying customers or to validate new ones. Some even provide fairly elaborate network management, because the billing data contains a lot of valuable systems data.

Most commercial billing packages will provide features and capabilities far in excess of the requirements of the average trunked radio operator, but for large public systems (statewide or nationwide), there is a reasonable argument that the requirements of a trunked radio operator are not too different to those of a cellular operator.

Most large trunked radio operators are unfortunately caught somewhere in the middle. The basic billing system provided with the system is a bit too small and basic, but a cellular billing system is still overkill and too expensive. The temptation is to write a computer program that does it all.

Those who take this path invariably underestimate the amount of work that is needed to develop such a system. All software has bugs, and in a billing system these bugs can be embarrassing at best and very costly at their worst.

Despite the cautions, the author has seen successfully developed in-house billing systems. The secret seems to be to use a flexible database system (Clipper, for example) and to have programmers available who are familiar with the development of commercial accounting packages.

The PC for any reasonable-sized customer base will need at least to be a high-end Pentium, and the hard drive will need to be around 9 Gigabyte, preferably with a second redundant drive. It is the storage capacity and access time that will ultimately determine the billing system capacity.

At the top of the scale, for very large nationwide or statewide trunked radio systems with tens of thousands of users a fully featured cellular billing system may be what is really required.

Most importantly, any billing system that is introduced will need to avoid double entry of subscriber data. This means that there must be a physical link between the billing package and the switch controller validation package, to allow the customer data to be simultaneously entered on the billing records and into the validation matrix. Where these two entries are separated, numerous inconsistencies will arise, including bills to disconnected customers and customers who are connected but not billed. With time, along with growth in the system, this problem will grow and will probably get completely out of hand in a very short time.

CHAPTER 14

MPT 1327 SYSTEMS

The MPT 1327 standard was released by the UK Department of Trade and Industry as a UK wide-open standard in 1986. As such, it became the first open standard trunked radio system, which later led to it becoming a de facto world standard. In the initial evaluation, many standards were examined including Mobitex, Radiocom 2000 (a trunk-like cellular system used in France), and most of the US standards. However, because the requirement was for a nonpropriety standard, most of the existing signaling schemes had to be ruled out.

A consistent problem with most then-existing standards was the relatively small number of subscribers that could be supported within the number range available within the signaling protocol. For the proposed nationwide system, the fewer restrictions on the number of users, the better.

In 1987 the first base station complying to the standard began operations from the Easton Square Talcum tower in London, and a year later the first multisite system went into service. By 1990 the MPT 1327 standard had become unofficially the European standard, and more recently it has become accepted worldwide in virtually every country except the United States. It is widely believed that politics has played a major role in keeping this standard out of the United States; however, there have recently been some new MPT 1327 systems installed there. Most European countries have a public MTP 1327 network, usually one or two national and a number of regional ones. The United Kingdom has the largest network, but it is also the oldest. However, in some countries the networks have only recently been established (e.g., Norway).

While MPT 1327 is the signaling protocol specification, the supporting standards include MPT 1347 (the base-station specification) and MPT 1343 (the mobile specification, including the air interface). Other relevant specifications are as follows:

- MPT 1317 is a signaling specification.
- MPT 1318 is basically an engineering memorandum on the trunking efficiency.

- MPT 1352 specifies testing procedures to verify compatibility of mobiles.
- MAP 27 is the standard for the interface between the mobile and other devices (data, intelligent control, etc.). MAP 27 is based on a 9600-baud rate.

Despite the clear intention that MPT 1327 was to be a common open standard, a number of manufacturers have adopted a "derived" version of the standard. In doing so, they have chosen not to implement all of the MPT 1327 functionality, and in some cases they have used reserved data bits to implement new functionality (e.g., duplex operation, passing of PSTN pulses, and transmit power control). While this new functionality is obviously advantageous to the user, it does mean that standardization is lost.

The United Kingdom has one national network and about 20 regional networks, with one of those regional networks recently being upgraded to a national one. Most regional networks are small, having between 1,000 and 3,000 users, while the national network has around 40,000 subscribers. The current growth is 10,000 per annum.

Across Asia there are national MPT 1327 networks in most countries, although most of them have only been introduced in the past few years and it is too early to draw any conclusions about the future, particularly because most of them have yet to complete their roll out plans.

Considering that the first multisite system began operation in 1987 (compared to cellular, which is generally accepted to have commenced service in 1983), MPT 1327 is still a relatively new technology. Like cellular, MPT 1327 had a lot of teething problems and can be considered to have matured as a technology only as recently as 1992.

Major manufacturers began producing second-generation equipment in 1993.

The advantage of MPT 1327 is that it is an open technology, with a large number of major players from the industry producing equipment. Added to that is a small army of "cottage industry" operators who are producing a vast array of value-added products for the technology.

MPT 1327 STANDARD

The MPT 1327 standard was conceived to provide a nationwide service across the United Kingdom. Unlike the other standards that evolved from the concept of a multichannel single-site system (see Fig. 14.1), MPT 1327 was required to efficiently provide multisite trunking from its inception and was envisaged as providing a nationwide service.

In many ways the concept of MPT 1327 resembles cellular radio without handoff. The implementation of the system by the various manufacturers, however, is diverse, and most have taken an approach based on a single central switch, with a single "interregional" switch to provide multiple switch operations (see Fig. 14.2). This approach is very much a compromise; and although it can allow quite large systems

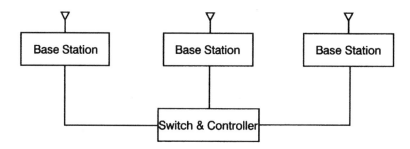

Figure 14.1   A basic MPT1327 network.

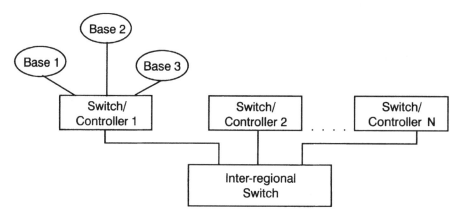

Figure 14.2   An interregional switch.

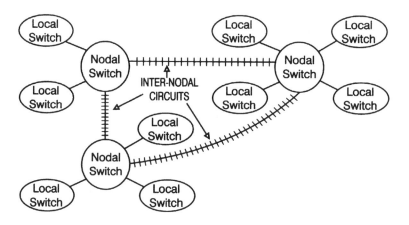

Figure 14.3   A nodal switch.

to be built, it does mean that they all have to have circuits back to a centralized controller. In countries where nationwide systems operate over large distances (like Australia, Indonesia, and the Philippines), the cost of the circuits back to the intersite controller can be daunting.

The other approach sometimes used is to have nodal switching. In this case all nodes are equal, and each node has access to all other nodes. This approach allows for large networks to be integrated over wide areas much more economically, and it more closely represents the way that cellular and PSTN switching is done. This concept can be seen in Figure 14.3.

Some manufacturers use dedicated nodal switches, while others use only one kind of switch, which can be made nodal, local, or both as required by the network design. This last option offers the greatest flexibility to the systems engineers.

WIDE-AREA COVERAGE

The dominant feature of MPT 1327, when compared to other systems, is its ability to work efficiently over wide areas. Most other systems, which evolved from single-site systems, effectively use parallel channels to cover wide areas. To see why this is not ideal for circuit efficiency, imagine a seven-site system where a wide-area call is placed to a roaming mobile. If the system is unable to determine which site the roamer is using and therefore switches the call to all sites in parallel, then seven channels are taken up by the one call, with most sites broadcasting the call to no avail. It can be seen that the larger the number of sites involved in a wide-area call using parallel sites, the less efficiently the channel resources are used.

MPT 1327 does not switch wide-area calls in this way. Instead it will determine where the called mobile is located and will connect only a voice channel at that site to the called party, and thus it only uses two channels for the one call.

Naturally, wide-area group calls (and broadcast calls) will require parallel operations from all sites in the wide area because it is not possible to simultaneously determine where all the members of the group are. For this reason, wide-area group calls need to be carefully monitored, either by restricting access to wide-area group calls or by implementing a billing regime that will charge the real cost for these calls.

FREQUENCIES

There are no formal standard frequencies for MPT 1327; and, generally, hardware is available in all "standard" mobile radio frequencies which today tends to mean 66–88 MHz and 136–520 MHz. Although not all of the frequency within bands conforms to the traditional "mobiles band," they are all in use in some countries.

Not all manufacturers make mobile products in all bands. The most commonly available bands are

66–88 MHz[*]

136–174 MHz

175–225 MHz

300 MHz (some countries only)

400–520 MHz

Notably excepted from the above bands is the 800-MHz trunking band used in North America. Although some manufacturers do produce hardware in that band, it is not widely available at the time of writing.

The networks are essentially full duplex, so that each channel consists of a pair of frequencies. There are no standards for RX/TX frequency separation, although 5-MHz and 10-MHz separation is widely encountered. It should be noted that although the MPT 1327 specification does not include duplex mobiles, most manufacturers can supply them. The efficiency of duplex operation will be very dependent on the TX/RX separation. The wider the separation, the smaller and more efficient the duplexers can be made to be.

The number of channels can be up to 1,024.

800-MHz MPT 1327

While traditionally MPT 1327 was operated in mobile bands below 520 MHz, there has been a demand for systems in 800 MHz. The reasons for this are availability of spectrum and the requirement for full duplex (the 800-MHz systems with 45-MHz spacing are ideal for handheld duplex—the high frequency makes the RF components small, and the wide frequency separation assists in simple duplex operation).

At the time of writing, only a few operational 800-MHz systems are known: One is located in Shanghai, China (supplied by Stanilite, which has been taken over by ADI, Australia), and the other is an airport system in Sri Lanka (Tait). The main reason for the paucity of hardware in the 800-MHz band has been the problem with the standard MPT 1327 protocol, which is not robust enough to operate at the higher frequencies (with its attendant higher Rayleigh fading). While this is not a problem at lower frequencies where the ability of the channels to carry voice reliably limits the range obtainable, if the system is simply moved to 800 MHz, then RF propagation on the control channel will limit the system range to less than the usable voice acceptability limits.

Studies by Stanilite revealed that at 800 MHz up to five errored bits could be expected in a single message. The solution was to incorporate a forward-error-correcting Hamming code that would permit up to a 5-bit burst error or any 2-bit error

---

[*]Note that the 66- to 88-MHz band is not currently highly favored due to the problems with interference and noise on these frequencies. Additionally, although the propagation in these bands can give wide coverage, it is unpredictable and sporadic. Because the propagation characteristics can vary seasonally as well as daily, this can lead the user to believe that the network is unreliable.

in a 64-bit segment to be corrected. When this code was implemented, Stanilite found that this restored the performance of the signaling at 800 MHz even in high-density downtown environments.

BASE STATIONS

In this text a base station means a base site including all equipment and infrastructure to support that site (note that sometimes the term "base station" is used to mean single RF channels, particularly in the United Kindom, but this use will not apply here).

A base station will consist of a control channel, voice channels, and supporting hardware and infrastructure. Figure 14.4 depicts a basic base station. Typically, there will be one control channel, which is capable of supporting about 24 voice channels (the upper limit is dependent on the paging data through the control channel). For very small bases, the control channel may be configured to revert to a voice channel when all voice channels have been allocated. Although this will increase traffic capacity, it will mean that while the paging channel is operating as a voice channel, no queuing can take place, and short and status messages that rely on the control channel cannot be sent.

Each channel has a supporting controller that enables it to connect local calls independently of the main base-station controller. When this is done, there will be no call screening; and any mobile frequency, regardless of whether or not it has a valid ID, can be connected to make local calls. This function is only activated when the base-site controller is inactive.

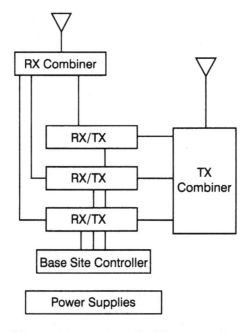

Figure 14.4 A typical MPT 1327 base station.

The base-site controller is a microprocessor that handles local calls and determines when to pass call requests onto the central controller (for intersite calls). It will be connected to each channel controller on a per-channel basis.

The RF channels will ordinarily be multiplexed into common antennas. Often separate send and receive antenna are used, although a diplexer (also known as a duplexer) can feed both TX and RX into a single antenna. There is about 3-dB disadvantage in using a diplexer, with the combining loss increasing with decreasing separation between the TX and RX frequencies.

CELL EXTENDER

For MPT 1327 systems, repeaters can be used (for additional detail see Chapter 8). All repeater systems that are protocol-independent will need to receive and transmit without frequency translation; and because of the potentials for oscillation, this will limit the service area.

ADI have an MPT 1327 repeater that intelligently repeats a base station "off-air" and translates the channel onto a new frequency. Frequency-dependent commands

Figure 14.5   An intelligent cell extender. (Photograph courtesy of RST.)

like "GOTO channel 10" are intercepted and re-sent as a command to go to the appropriate repeated channel.

The repeater is allocated only as many channels as are needed for traffic purposes and comes in a single rack with up to six channels. It is meant for operations in areas where it is not appropriate to install a new base site, which includes low-density areas like roads and outlying suburbs. A similar system from RST comes with a small rack-mounted system manager, which is designed to interface with any transceiver. This is shown in Figure 14.5, being used with Tait RF equipment.

SWITCHES

If there is one area that needs attention in MPT 1327 systems, it is the switches. It is hard not to draw the conclusion that most of the available systems had their switches designed by radio engineers. Although all of them will perform the basic intersite switching and switching to the PSTN, most are very limited in how they do it. The limitations may come as a limitation on ports per switch, PSTN capacity, PSTN signaling restrictions, diagnostics, routing capability, total base station sites per switch, or operational controls.

Like MPT 1327 itself, most manufacturers evolved their systems from basic one- or two-site systems, then onto many sites. It should be said at this stage that the limitations of the switching do not generally bother the smaller operators who probably have only two to five sites, one point of PSTN interconnection, and a few hundred to a few thousand users.

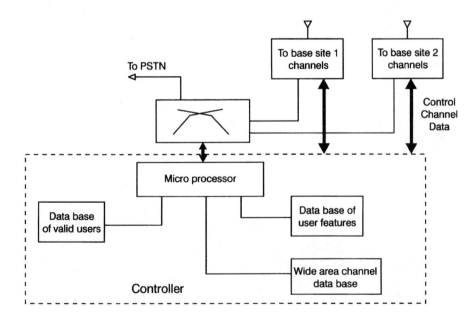

Figure 14.6   A systems controller.

It is the large operators who most need powerful switching, although through clever control of resources a powerful switch may also be a boon to the smallest of operators. This weakness in switching is not limited to MPT 1327 but is a characteristic of trunk systems in general. It is most evident in MPT 1327, because, with the nationwide systems that it can nominally handle, these limitations become frustrations for the operators.

THE SWITCH CONTROLLER

Often the control of the switch is integrated into the switch, and this will be most evident when the switch used is derived from a PABX or small PSTN switch. In trunking, it seems that a number of manufacturers have developed their own switches that are usually physically separate from the controller. Figure 14.6 shows the structure of a typical controller. This means that the switch will be (usually) a full-availability port-to-port switch that connects ports under direction from the

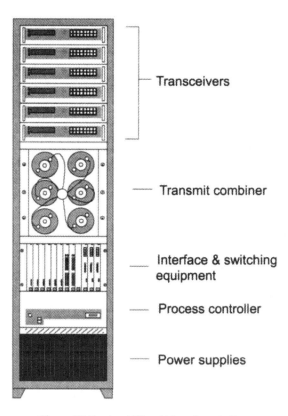

Figure 14.7 An ADI switch and controller.

controller. For example, it may be required to connect a wide-area call from one base-station site to another. This will involve switching together a channel from the first site to a channel from the second site. While this is physically done by the switch, it will have been the controller that initiated the paging to determine which site the called party was on, as well as whether channel resources were available, and that also initiated the command to the switch to connect.

Access to the controller needs to be restricted and controlled by a number of passwords to limit the access to legitimate users only.

An ADI switch and controller is shown in Figure 14.7.

CONTROL CHANNELS

Typically, one control channel can support about 24 voice channels. This limit varies with the traffic and its nature, but the ultimate limit is due to the number of messages sent on the control channel to support the voice channels. This means that setting up a lot of short calls, for example, will use up control channel resources more quickly than setting up a number of longer calls.

There is, however, no reason to restrict a base station to just one control channel, and a number of network suppliers allow multiple control channels on one site. This has the advantage of greater traffic efficiency, and even the ability to carry additional short message traffic.

At the other extreme, when very small sites are used there may well be a frequency shortage, and more efficient frequency use can be attained by time sharing the same control channel frequency at a number of sites. Note that this does not mean any savings in hardware because all sites still have a channel reserved for control. This operation is again only available from some manufacturers.

The control channels operate at 1,200 bauds and use fast frequency shift keying (FFSK). Each bit is represented by either one cycle of 1,200-Hz tone or one and a half cycles of 1,800 Hz. The 1,200-Hz signal is designated as 1, and the 1,800-Hz signal is designated as 0. The data are phase-continuous and transitions occur at zero signal level crossing, which is what distinguishes FFSK from the slower FSK.

It is worth noting that in most systems there is nothing to physically distinguish a control channel from any other channel, so that any channel can be converted into a control channel in the event of control channel failure.

Control channels can be set to revert to voice channel operation when required. Generally, they will revert after the queue has reached a certain depth. Although this can be an effective way of increasing capacity in very small systems, it is not recommended for even systems of moderate size. Once the control channel reverts to voice channel functionality, the short data message capability is lost, as is queuing and proper control of intersite calls.

The problem that all trunked radio systems have to contend with is that mobiles requesting a call may find that the calls clash (and thereby inhibit each other). The probability of a clash increases as the number of users on the system grows. MPT 1327 addresses this problem with its use of the dynamic frame-length slotted ALOHA technique.

The control channels are divided into time slots that are 128 bits (107 ms) long, each of which can contain one message. An ALOHA message indicates to the mobiles how many of the succeeding slots are to be allocated to call requests. In times of low traffic volumes, the ALOHA number is usually one. This means that the next slot is always available. As the traffic builds up, the probability of contention (call clashes) increases and the control channels look for these clashes. When they occur the ALOHA number is increased to two, and the mobiles will randomly choose which time slot to use. The ALOHA number be increased further to counter call contention.

It is conceivable that even the largest ALOHA number will not totally eliminate contention; in this case, partitioning is used. The calling mobiles are partitioned into groups, and in turn some of these groups are permitted to make a call request.

MESSAGES

A number of messages can be sent over the control channel. Some of the more frequently encountered ones are as follows:

ALOHA(ALH) A message sent to indicate that the base station is free to accept incoming calls.

AHOY(AHY) This is used to call a mobile.

REQUESTS(RQS) Mobiles use this to request a call.

ACKNOWLEDGE(ACK) Used to signify that the command has been received.

GO TO TRAFFIC CHANNEL(GTC) Used to send a mobile from the control to a traffic channel.

CLEAR(CLR) Used by the base to clear down a mobile call.

These codewords are arranged into time slots, which in turn are arranged into frames. A frame may consist of from one to 32 slots.

The call setup time, assuming no queuing, is ordinarily less than 0.5 s.

To see how these messages are used, consider a typical call that may proceed as in Figure 14.8. Here mobile 1 is calling mobile 2. It first waits for an ALOHA, which indicates that the next time slot is free. On that time slot it sends its call request. The response is an AHOY(AHY), which does three things. First, it acknowledges the call from mobile 1, next it inhibits all other mobiles from calling on the next time slot, and finally it requests mobile 2 to confirm its availability.

The call is now ready for completion, and the GTC (go to channel) instructs the mobiles to go to a speech channel. In the more complex situation where more calls are being handled simultaneously, the above procedures are interleaved.

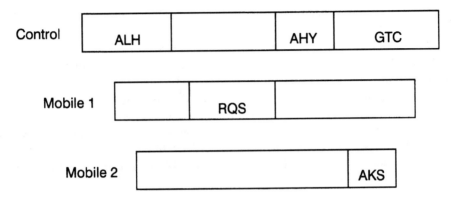

Figure 14.8   A typical simple, single-site call as seen by the control channel activity.

DATA

A very significant new development in MPT 1327 has been the release of the MAP 27 interface standard. This standard, which is currently being implemented, slowly, by all major mobile equipment manufacturers, defines a standard way of sending and receiving digital data messages on both the control and voice channels. There are three kinds of data messages that are defined:

1. STA—status message
2. SDA—short data messages
3. EDA—extended data messages

An STA is a status report, consisting of a number from 1 to 31, sent on the control channel. The message is only 5 bits in length. Typically, these are used to send "canned messages," such as assigning status 6, for example, to the message "out to lunch."

Short data (SDA) messages are sent on the control channel, while EDA messages can be sent either on the control channel or on the voice channels, depending on the length of the message.

An SDA message is in free format and may consist of up to 184 bits in length. It is sent on the control channel.

Although the STA and SDA messages are sent very efficiently, they do consume control channel overheads; and particularly on sites with a lot of voice traffic (nominally 15 channels and above), care should be taken that message loading does not compromise voice traffic by overloading the control channels. Some operators severely restrict access to the control channel for this reason, and others apply separate charges for its use.

EDAs (also known as transparent data messages and nonprescribed data calls) are sent on the voice channels and may be constrained by the same time limits as normal voice calls; alternatively, they may be set different time limits, if the system permits.

MAP 27 provides a uniform way to control a mobile through a standard RS232 port, so that any serial device such as a PC can be connected directly to the trunk network in a transparent manner. What is certain is that this facility will be attractive to the manufacturers of peripheral equipment, so that the value-added options available on MPT 1327 will continue to grow rapidly.

Data throughput as high as 6,000 bits/s on a 12.5-kHz channel has been achieved through MAP 27 data terminals.

CALL SEQUENCE

In the switch-on mode a mobile will begin to hunt for a control channel. When a suitable one is found the mobile will lock on and register. If mobiles are not engaged in a call, they monitor the control channel. Before the mobile can transmit, it needs to synchronize with the control channel. The control channel will invite call requests by transmitting an ALOHA call on the forward channel. In a four-slot time frame the "ALOHA" is sent out on every fourth time slot, while the mobile sends its request randomly on one of the remaining three. This mechanism reduces contention (clashes between mobiles).

Each base site will transmit its site identity code. The mobile will receive this and compare it to a look-up table that holds the identity of all sites on which it is valid. This prevents the mobile from locking onto a foreign system that will not provide it with access.

When not in use, the mobiles monitor the control channel (Fig. 14.9). A call is initiated by mobile A, by dialing the called party number (or by selecting it from a menu) over the control channel (Fig. 14.10).

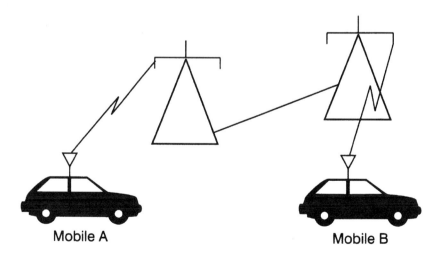

Figure 14.9   The mobile initially monitors the local control channel.

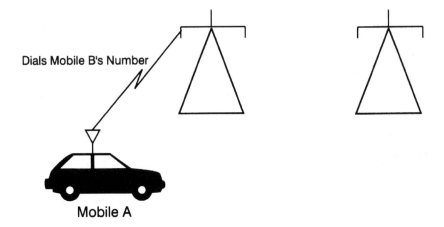

Dials Mobile B's Number

Mobile A

Figure 14.10   Mobile A initiates a call.

Having received the call request, the system will then page the called party on all sites (it may do this selectively, looking for the called party locally first) (Fig. 14.11). If the B party is available, it will acknowledge the call. When the acknowledgment is received, the system will instruct the mobiles to go to a voice channel. Note that if both mobiles are operating off the same site, then they will both be switched to the same channel.

Finally, the mobiles will switch to their respective voice channels for the duration of the call. If the call is not local, then an intersite switch will connect the two voice channels together and the call will proceed until terminated by one of the parties or until the system timer clears down the call.

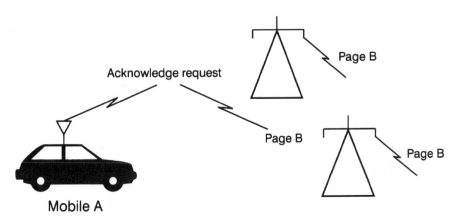

Acknowledge request

Page B

Page B

Page B

Mobile A

Figure 14.11   The call is acknowledged and the B party is paged.

MOBILE ID

The mobile ID is a seven-digit number consisting of a three-digit prefix, followed by a four-digit "identity." The first three digits can be in the range 000–127 (the size is constricted by the 8 bits used to specify the prefix), and the "identity" can be any number between 2 and 8100. Thus a valid mobile ID might be 078-3776.

A mobile may be active on the whole system, or its ID can be barred from access to certain sites. This enables a structured access charge based on the number of sites that each mobile is valid on.

In addition to a mobile ID, each mobile will have a built-in ESN (electronic serial number), which identifies the actual hardware of the mobile. This number is unique to each mobile, and it can be used as an additional verification of the caller's validity.

A TYPICAL MPT 1327 SYSTEM

Single-Site System

A number of suppliers arrange their channel hardware so that a single RF channel contains the necessary hardware to complete a trunked radio call. Without the associated base-station control hardware, there will of course be some limitations on the capabilities of the channel. Some typical limitations are as follows:

- All compatible mobiles operating in the correct frequency band can complete a call (i.e., there is no subscriber's validation list).
- Billing, operations, and fault data are not available.
- PSTN/PABX and landline access is not available.

Essentially, this single-channel mode is the default operational mode when the site controller fails. Use is often made of this facility for quick live demonstrations.

To avail of all the MPT 1327 features, a site controller will be added to the channels.

A single-site system can be configured for customized operation by the network management system. Parameters that can be controlled include the following:

- Call time duration. Both a maximum and minimum time can be set, and the system will dynamically vary the time, so that the minimum is approached as the system loads up.
- Maximum queue depth. This determines the maximum size of the queue.
- Maximum inactivity timer. Allows the system to drop calls after extended inactivity.
- Maximum time in the queue. This sets the time limit for a user to be held in the queue.

KX transceivers

Transmit and receive
combiners and isolators

Switch Matrix Frame

Process Controller

DC power supply modules
DC-DC converters or AC-DC
switch-mode power supplies

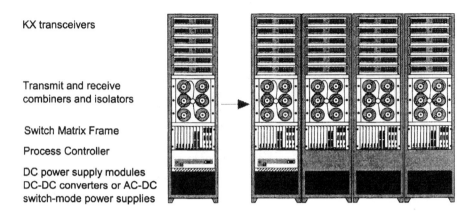

Figure 14.12 A large-site configuration from ADI.

- Option for the control channel to revert to traffic channel.
- Telephone traffic permitted.
- Group calls permitted.

Often these parameters (or combinations of them) can be set with a time-slot validity so that constraints can be relaxed in non-busy periods.

Larger sites are configured modularly from a single rack system as seen in Figure 14.12.

Multiple-Site System

A multiple-site system essentially consists of a number of single sites connected together via an intersite switch. When a call request is made, the site will first look to see if it can be handled locally and, if so, will connect the call. Where the called party is not on the same site, the call request will be sent to the intersite switch for processing.

Most systems are arranged so that base sites can have dedicated intersite channels plus dedicated local channels. These differ mainly in that the intersite channels have links back to the central intersite switch, while the local channels do not. This means that intersite channels can be used for local calls, but local channels do not have intersite abilities. A variant of this is where the base station has a local switching capability so that any voice channel can be switched to the local or intersite mode on demand. This leads to improved traffic efficiency.

SYSTEMS AVAILABLE

There is no standard network configuration required by MPT 1327, and a number of approaches are on the market. Most of them initially featured a single switch with a defined number of base stations under control. Limits were often also placed on the total number of channels, by the number of ports on the switch.

Recently, most manufacturers have moved to vastly increase the system capacities, and generally two modes have been adopted. The first is to expand in a star configuration by connecting a number of switches to a transit switch. For example, the original Fylde (a UK-based switch and base-station controller manufacturer) switch was capable of only 10 base sites, but its interregional switch enables the interconnection of 16 such switches to give a total of 160 base sites. This approach was also taken by Tait and Nokia.

The alternative is to make the system nodal. In a nodal system, any switch can work as a stand-alone or it can become a node (an interconnection point) for other switches. The Simoco TN10X is an example of a nodal switch. It can switch a number of local base sites and control them while at the same time acting as a node for other switches.

Rhode and Schwarz have an interesting system, which, like the Simoco system, is nodal, but it also has digital voice channels. It will be remembered that MPT 1327 not only leaves open the operating frequencies, but also does not specify the type of modulation. Rhode and Schwarz have taken advantage of this and added an FSK digital voice channel option. It operates at 4.8 kHz, with 4 kHz being used for voice (or data) and 800 Hz for overheads and error correction. The digital channels use the same bandwidth as the analog channels and have no TDMA splitting.

Dual-mode mobiles enable a blending of analog and digital channels. The advantages claimed for the digital channels includes better data performance and a wider coverage area.

OPERATIONAL PARAMETERS

Most systems have a number of time-tagged parameters that allow system management on a time-of-day basis. This means that call queue lengths, for example, can be tailored differently during the peak hour, as compared to off-peak hours.

Individual subscribers can also be validated on a time-of-day basis; thus, for example, "after-hours-only" or "weekend-only" users can be time-tagged and denied access outside the permitted hours (some operators offer this type of service for things like after-hours data transfer at reduced rates).

ACCESS LEVELS

Subscribers can be offered three access levels (normal, priority and emergency), which can be classified as local call area calling, PSTN access (allowed/disallowed), and data (allowed/disallowed).

DYNAMIC CALL TIMER

The call timer in some systems is dynamic. It has a maximum and minimum call length that varies as some function of the system load. A typical system may be controlled by the queue length. For example, suppose that the maximum queue length was set at 10. Assume also that the maximum call time is 90 s and that the minimum time is 50 s. Then the system can be set so that with a queue of 0, the call timer is set for 90 s, while for a larger queue it is reduced linearly at the rate of 4 s for each queued call. Thus when 10 users are in the queue, the call timer will be down to 50 s. This is the algorithm used by Fylde, but many others are possible. Fylde recommends that the queue depth be nominally set at 1.5× the number of channels and that the ratio of maximum to minimum call time also be 1.5×.

CALL HANDOFF

Under some conditions, call handoff is provided for. The most usual application of this is for local calls that because of channel availability were allocated an intersite channel on call setup, but which to conserve resources would be more economically carried on a local channel. Where this is implemented, calls are often handed off within seconds of call establishment.

B-PARTY QUEUING

B-party queuing is the mechanism whereby a B-party channel is held in reserve awaiting an A-party voice channel availability. Clearly, this only pertains to intersite calls, and it is essential to ensure the effectiveness of queuing. It can, however, cause a reduction is circuit efficiency (because a channel reserved but not carrying traffic is effectively temporarily out of use).

PREFIX AND ID SEGREGATION

Optionally, it is possible to assign all subscribers on a certain prefix to a particular dedicated channel, or alternatively assign a group of subscribers in a specified ID range to a specified group of channels. Although this will work against the basic traffic efficiency of a trunked network, it does permit certain users to have assured channel access (e.g., emergency services may want their own physical channels) or even the communal operation of a network with dedicated resources assigned to various users.

POOLED CHANNEL OPERATION

Where it is necessary to reuse frequencies on sites with overlapping coverage, it is possible to designate channels as pooled. These co-frequency channels will need to be "cleared for use" before being switched into service. In this way, frequency clashes can be avoided and calls queued until the co-channel is deactivated. Of course, use of this feature will reduce the efficiency of the use of network resources, and this feature would only be used where frequencies are at a premium.

CALL DIVERSION

Call diversion is available to either another mobile, a PSTN or PABX destination, or a dispatcher. Dispatcher positions can be hardwired from the main switch.

Third-party call diversion allows a suitably authorized third party (usually the dispatcher) to divert calls of a user to another address.

INCLUDE CALL

This enables a third party to be included into an ongoing call. If the third party is in the service area of one of the existing conversing parties, then it will be included on the same channel; otherwise it will be allocated as a new channel. This feature is not supported by all manufacturers.

FULL OFF-AIR CALL SETUP

Full off-air call setup enables calls to be entirely set up on the control channels. This has the advantage of not using speech channel time for ring tone, but has the disadvantage that it may well be that, upon answer, the called party is then queued until a free channel becomes available. This is particularly a problem when the calling party is on a PSTN/PABX line and is often left wondering what has happened after the ring tone ceased and no progress was noted.

The alternative is to operate a non-full-of-air call setup, which places the caller onto a voice channel for the duration of the ring tone (or in some cases without any ringtone). This is somewhat less efficient in terms of voice channel utilization, but it does give a near-instantaneous connect on called party answer. The choice is up to the users: Some emergency services prefer the instant access to the voice channel, while others are more comfortable to have a more "cellular"-like calling procedure. There is little agreement among manufacturers or operators on which mode is best, and both are widely used.

ENHANCED FUNCTIONALITY THROUGH NETWORK ADD-ONS

Being an open specification, MPT 1327 has allowed a large number of small companies to provide network enhancements. Radio Systems Technologies has a device called SmartBridge (shown in Figs. 14.13 and 14.14), which enables the network operator to link a conventional repeater network to the trunked network in an "off-air" mode. A trunked mobile is used as the link to the conventional network, and

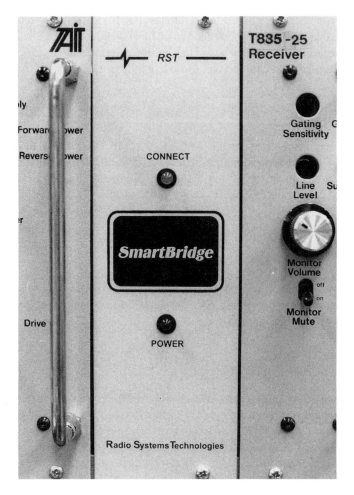

Figure 14.13 The SmartBridge system configured for Tait RF. (Photograph courtesy of RST.)

then the SmartBridge connects to the conventional repeater at the audio level, with a voice operated switch (VOX) controlling the PTT on the trunked radio. Access from the conventional system to the trunked network is via DTMF signalling. SmartBridge provides a path for a gentle migration from a conventional network to MPT 1327. SmartBridge is also available in a Smartnet version.

A product similar in concept, but different in function, comes from INDCOM, called LINELINK. LINELINK connects a telephone line or a telephone directly to a trunked mobile. It can be used in a number of ways. First, it can be used as an "off-air" link from the trunked radio network to a PABX. Second, it can provide a PSTN access to a network, where otherwise a physical link would be difficult or expensive (this is only relevant to small trunked systems). Finally, it can provide a "dispatch" type of access to an area like an office, where up to 10 DTMF standard telephones have access

Figure 14.14   The SmartBridge system to interface to any RF. (Photograph courtesy of RST.)

to the trunked network from a desktop "Batphone," which connects via the LINELINK to the trunked mobile and then to the network.

A similar product from Design Two Thousand, called a TACT TA-7T, is a trunked radio interface to a PABX (or PSTN). Designed to bypass the PSTN, when connecting between trunked radio users and a PABX, this device also bypasses PSTN charges. The TACT unit is called as though it were another mobile. Then using voice prompts, it instructs the trunked radio caller to dial the extension (or telephone number) via DTMF overdialing. Where DTMF is not available, it can automatically call an operator. Conversely, it can be used by a PABX extension to call any trunked mobile.

A number of suppliers offer GPS interfacing directly to MPT 1327 mobiles, and the dispatch and display console hardware to go with it. While most of these systems work readily with standard accuracy, at the time of writing, most have difficulty with adding differential correction in an economically practical way.

Interactive custom systems abound. One in the United Kingdom is an electricity meter reading system whereby the meter reader punches in the reading from a PC and gets a return data message, which is the customer's bill printed on the spot. The bill is delivered within minutes of the meter reading. Others have a range of SCADA applications transmitted by trunked radio.

MAP 27 is set to open an array of new opportunities for third-party suppliers once it becomes widely available.

All MPT 1327 mobiles have an electronic serial number (ESN) embedded in the hardware, which is quite separate from the mobile's ID.

MOBILES

Although all mobiles will generally comply with the MPT 1327 specifications, there are a wide variety of supplier options that are incorporated. Mobiles are available from a wide range of manufacturers, including Tait, Simoco, Motorola, Kenwood, Key, GME, and Nokia. Most provide both vehicle-mounted and a handheld version. Both the handheld and the mobile are often available in a dual-mode conventional/trunked, which is useful during the transition from a conventional network to a fully trunked one. Very few manufacturers cover all the frequency bands that are used around the world, so except for the most common bands, the range of suppliers may be limited.

Most mobiles and portables have talk-around; that is, mobiles can communicate directly to each other by bypassing the network.

Variations on the basic mobile may include encryption, scrolling, short message capability, extended alpha displays, RS232 ports provided as standard, and/or MAP 27 interfaces. These extended features are generally offered outside the MPT 1327 specification, which was deliberately designed not to constrain the manufacturers from offering enhancements. The Simoco alphanumeric head features a "personality key," which contains the mobile programming information for the mobile in much the same way as an SIM card does on a GSM mobile. This means that a user can operate from a fleet vehicle, taking the user ID, and number from vehicle to vehicle. All the user's functionality is contained, including access and priority levels and even the frequent caller list. Once the key is removed, the mobile can either be set to become inactive or default to a preset personality.

The screen itself plugs directly in place of the standard mobile head (so an upgrade is simple) and has a 4×20 character back-lit LCD. Double-height characters can be displayed.

An interesting variant of the handheld is the Tait T3040. It deliberately looks like a cellular phone and is the size of a medium cellular phone. Taking the cellular analogy a little further, this handheld can be worked in two modes: (1) as a normal two-way-style handheld and (2) at the press of a button the audio level drops and the unit can be held to the ear like a cellular phone. A small microphone is at the bottom of the unit to allow this operation. This is not merely a gimmick, because often it is undesirable to broadcast the incoming message to all of those in hearing range, and the feature is popular with users.

It is unlikely that trunked radios will ever get as small as the smallest cellular phones, simply because of their higher power requirements, typically 4–5 W. This higher power means that substantial heat sinks are needed, as well as more substantial batteries to achieve reasonable talk times.

However, a number of manufacturers now have handhelds that are similar in size to a medium-sized cellular phone. It may even be that some of the units have been deliberately designed to look and feel cellular-like.

MAP 27

One of the first suppliers to develop a MAP 27 interface was Tait. Many have been reluctant to do so because of uncertainties in the format, as well as doubts about user acceptance. MAP 27 is a data interface to MPT 1327 mobiles which also gives full access to the control functions of the mobile. So, for example, a mobile connected to a PC could be programmed to send a batch of data at a certain time each day, or to respond to incoming alarm data by alerting the operator.

The Tait "INFORM" text dispatch is a hardware and software dispatcher package using the MAP 27 interface and a mobile control head.

CHAPTER 15

SMARTNET

Smartnet is a Motorola propriety analog trunking system. The fundamental level of organization of the Smartnet system is the talkgroup, a grouping of users with a common interest. Generally, a mobile will be a member of one or more talk groups, each with its own ID.

The Smartnet system is built around the central site controller (see Fig. 15.1). All base-station channels are linked to the controller, while the control channel receivers are linked via the inbound recovery board.

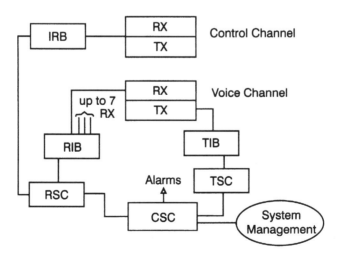

Figure 15.1 The Smartnet system. CSC, central site controller; RSC, receiver site controller; IRB, inbound recovery board; TSC, transmitter site controller; TIB, transmitter interface board; RIB, receiver interface board.

The CSC receives input from the RSC and TSC and makes the switching decisions. It also processes the calls, handles queuing and resource allocation, and performs diagnostics, alarm interfacing, and system management.

The CSC, RSC, and TSC are very similar hardware devices and are differentiated mainly by software.

SYSTEM CAPACITY

The system capacity of Smartnet is 48,000 IDs per system, with 4,000 talk groups.

Smartnet can have 20 channels per site, while Smartnet II is capable of 28 channels. Of these channels, up to four are nominally control channels, and they rotate as the active control channel on a daily basis. In the event of failure of an active control channel, another is assigned automatically.

REDUNDANCY

A high degree of redundancy is possible with Smartnet. A fully redundant central controller can be configured for immediate changeover. Additionally, the redundant controller can be used to try new software and updates, in a way that permits immediate switchover to the old systems if necessary.

A fully redundant base site can be set up, which duplicates the working base site. It would preferably be located nearby, but not co-sited, because the most likely cause of a total failure would be a lightning strike. Redundant sites can be employed more efficiently by arranging that two active sites cover the target area with overlap, but placing them so that with both working a significant fringe area is separately covered by each. Failure of one will then only amount in the loss of some of the fringe area coverage.

Failsoft is a feature that enables the mobiles to go to a conventional mode in the event of a repeater failure. The mobiles are preassigned a voice channel consistent with their talk groups. A subaudible tone identifies that the fail-safe mode is active, so that the users are aware of it, and also prevents this mode from being activated when the mobile simply drives out of range of a control channel.

Interference detected by a base-station channel will cause that channel to shut down. Similarly, a power output decrease in transmit channel will cause that channel to shut down.

RELIABILITY

A considerable degree of robustness is built into Smartnet. Because the operations are critically dependent on the control channel operation, up to four of the highest-frequency voice channels can be designated as standby control channels, and on control channel failure the next available channel will be allocated to the task.

Additionally, the four channels are rotated daily to spread the load (which is a 100% duty cycle) over a number of channels.

The transmitter power levels are constantly monitored, and a low TX power will result in the channel being shut down.

Self-diagnostic functions are carried out by the system central controller to check receiver and transmitter integrity.

DATA

Motorola has an associated data package called Intrac 2000, which enables remote data to be sent on the Smartnet system. The data are interspersed with voice, and they can be used for both monitoring and control.

The control channel signaling rate is 3,600 baud, and it sends data packets of 23 ms. Additional control information is sent over voice channels as subaudible and digital subaudible tones to provide misdirected radio protection, along with information updates.

Short messages such as ID and status messages are sent over the control channel, but text messages are sent over a data channel reserved for data transmissions.

MESSAGE TRUNKING

Although most Smartnet systems are set to assign a new voice channel for each PTT, an in-built hang time that can be set from 0.5 to 10 s will hold the current channel in use for an ongoing message (called *message trunking*). This is a compromise between the continuity offered by holding the channel in service until clear-down (which ensures continuity of service but uses more channel capacity) and immediate channel release on PTT (which maximizes channel occupancy, but can lead to many interrupted calls).

Messages can be queued, and this is ordinarily done on a first-in, first-out (FIFO) basis. There are five levels of priority on Smartnet I and eight levels of priority on Smartnet II, with an emergency call taking priority over all levels. The mobile is called back when a channel is free.

Automatic retry of unsuccessful call requests are initiated if the data are corrupted or fails to get through. Calls are randomly (in time) re-sent to avoid contention until the time-out condition is reached (typically 4 s).

The alternative, transmission trunking, which assigns a new speech channel on each PTT, is also an option (although it is less commonly used, it can increase channel efficiency if most messages are short).

FLEET CALL

A fleet call is one that allows a mobile to aggressively gain the attention of all members of fleet. The fleet call uses a good deal of network resources, and access to it is ordinarily restricted.

TABLE 15.1   The Smartnet Fleet Structure[a]

Fleet Type	Maximum Users	Sub Fleets	Maximum Fleets of This Type
A	16	3	255
B	64	7	32
C	128	7	16
D	512	15	2

[a]Smartnet II does not have these constraints.

Once placed, the fleet call directs all mobiles to the fleet call channels. Inactive mobiles can no longer place a call, and they will get a busy tone if they try. Call attempts during a fleet call are not queued, but simply ignored. Telephone calls in progress can continue but will be directed to the fleet call as soon as they are terminated. On Smartnet II systems a similar feature is the announcement call.

The system divides fleets up into categories that define the maximum number of users permitted in that fleet. As can be seen from Table 15.1, the larger the fleet size, the smaller the number of such fleets available. This means that while planning the fleet structure, it is important to allow for future expansion, but it is also important to ensure that the larger fleet allocations are only made to the biggest users. If a new fleet assignment is needed, the mobiles will need to be reprogrammed, unless dynamic regrouping is available.

PRIVATE CONVERSATION

Private conversation mode in Smartnet systems requires the user to switch to private call (PC) mode. The call is then placed, and it continues as a private conversation unless another caller also places a call to the same mobile. In this case, upon the first release of the PTT a conference call between the three parties will be set up. At the end of the call the originator must switch back to the group mode to receive group calls.

TALK GROUP SCAN

It is possible for a mobile to enter the talk-group scan mode, whereby it can monitor calls on a number of talk groups. It is also possible that the mobile can be programmed either to just listen or to be able to contribute to the conversation of the talk group.

MOBILE INHIBIT

A mobile can be inhibited off-air, in a way that prevents it from having transmit or receive capability until the operator restores it. This feature is particularly useful in the case of a lost or stolen unit. The inhibit command can be sent as an urgent request, with 1-s polling for 5 min, or regulated to one of several background activities of various duration and polling rates. The system will continuously monitor for the inhibited radio to ensure that the turn-off command is successful.

TELEPHONE CALLS

Telephone call capability depends on how many channels have telephone access. To call, the mobiles need to be suitably equipped. The caller presses the "phone" button and waits for a dial tone. Priority does not apply to telephone access, and calls are handled on a FIFO basis. On simplex calls a PTT tone is sent to indicate to the landline party that the channel is free.

Incoming calls require the caller to dial a system access number and then a four-digit mobile ID which is overdialed.

SYSTEM WATCH

A feature that allows the operator to follow the control channel activity in real time (and in plain English), called *system watch*, is available. It is particularly useful for following problem mobiles (e.g., ones behaving erratically) or deliberate misuse of such things as priority calling. This is a separate over the air monitoring system, with its own RF modem monitor and software.

ENCRYPTION

Encryption is available through Motorola's Securenet package. This involves adding the encryption package to the trunked repeater, and the call is diverted through the encryption equipment as required (ordinarily all channels can access the encryption path).

OPERATIONAL FEATURES

A "talk prohibit" tone, which is similar to the PSTN busy tone, is provided on the mobiles; it is heard when a user presses the PTT but the system is busy, the mobile is out of range, or the system is out of service. A "busy" light is also provided.

Callback on queuing is provided to alert a queued user that a channel is now available for the call.

Automatic retry causes the mobile to resend an unacknowledged call after about 4 s. This provides repeats in the event that the original call failed for some propagation reason. The retry procedure is randomized to avoid contention and continues for up to 16 tries if the call remains unacknowledged.

On fleet calls, misdirected mobile traffic (mobiles that have wrongly been sent to a group call) is all but eliminated by the continuous send of a subaudible fleet tone ID, which will redirect any errant mobile.

To ensure that even newcomers are directed to an ongoing group call, the group call has continuous assignment updating that constantly directs newly switched on mobiles to the group call in progress.

CONSOLE CROSSPATCH

Linking conventional or incompatible mobile networks into Smartnet can be done using a console crosspatch.

PAGING

Paging can be provided on Smartnet and can be placed using DTMF or a manual dispatch. When a page is initiated, it is directed to a specified channel and sent in a nontrunked format.

SMARTZONE

Smartzone can be viewed as an enhanced version of Smartnet. It initially allowed 25 sites to be controllers (compared to 10 by Smartnet), but later systems have even more capacity. Central to the Smartzone philosophy is that user talk groups will generally operate in local areas. Upon switch-on or changeover to a new site, the mobile registers its ID with the zone controller. The zone controller keeps a "talk-group registration table" and at any time knows where all members of a talk group are. Calls from one talk group are then sent only to all the sites that have registered users in that group.

Because of the zonal control, unlike Smartnet, it does not necessarily follow that all sites will need the same number of wide-area channels (except in the case where all talk groups are active in all zones), and Smartzone allows for the allocation of wide-area channels on a traffic demand basis. Motorola calls this *variable density trunking*.

Smartzone can coordinate frequency assignments at sites in a way to minimize adjacent site interference.

Members of a talk group can be tracked as they roam because each mobile sends its talk-group affiliation together with its ID.

UNIX-BASED FAULT-TOLERANT CPU

Smartzone uses UNIX-based fault-tolerant CPUs. These consist of redundant processors and redundant I/O subsystems. Failures are detected by dedicated hardware, and the failed components are switched out of operation. A Windows-based CPU should soon be available.

STARTNET/SMARTWORKS

Startnet and Smartworks are single-site versions of Smartnet. While they are RF-compatible with Smartnet, the controllers are different. Startnet can have up to five channels, while Smartworks can have seven (with up to three telephone interconnect channels). In both cases the control channel can be either dedicated or allowed to become the voice channel of last choice.

The number range capacity is up to 16,000 users and up to 2,000 talk groups.

Lost or stolen mobiles can be disabled from the system with the "selective radio inhibit" function.

The "automatic callback scan" ensures that once a mobile is called it will be part of the talk group by simply pressing the PTT.

The "failsoft operation" ensures that in the event of a system failure the mobiles can continue to operate in the conventional mode.

Smartworks is effectively a cut-down version of Smartnet and uses the same cards, but the limitation of seven channels is imposed by its physical design.

DYNAMIC REGROUPING

The system enables talk groups to be reconfigured, which can be particularly useful in emergency situations where the response team is made up of users who have been drawn from disparate services. Because most Smartnet mobiles have a limited addressing capability, this feature is essential to ensure that anyone can access any mobile in a emergency.

SHARENET

For very small operations a system called Sharenet enables a number of small users to collectively purchase a system, and yet still have their own dedicated voice channels while sharing control channels, central controllers, and antennas. Partitioning of channels can be either hard (each user accesses only the assigned channels) or soft (the user first accesses assigned channels). Up to eight individual users can be assigned to Sharenet.

EMERGENCY CALLS

An emergency call button is on all mobiles. When pressed, the call can be accorded the highest priority and sent either to an address in the caller's own talk group or to a talk group designated to handle such calls.

CONTROL STATIONS

Control stations are best thought of as operator or dispatch positions. They may be either RF stations (effectively a dispatch position operating via a mobile) or hard-wired consoles.

Mobile-based dispatchers are best for only a few users (although up to six can be accommodated). Hard-wired consoles will often require leased lines to connect them, which can be expensive, but they do provide enhanced access and can accommodate a large number of dispatchers (originally up to 24, but the limits have recently been removed).

TRUNKING TERMINALS

These are similar to control stations, but they have a deeper level of network access. Features such as dynamic regrouping, selective radio inhibit, status, and messaging are all available, only through a trunked terminal.

MOBILES

A feature of the Motorola trunked radios is that they are all dual-mode and can be used outside the trunked area for conventional calls. Often conventional repeaters are used for regional coverage at the boundaries of the trunked radio systems. Alternatively, stand-alone trunked systems (that is, they are not linked to the main system) are sometimes used.

The mobiles are also capable of talk-around, which enables simplex calls mobile to mobile. These calls are independent of the trunked network.

Motorola Syntor X mobiles, as well as an earlier model, were made so that the basic RF part is the same for all units, and the personality is contained in a code plug located in the control head. This made "loaner" mobiles easy and effective, because the whole RF package can be changed over without changing the mobiles characteristics.

The priority levels available on all mobiles were as follows:

1. Emergency call
2. Tactical priority
3. Command priority
4. Operational priority
5. Normal

The emergency call can be assigned to a reserved emergency channel if none exists. In the case where there is no reserved emergency channel, the call causes the system to become a transmission trunked system, with the emergency call receiving highest priority. As an alternative to this, ruthless preemption is possible, which takes over a channel in use from an existing low-priority user.

The other priorities are controlled by the dispatcher and depend on the gravity of the situation. The exception is operational priority, which is ordinarily assigned to public safety and emergency response services for routine operations. All other units will be assigned normal priority.

WIDE-AREA SERVICES

Receiver Voting

When more than one site is required to service the coverage area, a number of options are available. The cheapest option, which can only be used to improve areas of marginal coverage or extend handheld range, is to establish a second site that is receive only. This can be especially effective if the receive site is a quiet one (that is, with few or no local transmitters).

The system will have a remote link to the receiver site, along with a voting system that will determine which receiver is best situated to handle the call. Figure 15.2 illustrates this.

Voting comparators called Spectra-Tac and Digitac are available.

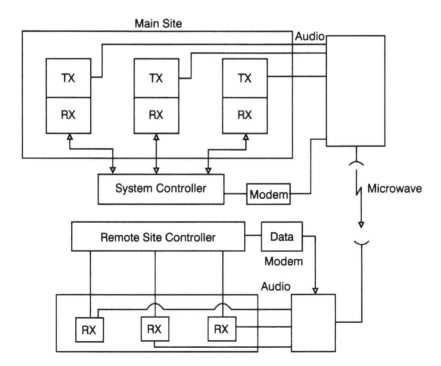

Figure 15.2   A remote receiver-site operation.

Simulcast

Simulcast is a reliable but expensive way of ensuring wide-area coverage. In simulcast a number of sites are trunked in parallel, using the same frequencies at each site as shown in Figure 15.3. It is critical in simulcast that the frequencies be exactly matched, and it is also important that in areas of equal field strength (from one or more transmitters) the audio delay be equalized.

To accomplish the frequency stability, a rubidium standard was originally used at each site to control the synthesizer clocks. This has since been replaced by GPS clocks. Audio levels and delays are equalized by an audio processor. The links between the sites need to be microwave, optical fiber, or other wide-band media.

Although each site is always simultaneously transmitting during a call, the receivers vote for the best signal, and the network audio is derived from the best site.

Improved simulcast performance is obtained by separating out the voice audio and the data at the receiver site.

Automatic Multiple-Site Select (AMSS)

Where spectrum is not at a premium, AMSS can be used which effectively parallels sites in a way similar to simulcast except that the various sites use different frequencies. AMSS mobiles need to be ordered as such and can have up to eight different control channels.

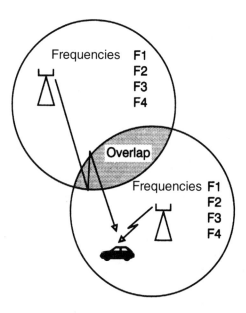

Figure 15.3 A simulcast operation.

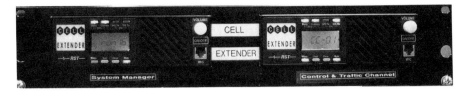

Figure 15.4 SmartBridge can be used to link conventional and Smartzone networks. (Photograph courtesy of RST.)

Using AMSS, calls can be initiated as local or wide-area, the option being chosen by the originator with the fleet selector.

REPEATERS

A repeater, called a PAC-TR, is a repeater that is essentially a trunked radio. This repeater is designed for in-building coverage, and it is accessed by the mobile in the conventional mode. Because the portable is operating as a conventional mobile, it does not have access to all the trunked radio features.

A similar configuration is available to allow a vehicle-mounted mobile to repeat to handheld transceivers operating in the vicinity.

SYSTEM-TO-SYSTEM INTERCONNECT

Products exist to connect a Smartzone system to a conventional network. Radio Systems Technologies provide a rack-mounted package called a SmartBridge, which links the two networks using any RF equipment. This has applications where a gentle transition is required to phase-out an existing conventional network, as well as in low-density rural areas where the cost of a full trunked system may be hard to justify. The RST SmartBridge is shown in Figure 15.4.

CHAPTER 16

LOGIC TRUNKED RADIO

Logic trunked radio (LTR) was developed by E. F. Johnson as a simple multichannel mobile radio system. These mobiles are marketed as "LTR." They do not use a control channel.

Although LTR mobiles are widely available (manufacturers include Uniden, King, Standard, and Repco in addition to E. F. Johnson), it is possible to convert conventional synthesized mobile radios to the LTR standard by adding a suitable control board.

The base stations consist of individually controlled channels, and each channel has an associated controller. Each channel can have 250 IDs that may be assigned to talk groups or to individual mobiles. Up to 20 channels can be provided per base station. Generally, there is no centralized control.

Users are divided into systems (there are mobiles available that can access a maximum of 10 systems, while some can access all 250), with each system containing a number of user groups. Users in one system cannot talk to users in another.

Within each system there are typically 10 user groups. Some mobiles are not equipped with a group select facility, and such mobiles can only have one ID within each system. The calling groups are defined by subaudible tones. Individual calling can be activated for some mobiles.

Where interconnection is permitted, the repeater is programmed with those IDs that are permitted interconnect access. These IDs, known as *repeater interconnect codes* (RIC ID), can include some or all of the 250 possible codes available on the home repeater.

Most LTR mobiles are in the U.S. trunking band (806- to 824-MHz mobile TX and 851- to 869-MHz mobile RX), although units are also available in the 900-MHz band. Because there is 45 MHz between the send and the receive frequencies, cellular-type duplexers can be used in the mobiles to give full duplex operation. These frequencies are directly adjacent to the AMPS/TDMA/CDMA cellular bands, which start at the 824-MHz mobile TX as depicted in Figure 16.1. The band occupies 18 MHz, and most

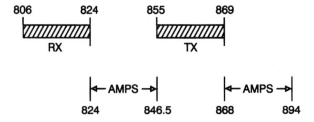

Figure 16.1 The LTR band is adjacent to the AMPS band.

mobiles can be used over any part of the band. This allows good frequency spreading for efficient duplexing (at the RF level).

MOBILES

The mobiles frequently have a talk-around capability that allows them to talk mobile-to-mobile, in the simplex mode, when they are out of range of a repeater.

Most mobiles have a "system selector" switch that enables them to change their home channel, and all have a group selector switch that selects the called party.

Mobiles can have one or more IDs, each with its assigned priority. If the mobile is receiving a low-priority call and a higher one is initiated, the new call will drop the old one. However, this feature only works if the mobile is on its home channel at the time of initiation of the high-priority call.

Access priority in a busy system is not a feature of LTR systems, and all mobiles have an equal, first-come, first-served access priority.

All LTR mobiles can be programmed for conventional operating as required.

A clear-to-talk tone is available in some mobiles. This tone sounds after the PTT has been pressed, to indicate that the channel is available. It can optionally be disabled.

A wide range of mobile types and features are available. The TX power can vary from 1 W to the special high-power versions delivering 35 W. Be aware that the coverage will be quite different for the mobiles spanning this wide power range (because there is 15 dB of variation in the transmit power levels). Also, system design will to some extent be constrained by the performance of the lowest-powered mobiles being used.

Both full-duplex and two-channel simplex mobiles are available. One of the smallest offerings in full duplex is the Viking CX from E. F. Johnson. This 0.6/1-W model is similar in styling and size to the early Motorola MICROTAC AMPS phones. It even has a similar keypad and the familiar flip microphone.

Free System Ringback

On telephone calls, an attempt to access a busy system will automatically activate the "free system ringback," whereby the system will signal back to the caller once the system is free. This facility is not available on mobile-to-mobile calls.

Other LTR Mobile Features

Missed calls (ones that came in while the mobile was unattended) can be registered on some mobiles, with only selected IDs being displayed.

A horn alert feature and time-out timer are available in most LTR transceivers.

Conventional Mode Features

In the conventional mode, most LTR mobiles can revert to a subset of the LTR features. These include subaudible (CTCSS) or digital squelch, automatic transmit disable on busy, talk-around indicator, horn alert, and time-out timer.

Voice Encryption

Voice encryption is available on some LTR models and can have up to 5 billion possible codes. Encryption can be used for both interconnect and mobile-to-mobile calls. It cannot, however, be assumed that encryption can be fitted (or retrofitted) to all mobile models, so it would be wise to check on this option with the supplier if encryption is a consideration.

BASE STATIONS

The base station can consist of up to 20 channels, each of which are connected together by a high-speed bus with a coaxial cable, which may be up to 160 m long. One of the channels is designated to be the bus management repeater, and it coordinates the interchannel traffic. This structure is seen in Figure 16.2.

Failure of a home channel disables all the mobiles that ordinarily home on that channel. While mobiles with a "system selector" can then change to a new home channel, this is only possible if that mobile has been validated on the new home

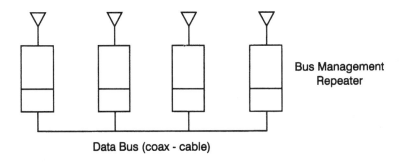

Figure 16.2   The structure of an LTR base station.

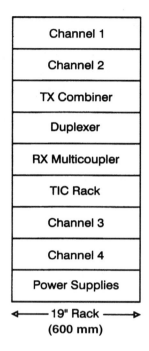

Figure 16.3   An LTR base-station rack.

channel. To provide alternative home channels for mobiles, will use up home addresses.

Signaling is done at the subaudible frequency 150 Hz, and it is overlaid on the voice transmissions. This allows calls to mobiles recently signed on to be initiated at switch-on, even if the mobile missed the first part of the call.

Base-station repeaters are available in a range of TX powers from 25 to 175 W. A typical base station may be configured as seen in Figure 16.3.

INTERCONNECTION

The simplest but most expensive way to provide telephone interconnect on an LTR system is to have each channel connected to its own PSTN line. The alternative to this is to have a block of ID codes assigned to interconnection. In this way, each mobile that can be permitted telephone access has a code reserved for these calls. The system works best if these IDs are assigned in a block across all home channels.

Each assigned ID code can have associated restrictions on local or long-distance calling.

The interconnection feature is a separate capability and is installed in a repeater card rack, on the basis of one card per repeater channel that has PSTN access. To dial a PSTN number, most mobiles need to be fitted with an optional keypad, although a few models have preprogrammed telephone number capability.

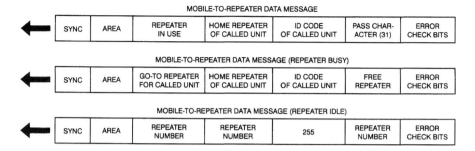

Figure 16.4 The signaling format.

Telephone calls can be full duplex or half-duplex (two-channel simplex), depending on the capabilities of the mobile. Toll call restrictions can be programmed into the interconnect cards.

The mobile telephone call request is steered to the repeater with an interconnect capability by the ID code used when making the call request. To enable future telephone access, each repeater should be allocated some "reserve" PSTN IDs.

SIGNALING

The mobile and base sites maintain a continuous message exchange during calls. The data are sent at 300 bits per second.

The signaling between the base stations and the mobiles follows the message format shown in Figure 16.4.

The two sync bits are used to begin the messages and are followed by an area identifier, which is used to prevent an adjacent system capturing the call request. This bit is ordinarily set to "0" or to "1" if there is an adjacent system that could interfere.

The repeater-in-use or repeater-go-to specifies the repeater number (1–20) being addressed. Again the main function of this data slot is to prevent false repeater seizures.

ID code contains the ID (1–5000) of the mobile or group being called.

The home repeater is the repeater number (1–20) of the mobile.

The free repeater function either identifies that the repeater is free (own repeater number transmitted) or identifies the number of an alternative repeater that is free and can be used to originate calls. If all repeaters are busy, then this slot will contain "0."

ATTEMPTING A CALL WHILE OUT OF RANGE

A mobile out of range that attempts a call will fail to receive a handshake and thus will try again. After a few repeated call attempts the mobile will switch to out-of-range tones (alternating high and low tones) and cease to attempt to contact the repeater.

DATA

Data can be sent on a voice channel at the rate of 300 baud. Signaling is done via subaudible tones, which limits the throughput rate but does provide for signaling even when the voice channel is in use.

HOME CHANNEL CONCEPT

Because there is no control channel, there needs to be a way for an incoming caller to find the called mobile. This is achieved by making the mobile "home" onto one particular channel whenever it is in the standby state. Up to 250 ID codes can be assigned to one home repeater channel.

The communication between the mobile and the home channel follows this sequence:

1. While the home channel is unoccupied, it sends a message every 10 s to the mobiles, instructing them to initiate calls on the home channel.
2. The home repeater, if free, is the first choice for originated calls. When the home channel is busy, the mobiles in the talk group are assigned another channel for originating calls (if one is free) or told that the system is busy.
3. As soon as there is a change from home channel busy to free (or vice versa), the mobiles are notified.

Incoming calls consist of (a) a data burst that identifies the mobile(s) called and (b) the channel assignment to receive the call.

Originating Calls

Calls are originated by pressing the PTT. The mobile will then go to the last assigned originating channel. If it is free, then a "clear-to-talk" signal will be sent and communications can begin. If the channel is busy, then a busy signal will be sent.

In LTR systems the control channels are distributed, meaning that any free voice channel can act as a control channel for an originating call. When compared to a dedicated control channel system, a distributed control channel is more efficient for small groups of channels because all channels are voice channels.

The distributed control channel means that there is no real queuing in LTR systems. However, the caller can hold down the PTT, even while a busy tone is being received, and in that way be assured of first access to the channel when it becomes free.

In a conversation between two mobiles, the above originating call sequence is repeated for *each* PTT. This means that the channel in use is not reserved for the reply (or subsequent use by the originator). Transmission trunking is the term used to describe this access mode.

Hang Time

Most transmission trunking systems allow the operator to select a hang time, so that calls stay on the same channel for a designated time after the release of the PTT. This ensures continuity of the call but will reduce circuit efficiency. LTR has a hang-time function on interconnect calls but not on mobile-to-mobile calls.

MULTISITE OPERATION

The original LTR specification did not envision multisite operation, and it is not an option available from most vendors. However, if it is required, it can be done using third-party equipment. Americom offers such equipment, whereby multisite calls are recognized by the caller ID and are switched accordingly.

NARROW-BAND TRUNKED RADIO

A narrow-band 220-MHz trunking system based on the LTR protocol was released in the United States in 1994. It uses 5-kHz channels and a form of SSB modulation which uses pilot tones, transparent tone in band (TTIB), to synchronize the SSB frequencies. It is claimed that data rates of up to 9,600 baud can be achieved with suitable modems.

CHAPTER 17

TETRA

Trans-European trunked radio (TETRA) is a trunked radio equivalent to GSM, in that where GSM was conceptually a radio extension of ISDN, so TETRA is a radio extension of ISPBX (integrated services private branch exchange). It is a digital standard that has been conceived to take over from MPT 1327, the de facto European analog standard, by the European Telecommunications Standards Institute (ETSI).

The main objectives were to make the hardware truly compatible from the different manufacturers (something MPT 1327 only partially achieved) and to double the spectral efficiency (from the 12.5-kHz analog bandwidth). Considerable effort has gone into ensuring compatibility of interfaces at all levels. Interoperability is assured; and in Europe at least, where the spectrum is also standardized, cross-border operations are foreseen.

The modulation method is TDMA, based on a 25-kHz carrier that can have up to four time slots with a total throughput of 28.8 kbits/s. These four channels are derived from a single transmitter, without the need for combiners or splitters. Both the mobiles and the handhelds can operate in the full duplex mode without the need for a duplex filter.

Unlike APCO 25, TETRA has no backward compatibility, and the interim plan is to establish "gateways" through which analog and digital systems can be linked. This rather unsatisfactory transitional method will probably slow the introduction of the technology considerably.

FEATURES

TETRA will provide the usual trunked radio features including the following:

- Broadcast calls
- Direct mobile-to-mobile calls

199

- Mobile used as a repeater
- Group calls (with dynamic assignment)
- Priority calls (up to eight levels)
- Encrypted speech
- Circuit mode data
- Short messages
- Conference calls
- Call diversion
- Mail box
- Automatic callback
- Include call
- Discreet listening
- Calling number ID
- Call me back
- Telephone access

For high-security applications, some channels in the system can be removed from the general pool and assigned full-time to specified users.

Stolen mobiles can be disabled from the network, to prevent unauthorized usage.

The target call setup time is 300 ms.

Among the claims for TETRA are the possibility that a multimode TETRA/GSM/DCS1800/ERMES terminal could be made available.

MODULATION

The modulation chosen is $\pi/4$ digital quadrature phase shift keying (DQPSK), which permits a 36-kbit/s transmission rate on a 25-kHz channel. The modulation scheme has four symbols (being the phases $\pi/4$, $\pi/2$, $3\pi/4$, and 2π) that transmit two bits per symbol period, at 18 kBaud, for the gross rate of 36 kbits/s. The decoder looks for phase changes, rather than absolute phase states, and so a very simple decoder can be used.

The signaling rate allows a 19.2-kbit/s data throughput after code redundancy is added. Three rates of circuit mode data are available, with the data rate depending on the level of protection. For unprotected data the throughput is 7.2 kbits/s per 6.25 kHz of spectrum. This rate is halved for the first level of protection, and halved again for high protection. When high protection is applied, the throughput is reduced to 9.6 kbits/s for the full 25-kHz bandwidth. The corresponding unprotected data rate is 28.8 kbits.

The full-rate speech CODEC has a basic rate of 4.567 kbits/s, with 2.633 kbits/s of error protection, using advanced code excited linear prediction (ACELP). The CELP algorithm is a generic class of analysis by synthesis. It works like this. There is a codebook of excitation sequences (a library of noise fragments), which are usually

Gaussian white noise. If these library signals are passed through suitable set of filters, they can produce voice-like sound fragments. By using a standard filtering arrangement, the transmitter passes the library bits through the filters and selects the one that most nearly matches the analog voice signal. Once the selection has been made, the CODEC then sends the index number of the noise fragment of best fit, which can then be used by the receiver to reconstruct that sound. Figure 17.1 shows a simplified CELP coder.

As you might imagine, the CODEC has to work very fast to sort out the best-fitting code fragments, and a measure of this is the number of generic operations (GOps) that need to be performed. For a TETRA call, this works out to be about 13.6 MOp/s (mega operations per second).

The half-rate CODEC has been extensively discussed; but because there are still some reservations about the full-rate CODEC, the half-rate version had not been developed at the time of writing.

Notice that the 25-kHz channel is divided into four 7.2-kbit/s TDMA speech channels. When only one channel is occupied, the data rates are one-quarter of the above. Note that it is possible to combine either one, two, three, or all four of the time slots into a single send/receive data stream.

A 12.5-kHz version is also proposed. The transmission mode can be either voice + data or a packet data optimized mode (PDO).

The time slots will be multiplexed as shown in Figure 17.2. Each carrier is divided into time slots of 14.17 ms, and each slot corresponds to 510-bit periods (including guard band and ramping times). In turn, four timeslots are arranged into frames (56.7 ms), and 18 frames are arranged into a multiframe (1.02 s). Every 18th frame is exclusively used to control signaling and is known as a control frame. Finally, 60 multiframes make one hyperframe. The frame structure is shown in Figure 17.3.

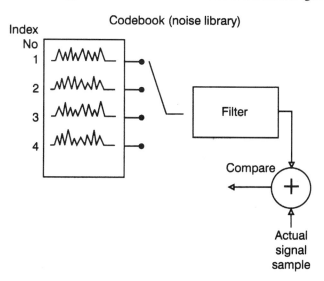

Figure 17.1 A simplified CELP coder.

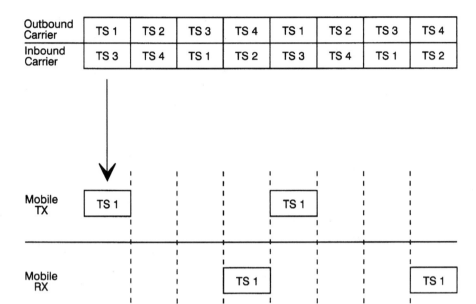

Figure 17.2 The TETRA time-slot structure.

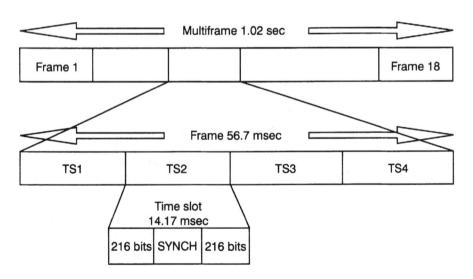

Figure 17.3 The TETRA frame structure.

ERROR CORRECTION

Like all digital systems, TETRA uses error correction techniques. Forward error correction (FEC) is used to correct minor errors in the transmitted data. Because mobiles are subject to Rayleigh fading, which in turn causes burst errors, bit interleaving is used. When the errors are too massive for correction, an automatic repeat request (ARQ) is sent to request a resend. In speech coding, too frequent use of ARQ will cause discontinuous speech and cannot be tolerated.

FREQUENCY

The bands allocated for TETRA are 380–400 MHz for emergency services, parts of 410–430 MHz, 450–470 MHz, and parts of 870–933 MHz for civil applications. The duplex spacing will be 10 MHz except in the 900-MHz band, where it will be 45 MHz.

Because timing is critical in TDMA systems, a frequency correction channel (FCCH) is sent periodically to enable the mobile to check its frequency and to measure any offset against the station frequency and thereby derive a frequency offset factor. Baseband frequency checks against the local oscillator are also done.

ROAMING AND HANDOVER

An intersystem interface (ISI) is to be provided to allow roaming between different systems. Handover between cells is provided.

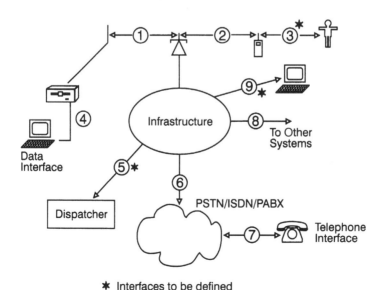

Figure 17.4   The simplified TETRA interfacing.

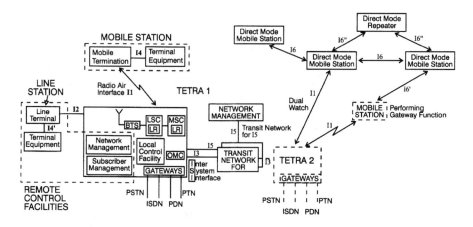

Figure 17.5 The TETRA reference standard and defined interfaces. *Abbreviations*: BTS, base transceiver station; ISDN, integrated services digital network; LR, location register; LSC, local switching center; MSC, main switching center; OMC, operations and maintenance center; PDN, packet data network; PSTN, public switched telecommunications network; PTN, private telephone network. *Defined interfaces*: I1, radio air interface; I2, line station interface; I3, intersystem interface; I4, terminal equipment interface; I5, network management interface; I6, direct mode radio air interface; I7, direct mode via gateway.

DEFINED INTERFACES

Six different standard interfaces have been defined, and three more are proposed. These are shown in a simplified structure in Figure 17.4 and in more detail in Figure 17.5.

MOBILE REPEATER

Motorola's Smartnet allows limited use of a mobile as a repeater, but TETRA allows almost transparent usage. This provides for mobile-to-mobile coverage without using the repeater network, and it also provides extended coverage from mobile to handheld. The range in this mode is expected to be from 400 m to 2 km. Features to be included are encryption, short data, and TX identification.

DIRECT-MODE MOBILES

The direct mode enables mobiles to connect to each other without going through the repeater. This can allow the mobiles to be used for local transmissions, even when they are outside the network coverage areas. Mobiles can be set for direct mode only, trunked or direct under manual switching or dual watch mode, where the mobile monitors both modes. The direct mode uses all the standard security features such as encryption and authentication.

NETWORK CONFIGURATION

TETRA provides interfacing between TETRA networks via a transit network. This enables roaming, and also direct interconnection between networks, in different areas (even when they may not be frequency-compatible).

It is of interest that the mobile radio can perform a gateway function and act as a repeater. In fact, in theory at least, a network could consist of a single base site and a number of mobile repeaters, located so that the required area was covered.

RF SPECIFICATIONS

The base-station power is 25 W, while the mobile is 10 W and the handheld 1–3 W with power control.

The receiver sensitivity will be a limiting factor; as like all TDMA systems, the range of any one site is generally less than the analog counterpart. The receive sensitivities are as in Table 17.1.

When compared to a conventional mobile with 25 W ERP which is transmitting to a base station with a sensitivity of –120 dBm, it can be seen that a TETRA base will have a path gain disadvantage of 9 dB, if the average base sensitivity is used. Assuming that the signal strength falls off as the inverse forth power of distance, the relative range is

$$(D_c/D_t)^4 = 10^{0.9}$$

where D_c is distance covered with a conventional system and D_t is distance covered with a TETRA system (Note that 9 dB = $10^{0.9}$ expressed as a power gain). This gives the conventional system a relative range of 1.68× TETRA (alternatively, TETRA has 60% the range of a conventional base).

The relatively poor range can in part be compensated by the use of diversity receivers; and in most applications, except where the site is in place only for local traffic purposes, it would be advisable to use diversity. A gain of around 3 dB can be expected with diversity use.

With the expected wide usage of handhelds, TETRA provides receive-only base stations to improve coverage.

TABLE 17.1   Receiver Sensitivity

RX Type	Stationary (dBm)	Moving (dBm)
Base	–115	–106
Mobile	–113	–104
Handheld	–112	–103

QUASI-SYNCHRONOUS TRANSMISSION

In order to improve spectrum utilization, especially in low traffic regions, the TETRA network can operate in a quasi-synchronous mode. This permits use of the same frequencies from different sites operating effectively in parallel. When using this it is necessary to identify the area where the field strength will be within ±6 dB, from two adjacent sites. In this overlap area the arrival of the two signals needs to be kept within a difference of less than one-fourth of a symbol period, which is 13.9 μs or equivalent to 4 km. Deliberate time delays can be set at the base sites to obtain the desired delay. Note that the TETRA modulation uses two bits per symbol, and a bit rate of 36 kbits/s, so the symbol rate is 18 kbits/s, which gives a symbol period of 55.56 μs.

Provided that the symbol delay is kept within the one-fourth symbol period, the in-built equalizer will be able to handle the resultant multipath. Three classes of equalizer are available for different multipath environments. The class A equalizer is for urban and mountainous areas (extreme multipath), class B for typical built up and urban areas, and class Q for mobiles. Note that all of the classes are fine in rural areas.

TIME-SHARED TRANSMISSION

In very low density areas the TETRA system allows time-shared control channels (so that one frequency can serve the control channel requirements of a number of sites) and traffic channel assignment on demand.

RECEIVE-ONLY SITES

Handheld coverage is a limiting factor on all trunk systems. This is because of the low output power of the handheld compared to the mobile or the base. Often a handheld can receive but cannot talk back. Local receive-only sites can relay the handheld transmissions and thereby extend coverage relatively cheaply.

POWER CONTROL

The power control for a mobile is stepped upwards from 30 mW, in 5-dB steps to the maximum power level of the class of mobile. This provides battery savings as well as reduced interference to co-channel and adjacent channel cells. In most cases the power control is directed by the base-station. The base station transmitters themselves have no adaptive power control.

CHAPTER 18

APCO-25

The APCO-25 standard is being driven by the Association of Public Safety Communications Officers Inc., which covers emergency services such as fire, ambulance, police, and so on. The project was initiated in 1989 and is basically a U.S. and Canadian standard. The name derives from an earlier body founded in 1935, the Association of Police Communications Officers, which was formed to lobby the FCC on behalf of the law enforcement community. The number 25 refers to project 25 set up as a work group in 1989 to determine a standard for a digital mobile network. The objective of this group was to achieve the following:

- Maximize channel efficiency and achieve a transition from 25-kHz channels to 12.5-kHz channels (later extended to include 6.25-kHz spacing).
- Ensure a competitive product through the development of an open standard.
- Deliver effective and reliable interagency communications.
- Provide a smooth transition from digital to analog services.
- Ensure that the equipment is as user-friendly as possible.

The objective of the standard is to define future digital trunked systems that will be based on an open standard and will satisfy the needs for radio communications into the next decade. The APCO-25 steering committee has 11 members. Four are from APCO, four are from the National Association of State Telecommunications directors (NASTD), and one each is from the Department of Defence, National Communications Systems, and National Telecommunications and Information Administration (NTIA).

A part of the original charter was to design the system around proven standards. This led to the choice of FDMA and, interestingly, to the exclusion of CDMA which was judged to be insufficiently proven at the time to be considered. On similar grounds, TDMA was rejected.

The primary objective was to define six open interfaces:

- Data host
- Data port
- Telephone interconnect
- Network management
- Intersystem interface
- Common air interface (CAI)

Two additional interfaces were defined during development:

- Console
- Fixed station

These layers are shown diagrammatically in the APCO-25 reference model in Figure 18.1. Additionally, the standard requires backward compatibility with the manufacturers analog trunked systems. This allows for the phased introduction of the new standards.

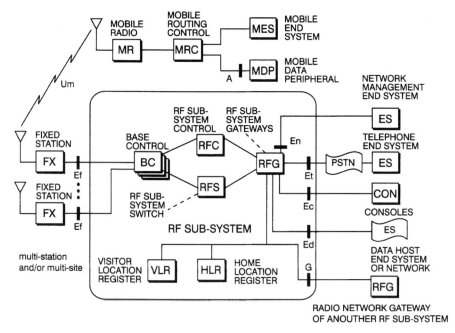

Figure 18.1 The APCO 25-reference model. *Defined Interfaces*: Um, common air interface; A, data port; Ed, data host; Et, interconnect; En, network management; G, intersystem interface; Ec, console; Ef, fixed station.

The CAI will ensure that a nonpropriety system can be implemented so that mobiles can be sourced from a wide range of suppliers.

CALLING FEATURES

- Priority calling (including emergency calls)
- Encryption (all voice and data calls can be encrypted using the data encryption standard (DES), with dynamic rekeying over the air)
- Call alert
- User ID (the terminal displays the users ID)
- Group calling
- Over-the-air programming (OTAP) (subscriber terminals can be reprogrammed over the air at any time)
- Affiliation (subscribers can change groups dynamically)
- Call restriction
- User programming (certain mobiles can be given the ability to reprogram themselves)

The frequency bands currently under consideration are 138–174 MHz, 380–512 MHz, and 800–900 MHz. The duplex separation is 45 MHz for the 800-MHz band and will be band-dependent in other frequencies. The initial systems will use 12.5-kHz channel spacing with a plan to migrate to 6.25 kHz around the year 2005.

Transmit powers range from a maximum of 350 W for a base site and from 10 to 110 W for mobiles and from 1 to 5 W for a handheld.

Protected data rates of 9,600 baud is available, giving a data throughput rate of 6,100 baud.

The CODEC, using IMBE (improved multiband excitation), operates at 4.4 bits/s, and at an error protection data rate of 2.8 bits/s. Thus the net data rate is 7,200 bits/s. The modulation uses 12.5 kHz of bandwidth, with a four-level FM modulation (C4FM). A half-rate, 6.25-kHz modulator using QPSK-c modulation is proposed for the future. The half-rate modulator will need a linear RF amplifier.

Channel coding is BCH for network ID, Trelliss codes for data, and Hamming codes for embedded signals.

The hardware will be capable of carrying conventional traffic with CTCSS signaling; also, digitally based conventional mobile calls can be carried. Talk-around between mobiles (independently of the network) is featured.

ADI SYSTEM

One of the first manufacturers to produce an operational system was Stanilite (now ADI). Their base station, seen in Figure 18.2, can be configured with up to 23 full-duplex channels, one or more of which can serve as control channels. A fully

Figure 18.2   An ADI APCO-25 base station. (Photograph courtesy of ADI.)

loaded rack contains eight channels and weighs 350 kg. All channels can be switched to the PSTN at any level up to T1/E1.

The system is nodal in design, and the Stanilite node controller is based on a UNIX software hub. The node controller looks after base-station, switch, and interface control and supervision.

The home location visitor register (HLR) and the visitor location register (VLR) are contained within the node controller, but on larger systems they may be separated. This contains

- VLR number
- HLR number

- Subscriber's service package
- Location/segment numbers
- ESN
- Authentication key
- Diversion on busy number
- Last register status
- Bad-user status

A network management facility (NMF) that is Windows-based holds data bases containing subscriber's information, terminal equipment details, and network resource data.

Network functionality of the NMF includes fault and alarm management, accounting, operations and maintenance, and system configuration.

A dispatch terminal is available.

The associated mobile is shown in Figure 18.3. It allows 8 h of normal operation (defined to be standby:receive:transmit % split of 90:5:5) between battery charges. A separate socket for data is provided. All APCO-25 mobiles can be programmed either in the field or over the air. The over-the-air package enables the following to be changed at any time:

Figure 18.3   A prototype handheld mobile. (Photograph courtesy of ADI.)

- System, group, and ID numbers
- Encryption capability
- Control of power levels
- Priority levels
- Assigning talk-group numbers
- Analog channel selection

APCO-16

APCO-16, an earlier specification from the same body, is a functional specification only, leaving manufacturers free to implement their own modulation and signaling techniques, with the result that compliant hardware from different suppliers need not be compatible with each other. APCO-16 was used as a de facto standard by US manufacturers to develop proprietary systems that in a way worked against the original aims, by making neighboring services and even some local service facilities incompatible with each other. In a way, APCO-25 seeks to redress the shortcomings of this earlier standard.

CHAPTER 19

---

iDEN

Formerly known as MIRS (Motorola integrated radio system), iDEN is a digital narrow-band TDMA (squeezing six channels into 25 kHz), using a multilevel linear modulation scheme called M-16QAM. This particular modulation method has proven quite robust and has the ability to tolerate severe differential path delays, without needing an adaptive equalizer for timing corrections. Adaptive equalizers are used in GSM and IS 54 to reduce intersymbol interference. However, these equalizers, at least in 1994, when the system was being specified, added significant cost complexity and power consumption to the mobile, and for these reasons it was avoided.

To avoid using an adaptive equalizer, the iDEN technology sends the signal out over several 25-kHz channels at a relatively slow symbol rate, which means that the spread caused by multipath does not cause serious intersymbol interference spreading.

The spectral efficiency of iDEN comes from its modulation method. It should be noted that because most modern trunk systems operate at 12.5 kHz per channel, the actual capacity gain over analog is only threefold. This, however, is still significant, given that frequency reuse is very limited in trunked systems.

TRANSITIONING FROM ANALOG

The original concept was that analog operators in the 800 MHz band would migrate gracefully from analog to iDEN, recovering a few channels at a time (and gaining six digital channels per analog channel). It is claimed that no guard band between the digital and analog channels is required.

HYBRID CELLULAR/TRUNK?

iDEN has all the features of a trunked radio system plus most of the features of cellular. Overall, it is more cellular-like than trunk-like. Facilities provided include:

- Short alphanumeric messaging
- Paging capabilities
- Automatic redial
- On-hook dialing
- Call forwarding
- Voice mail
- Busy transfer
- Call restriction

While all of these features are available on other systems (such as MPT 1327), it does seem that the iDEN system is effectively a trunked radio system that was envisaged as being directly competitive with cellular. This is particularly evident in the size and capabilities of the switch.

The implementation of iDEN leaves no doubt as to its origins (which were GSM). The system was initially offered based around Northern Telecom's DMS-100 central office switch. The basic structure of iDEN also looks a lot like GSM as can be seen in Figure 19.1.

The essential switching structure of GSM has been retained, while the Achilles heel of GSM, the TDMA modulator, has been replaced by what seems to be a more mobile-friendly 16 QAM system.

However, a new complexity has been added. Dispatch calls are not routed in the same way as PSTN calls. A separate packet switched network is overlaid onto the system to provide a more efficient way of handling short-duration calls. While they use the same RF resources (channels, antennas, etc.), they are switched by an independent packet switch.

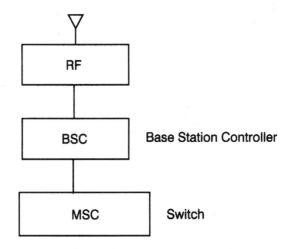

Figure 19.1   The structure of iDEN.

DESIGN

The design concept of iDEN is somewhat like that of cellular. The thrust seems to be to use a large number of smaller cells to enhance frequency reuse and thereby increase capacity. While this can be achieved, it is at the cost of economical wide-area group calling. The more sites that are used to cover an area, the more channels there are that will be tied up in a group call.

The small-cell approach also increases the switching complexity (and hence the switch cost) because in iDEN, site-to-site handoff of calls is allowed. The handoff facility (which operates like cellular handoff) is very demanding of switch processing time.

PERFORMANCE

In the United States the major user of iDEN is Nextel, which has underway a nationwide roll-out. Service began in California, with mobiles costing around $850. Monthly fees are $27 (1998) for a basic trunked service which includes 75 min of free airtime. Airtime is calculated in seconds of talk time, and it is measured only when the PTT is pressed. Telephone access is available at $40 per month (1998). The other main user is Onecomm, who offers service in Colorado.

At the time of writing there were systems in operation in most major centers of the United States, and many other countries have announced proposed systems, are doing so either with supplier credit (i.e., Motorola funding its own roll-out) or as a joint venture with Motorola.

Japan Shared Mobile Radio (JSMR), who as one of the main analog operators has almost 200,000 customers, operate an iDEN system in Japan in JV with Motorola. This system is initially stalled because permission is needed from the Ministry of Post and Telecommunications before the paging and "cellular alternative" services can be offered. Service is charged at 3,700 yen (approximately $45) per month.

Another major system outside the United States is in Beijing, where it is operated as CATCH.

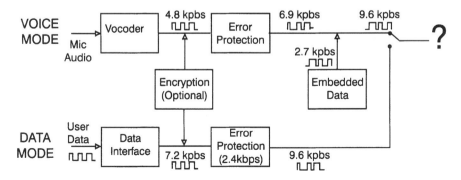

Figure 19.2   The digital channel structure.

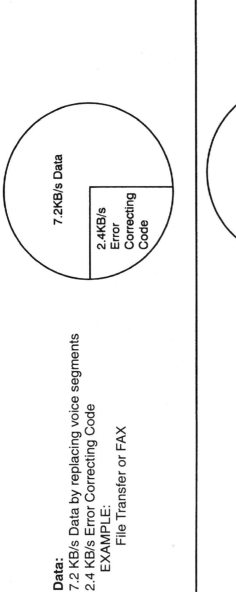

Data:
7.2 KB/s Data by replacing voice segments
2.4 KB/s Error Correcting Code
EXAMPLE:
 File Transfer or FAX

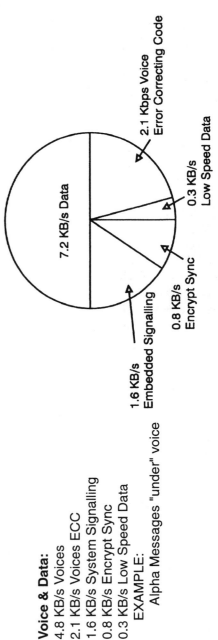

Voice & Data:
4.8 KB/s Voices
2.1 KB/s Voices ECC
1.6 KB/s System Signalling
0.8 KB/s Encrypt Sync
0.3 KB/s Low Speed Data
EXAMPLE:
 Alpha Messages "under" voice

Figure 19.3 The two data modes used in iDEN.

The performance of iDEN in terms of coverage is claimed to be better than that of an 800-MHz analog by about 30%. Voice quality is subjectively claimed to be consistently better than IS 54 TDMA. Such claims of more range and better voice quality for digital systems are often made, but they are rarely justified; users have reported less-than-agreeable voice quality.

The modulation method is quite different to TDMA, although the vector sum excited linear predictive coding (VSELP) codec operating at 7.2 kbits is similar to that used in the IS 54 standard. Error correction accounts for 2.4 kbits/s, while the digitized voice data stream is 4.8 kbits, as shown in Figure 19.2.

The voice quality is fairly consistent within the coverage area, and like most digital systems the quality degrades rapidly at the boundaries of coverage, and the link fails soon after.

DATA

iDEN provides for two data modes: The first is effectively equivalent to using a voice channel for data, while the second permits slow-speed data to be carried on top of a voice signal. These two modes are seen in Figure 19.3.

MOBILES

The first handhelds weighed 450 g. They featured paging and data communications through a built-in RS232 port.

CHAPTER 20

OTHER TRUNKED SYSTEMS

RADIO TELEPHONE EXCHANGE (RTX)

RTX is a trunked radio system that has been designed to upgrade conventional mobiles to trunking. The mobiles need to be synthesized, and the control module for RTX is added. Up to 1,000 mobiles can be connected to a system, and additionally 1,000 pager numbers are allowed. It is essentially a single-site system, although parallel multisite operation is available.

It allows full duplex, half duplex, and paging operating together. The telephone interface is via two- or four-wire connection. DID (direct in-dial) is possible by using a DID module that allows two, three, four, or five digits to be forwarded from the PSTN.

An activity timer allows either 3, 5, or 9 min, or alternatively no time limit on calls.

Each channel has its own controller, and there are no control channels. Paging is sent out on any available free voice channel.

Signaling is via CTCSS tones, or two-tone sequential.

Incoming Calls

Incoming calls are sent on a free channel that first sends out a "collect tone" to inform all the mobiles in the scanning mode that incoming call data are coming. The mobile or mobiles that are being called will stay on that channel for further instructions, while the rest will recommence scanning for a new free channel.

THE CES LANCER

The Lancer system is a derivative of RTX, and initially had a capacity of 2,000 subscribers but has recently been extended to 3,000. Like RTX, the philosophy of

Lancer is to use conventional synthesized mobiles and upgrade them to trunked radio with an additional control module. The interface module fitted to the radios is known as a Lancer interface-to-network controller (LINC). Like RTX, Lancer uses DTMF signaling.

All connections on the Lancer network require the automatic transmission of the mobile ANI. This is used for both call validation and billing. Although the Lancer network is a very simple one, it does provide a host of features including the following:

- Group calls
- Conference calls
- Call accounting
- Validated ANI
- Dial click regeneration
- Activity timers
- Full- and half-duplex operations
- Masking tones
- Stations ID
- Toll restrictions
- Priority override

Operations

In operation the Lancer trunked system to the user is much like a standard patched telephone operation. This is particularly evident in calling procedures.

Incoming calls from the landline are established first by ringing the telephone number assigned to the network and then waiting for a triple beep or dial tone. On hearing this the caller has 10 s to begin to overdial the called parties four-digit ID. Provided that the ID is valid, ring tone follows and the mobile has from 6 to 36 s to answer (depending on the programming). Disconnection is by pressing "#" (to which the mobile appends the disconnect sequence).

Mobile-to-landline calls are established by pressing "*" and then waiting for dial tone. The caller then has 10 s to begin dialing the required mobile. Both the system and the individual caller may be subject to toll call restrictions. Call timing is 1 min, 3 min, 9 min, or no limit.

Group calling is identical to individual calling, and the group call ID is a second number programmed in the memory. There are some small operational differences once a group call is made.

Paging is possible using one of four call formats:

- Motorola direct code
- Motorola general code

- GE direct code
- GE general code

Conference calling allows mobiles to add other mobiles or one (only) landline to the call in progress.

CTCSS

CTCSS is an optional feature on Lancer. When used, it should be noted that mobile-to-mobile calling between different CTCSS groups is not possible.

Call Accounting

The Lancer network controller has an in-built call processing capacity, and it comes with four options:

- Print every call
- Print landline calls only
- Store every call
- Store landline calls only

Because only one serial port is provided, connecting a local printer for calls will mean that remote control via the RS 232 port will no longer be possible.

The call record contains the calling party ANI, the called party, the time the call started, time completed, and duration. These details can be stored, and about 1,000 call records can be held in ASCII format. The records can be downloaded to a remote site via the RS 232 port using XTALK.PROCOMM and an equivalent communications program.

Deadbeat Disable

This feature permits the disabling of a mobile over the air without totally deleting the mobile from the system. It is available from the Lancer network controller, and it can be accessed by a landline.

Priority Level

There are six priority levels. Because high-priority users can disconnect the call of any lower-level user, allocation of these levels needs to be done with some care. If a high-priority user is seeking a channel, the mobile will first scan all available channels; if none are free, it will continue scanning and take over the next-lower-priority user it encounters.

Half-Duplex Repeater Audio

There are three options for the repeater audio operation during landline calls, when the mobile is talking to the landline. First, the audio can be repeated so that both sides of the call are clearly heard, and of course there is no privacy. Next the audio line can be disconnected so that only half the conversation is heard, making it more difficult to follow the conversation. Finally, an annoying 800-Hz tone can be sent during the mobile-to-landline period to discourage eavesdroppers.

Toll Restrictions

The toll restrictions that are available are quite limited and have clearly anticipated only the US market. Toll calls may be restricted on any first or second digit dialed, and up to 10 area codes may be included which bypass these restrictions. A limitation is that the bypass area codes are three-digit codes only.

Scan Times

The scanning time of the mobile is about 230 ms per channel, so the total time to scan all channels increases as the system size increases. This is a consideration in group and system calls. It also limits the response time of the network.

SmarTrunkII

The SmarTrunkII system evolved from the Selectone Corporation product range. SmarTrunkII is an LTR-like trunked radio system manufactured by SmarTrunk Systems Inc.

The SmarTrunkII system is available in most frequencies and is compatible with radios manufactured by Alinco, Kenwood, Motorola, Standard, and Yaesu. The mobiles are converted to trunked radio operation by the insertion of a proprietary module.

This analog system uses digital signaling for increased throughput on the signaling channels. Full- and half-duplex operation is possible.

Essentially a single-site system, SmarTrunkII can accommodate 16 channels maximum and has up to 4,096 user IDs.

It differs from LTR in a few ways. In LTR the controller assigns mobiles to a voice channel, while in SmarTrunk the mobile scans for and selects a free channel. Because of the digital signaling, it is possible for any mobile to call any other directly, which is something that can be quite difficult with LTR unless the callers are physically assigned to the same groups.

Telephone interconnect access is a little awkward with this system and requires incoming calls to be directed first to the controller, which then prompts the caller to overdial the mobile number. Calls are answered by pressing "*" and disconnected by pressing "#". Outgoing calls also require some extra dialing as a telephone number is identified by adding "1*" to the end of the dialled string of digits. In operation, the SmarTrunkII telephone interconnection is really no different from a telephone patch, long familiar in conventional mobile radio.

A mobile kill function is available.

In the WLL (wireless local loop) configuration, SmarTrunkII provides access for up to 1,100 subscribers.

For each call generated, a call record that shows the subscriber's number, the call type, the number called, call time, date, and duration is made. This can be downloaded from each controller (one per channel) over a built-in 1,200-baud modem.

A base station consists of up to 16 channels, each having its own independent controller. PSTN access is provided on a per-channel basis, and ordinarily only some of the channels will have such access. Figure 20.1 shows a typical system that is used for both mobile-to-mobile communications and for WLL.

SmarTrunkII is a scan-based system, and each of the 16 channels is scanned by the mobiles, until a vacant one is found and locked onto. Channel signaling is passed on the speech channel at the beginning and end of each call.

220-MEGAHERTZ NARROW-BAND TRUNKED RADIO

Also a derivative of LTR, the 220-MHz-band, narrow-band radio is essentially an LTR system with a narrow-band SSB RF. The bandwidth used is a mere 5 kHz, and the usual shortcoming of SSB, namely the "Donald Duck" effect caused by frequency

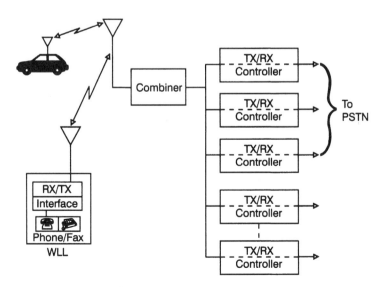

Figure 20.1   A typical SmarTrunkII system.

errors between the send and receive tuning, is eliminated by the use of pilot tones that are used to lock the receiver to the transmitter; these tones are known as a transparent tone inband (TTIB).

Being SSB, this system should be able to achieve sensitivity figures comparable to the conventional FM mobiles, even though the operations is on a much narrower bandwidth.

Data can be sent at 2,400 bits/s, but it is reported that modems for 9,600 bits/s and even 19.2 kbits/s are proposed. It is not clear exactly how these will work.

The controllers are identical to those used for LTR, except that an SSB interface is fitted.

ENHANCED DIGITAL ACCESS COMMUNICATIONS (EDACS)

EDACS is a propriety trunk radio system from Ericsson, featuring distributed processing for enhanced reliability. A typical EDACS base station consists of a base-site controller plus channels, each of which has their own trunking card as seen in Figure 20.2.

The trunking card on each channel can on its own provide basic trunking services, while the controller adds features including the following:

- Eight levels of priority
- System diagnostics
- Telephone interconnect
- Unit enable/disable
- Logging of mobiles
- Dynamic regrouping

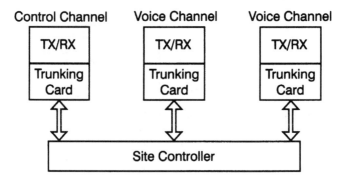

Figure 20.2 An EDACS base site.

One channel per base is designated the control channel, although in the event of a failure a voice channel will be reassigned the control channel role.

The base stations are connected to a central controller; but should the link to the central switch be lost, then local stand-alone operation of the base stations continues.

The system operates at 9,600 baud for a 25-kHz channel and 4,800 baud for 12.5-kHz channels. The two baud rates 9.6 bits and 4.8 kbits are used for the CODEC and all other services. As inband signaling is available on the voice channels the voice channel frame is shared as follows: VOCODER information 45%, error detection and correction 25%, signaling 20%, synchronization and framing 10%.

EDACS claims call setup times of less than 250 ms and cleardown in less than 150 ms. Status messaging is available.

The frequencies that are available are 150, 450, 800, and 900 MHz.

It is intended to provide an evolutionary path to TETRA. The original EDACS in 1991 was a frequency division analog and digital channels system that evolved to a three-time-slot TDMA system by 1994.

SmartLink

The SmartLink system is relatively new and is a multiprotocol system that can simultaneously handle LTR, GE Marc V and VE, and EDACS. Operating in the US frequency bands of 150 MHz, 220 MHz, 450 MHz, and 900 MHz, the system offers interconnection of calls between the protocols and even between mobiles using the different frequencies.

The system comes in three size options: a standalone site, a system with 50 channels and 32 sites, or a system with 640 channels with 32 sites. The network is controlled by a PC, which is connected directly to the main switch. The main switch, in turn, connects to local switches located at each site.

Wide-area operations are possible, but require manual registering on each new site. Handoff between sites is available: If the user manually switches to a new site while a call is in progress, the call will hold up and be transferred.

Interconnection is at the two-wire loop level, with DID/DOD. The interconnection is not restricted to certain user channels but is available to all the channels at the site.

Users can be restricted to access based on the time of day or day of the week.

The billing records are contained in an ASCII file.

The size of the equipment is 19 in. (600 mm) and is rack-mounted.

GENERAL ELECTRIC MARC V SYSTEM

The GE MARC V trunked radio system is one where most of the logic is in a module in the mobile. To make a call, the mobile identifies an idle channel and sends a tone to request the channel. If it is free, then the channel will prompt the mobile to send the ID of the called party.

The called mobiles are meanwhile scanning all channels for their ID. When they recognize a call, they lock onto it.

The mobile radio basically has three modes: idle, wait, and ready. In the idle mode the mobile is scanning the base station channels for its called ID, in the wait mode a call is being set up, and in the ready mode it is in conversation.

Because inband signaling is used, the mobiles have filters to notch out the signaling.

CHAPTER 21

ENHANCED PMR

It would be impractical to cover all the numerous variants of PMR enhancements that exist. Instead the approach taken in this chapter will be to explore the building blocks of an enhanced PMR system and to understand the potentials. As mentioned in Chapter 1, PMR enhancements may be minor, or they may be so extensive that the resultant network has all the features of a powerful trunked network (in which case the only thing that distinguishes this network from a trunked system is the fact that it is an one-of network).

Mobiles themselves can be broadly divided into two groups: those that have essentially manual operation and those that are processor-controlled. There are severe limits on things that can be done to enhance a network of manual mobiles, but the possibilities of processor-controlled units are extensive.

Processor-controlled units can generally scan a wide number of channels, be programmed to perform complex search sequences, perform handovers, and memorize a significant amount of end-user code. In fact, it is often the case that there is very little to distinguish a powerful PMR radio from a trunked radio other than the software code.

SELCALL

Selcall, or selective calling, is a signaling system that is widely used in PMR. In fact, it has become so common that it has found its way into most CB equipment. Primarily used for privacy, mobiles can be called as a group on a selected selcall code (or addressed individually by a unique selcall code). Selcall enables a number of unrelated users to share a repeater without having an open channel that broadcasts information indiscriminately. If one user group has seized the repeater, then all others are blocked out for the duration of that call.

In most countries today, some kind of selcall is mandatory for repeater operation. Selcall can significantly reduce false triggering of a repeater, which happens because the operating frequencies of repeaters are reused throughout the country. This means that it is possible for a mobile to be within range of more than one repeater at any given time. If such a mobile initiates a call, it will trigger all repeaters in range and rebroadcast the call from all of them. If selcall is used, the repeater will only trigger if the selcall sequences match. Of course, false triggering can still happen if the selcalls happen to match, but this will occur rarely.

Selcall can take a number of forms. The earliest type of selcall was an inband tone that was received by the repeater and then notched out before being re-sent to the mobile. Frequencies of around 1.5 kHz were widely used. Later came the subaudible tones. These have the advantage of being out of band and thus do not affect the speech path. The subaudible tones are detected and processed before the audio path, which will have a bandpass characteristic that will reject the subaudible tones.

By using sequential tone signaling, a large number of selcall IDs can be sent. A common format is five-tone sequential signaling that permits 100,000 codes. The tone can be encoded automatically or dialed using a keypad similar to a telephone dial. With this number of codes available, even very large fleets can be accommodated on a single system.

DTMF

Dual-tone multifrequency (DTMF) is a format originally designed for tone dialing telephones. It has been widely adopted for mobile use, particularly for use with

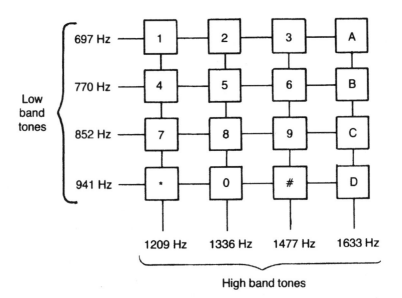

Figure 21.1   DTMF tone structure.

telephone interconnect devices. DTMF is far from an ideal signaling system for mobile radio. The standard DTMF decoder being designed for wireline telephone networks has a number of anti-falsing features that can be problematic in the mobile environment. Because the signaling is inband, it is important that the decoder does not mistake speech or noise as a valid signal. To ensure correct operation, some characteristics of the signal other than the frequency need to be examined. First, the decoder requires a minimum continuous signal time, while the actual time is determined manually (the tone length varies with the key stroke and the longer the key is held down, the longer the tone is sent). The mobile environment is subject to fading, and a long key stroke interrupted by a fade can be interpreted as the digit being sent twice. Another problem is a factor called *twist*. The decoder looks to see that the two tones are within a predetermined level of one another, and it rejects the signal if the test is not met. The mobile radio environment is harsh on this test too. The DTMF tone structure is seen in Figure 21.1.

TELEPHONE INTERCONNECT

Adding telephone interconnection to a mobile network can be as simple as running a few wires between the repeater and the telephone interconnect device and two more between the interconnect and the phone line. Although there are a wide variety of telephone interconnect manufacturers, the equipment can broadly be divided in to three types. The first is simplex operation, where the mobile will use a PTT microphone. The next is duplex, where there is no PTT and the call will proceed much like a regular telephone conversation. Finally, there are interconnects that are meant to work in the short-wave bands and perhaps even with SSB. This last category must operate very differently from the first two because DTMF dialing cannot be used over SSB (due to the frequency changes); and because the noise levels are much higher, they require a more elaborate VOX operation.

A telephone interconnect can be added to almost any mobile system, and it can be of great utility. However, once the interconnect is added, there are some new complications.

Charging for telephone calls can be simple if they are all charged at the one rate. As an example, the interconnection may allow local calls only, in which case there can be a fixed charge per call or standard charge rate per minute. This can be done with simple software, and some up-market interconnects actually come with a simple billing package.

Once long-distance calling is permitted, a much more powerful billing system (which will be costly) will probably have to be used. Alternatively, some operators make the billing simple by having only a few charge zones (e.g., local, statewide, and national).

International calling can be more problematic because there is a very wide range of charges that apply, particularly to some of the smaller countries who see international calls as a source of state revenue.

Finally, some systems with telephone interconnection have little or no protection from fraudulent users. There have even been cases where, in rural areas, some mobile users have put an interconnect onto a CB radio and included no security (for example, an access code). Many radios today have built-in DTMF keypads; and if some security is not put in place, some big telephone bills can be expected.

There can be no doubt that for the end-user, telephone interconnection can be most useful. But there are some downsides for the user as well. Most PMR systems are still press-to-talk. Telephone users used to full duplex can find it hard to get used to the need to "talk in turns." The mobile can usually take control of the circuit by pressing the PTT, while the landline telephone user must rely on VOX (voice operated circuits) for control. The VOX operation can itself present problems if the line is noisy and the noise begins to take control of the circuitry.

Another problem with interconnection is that people usually talk longer on the telephone than they would over a mobile. This ties up more airtime, and even more is tied up with dialing and waiting for the answer to ring tone. The simple fact is that adding an interconnect will increase the average airtime per user and thereby reduce the number of users per channel.

Operators have taken a variety of approaches to this problem. One approach is to disallow telephone interconnection. Another is to allow it, but to increase the monthly charges for those who use it and place call timers on the call. Alternatively, one can ask that the mobile users pay for and install their own telephone line and take responsibility for call charges that they incur. Others simply see telephone traffic as another revenue stream and charge by the minute (in the same way that cellular companies do).

Telephone interconnect will dramatically bring home the advantages of multichannel operations, and PMR operators can use relatively simple techniques to allow processor-controlled mobiles to access a pool of channels.

DATA

Traditionally in PMR, instructions were sent and received as voice commands. Today, many mobiles come with a built-in display that is capable of displaying (and perhaps storing and scrolling) short messages. Alternatively, there are a number of manufacturers who make data heads that can store and send data from any mobile. In either case the user effectively has a two-way pager. A simple instruction sent as data—for example, "pickup required at warehouse No. 4"—can be sent in a few seconds. Additionally, the acknowledgment can be a simple digital response. Taking advantage of this, some companies now have the next day's work details logged into a computer, which the employee can access at the beginning of the shift in the morning and then acknowledge completion over the air in real time.

PMR is ideally suited for short data bursts, the limited bandwidth, and the high overhead in error correction code, making it far less suited to large blocks of data.

DISPATCH

While the traditional dispatch operator sat at a console and issued instructions manually, today's dispatch systems are usually computerized and can handle vast amounts of data. The radio network, in addition to monitoring mobile traffic, can also monitor remote alarms, status levels of equipment, position location information from GPS systems, and more as background tasks, while the conventional dispatch functions are ongoing.

WIDE AREA

Although basic PMR systems are frequently single-site, there is often a need to cover a wider area, and this can be done in two basic ways. The first way is to parallel the sites that are needed to cover the area at the audio level and have the mobile select the site with the strongest signal, by scanning the alternative repeaters (which will be on different frequencies) sequentially. In a more advanced version of this, the mobile may even handoff by scanning again at a later time, should the signal drop below a certain threshold. This method is simple and reliable but is wasteful of spectrum.

Where spectrum is in short supply, it is possible to repeat the signal on the same frequency provided that the transmitters have a frequency offset (typically of around 100 Hz) and the modulated signal is sychronized in the area of overlap between repeaters. In this instance the only overlap of significance is where the signal from two repeaters is within ±6 dB of each other. Whenever one signal is dominant by more than 6 dB, interference will be minimal; but as the levels converge, destructive addition of out-of-phase signals can occur. To ensure that the signals in the overlapped area arrive in phase, it is necessary to measure the overall audio delay in the system of both transmitters. In particular, this includes any delays in the audio links between the transmitters. Once the delays are known, it is a relatively simple matter to calculate the net delay to the overlap area (adding the time that the signal takes to travel at the speed of light to the area to any system delays) and then to add in a delay to one of the transmitters to equalize these transmission delays. This system is synchronized at the baseband level but not at the carrier frequency and is known as quasi-synchronous.

Sychronization (at the audio level) is relatively easy to do, but it can have problems. Because a significant part of the overall delay is in the base-station links, any changes to those links (sometimes even including the replacement of microwave equipment with a spare of the same kind) can alter the overall delays and lead to coverage problems. It is also essential to have a high-precision frequency reference at the base sites to ensure that the offset is maintained and that the station oscillators do not drift to the same frequency. Alternatively a "wobulator" can be used to randomly vary the carrier frequency at a rate of 100–200 Hz. This again prevents interference at the carrier frequency.

A hybrid way to conserve frequency lies somewhere between the two option above. By equipping the base station with a scanning receiver and having all base stations receive on the same frequency, they can report to a central controller the mobile received level at each site. The central controller can then determine which base site is the most appropriate to be used. Different transmit frequencies are used at overlapping sites so that sychronization is not needed. Such a wide-area system will use less spectrum than the first option but more than a sychronized system. An early version of this system used "race voting." In the days when central controllers were not necessarily inexpensive, the base station could still determine which base site was the most appropriate by having a simple integrator on the signal level monitor. The integrator would reach a predetermined threshold quicker for stronger signals. Once the threshold was reached, the base site would seize a signal line (place a voltage on it) to indicate to all others that it had taken control. To ensure updating, a "race vote" could be repeated each time the user operated the PTT switch.

A TAXI DISPATCH SYSTEM

As an example of just how big an enhanced PMR system can get, Sigtec has installed a dispatch system for 80% of Sydney's taxis (in total 3,400) which operates as a shared system for six different competing cab companies.

The system uses GPS for positioning, but Sigtec found that the accuracy available from the GPS was insufficient, particularly in the center of Sydney, where the high-rise buildings tend to block access to satellites. To overcome this, they incorporated a dead reckoning system based on inertial navigation and distance information from the vehicles odometer, which is activated when the GPS satellite data are not available. This system, called mobiTAG, is a stand-alone unit that can work over any radio system (including trunked radio).

A propriety differential system is used to increase the accuracy so that the design goal of locating the taxi to an accuracy of 20 m for 95% of the time is met. Sigtec claim that in most cases they can locate a taxi to an accuracy of 5 m.

Naturally a large number of resources are needed to operate such a system and 44 RF channels are used, together with a structured operating system.

CHAPTER 22

EQUIPMENT SHELTERS

For many trunked radio applications, there will not be much choice of shelter types or construction. However for some large systems and perhaps some new smaller ones, there may be a need to design the buildings from scratch. The following text provides some guidelines.

Equipment shelters should be designed to maximize flexibility for expansion, to minimize operational costs, and to provide a clean, safe working environment for the staff. There are many different ways to achieve these objectives. This chapter presents some of the factors to be considered when designing shelters for a large network; smaller operators can scale this down considerably.

BASIC CONSIDERATIONS

Planning for expansion is particularly important because it seems that no matter how large an equipment shelter may be, soon after installation it will be found to be too small. Even when no expansions are foreseen, it is wise to build the shelter so that at least one wall can be removed to extend the building.

When transportable huts are used as equipment shelters, they should be placed on the site in a way that allows for additional huts to be added. Too often, transportable huts are placed centrally on the site, mainly for cosmetic reasons. This can make future expansion awkward.

Switch rooms in particular should be built with ease of expansion in mind. It is expensive to relocate a switch once it is placed in service.

Operators, particularly new operators, should not underestimate the amount of space that may be needed for other services. In particular, space must be found for microwave, power, and, sometimes, billing equipment.

Shielding

In an ideal world, shielding of the shelter would not be a consideration, but with the proliferation of other RF services, interference to the equipment through the shelter wall is always a possibility. Because of this, preference should be given to construction techniques that minimize interference. In areas where high-powered RF sources are known to exist, such as AM, FM, or SW broadcast transmitters, as well as on shared sites, only metallic-clad buildings should be considered. In other areas a preference for metallic cladding should exist provided that this does not compromise structural integrity or—to a significant extent—cost.

Concrete shelters, with their steel reinforcing, can double as effective shields, particularly if care is taken to bond the reinforcing bars together. This may need to be allowed for at the design stage.

Fiberglass shelters can be lined with conductive tape and filled with metallic-backed insulating material. Lining of the internal walls with metal sheets should be considered if there is any suspicion that interference will be a problem.

SWITCH BUILDING DESCRIPTION

The building will generally have only one floor and include the following areas:

- Equipment room
- Control and (perhaps) billing room
- Battery and power room
- Emergency-plant room
- Uncrating area
- Storeroom
- Air-conditioning plant
- Toilets
- Staff facilities
- Cleaner's/janitor's room

Equipment Room

The equipment room houses the switch and microwave equipment. Because the equipment is often bulky, good access for delivery vehicles and movement of equipment is essential. The area must be air-conditioned and the humidity regulated.

Because demand is difficult to forecast, the room should be designed to expand in at least one direction (at least one wall should not be structurally supportive). Expansion can be facilitated by the use of steel-framed, non-load-bearing walls, which can readily be removed when additional area is required. The position of the equipment room with respect to the site should allow for this expansion.

An equipment room allowing for three suites of equipment should be about 5×5 meters. The actual dimensions will depend on the actual equipment purchased.

Floor tolerances for the equipment room should be precise. In the largest dimension of any floor area, the level should not vary by ± 12 mm; in any 3-m length, the floor should not vary by more than ± 5 mm in any 300-mm section.

Control and Billing Room

A room about 5×5 m could be included to house the control and billing equipment. If necessary, billing functions can be handled remotely, and there are often good reasons for doing so. Real-time billing (hot billing), however, requires a data link between the switch and the billing computer.

Battery and Power Room

The battery and power room should be about 3×3 m—large enough to accommodate two battery stands and the rectifiers. High ohmic distribution should be employed. The room should be fitted with a small washbasin and a handheld hose spray attachment, as well as an exhaust fan (if wet cells are used).

When lead-acid batteries are used, the battery and power room should be physically isolated from the switch room. Today, however, sealed rather than lead-acid batteries are usually used, and these can be located in the switch room. In fact, with sealed batteries. it is common to place the batteries and rectifiers very close to the equipment to reduce copper costs (smaller bus bars), which can result in overall savings of 30%.

Emergency-Plant Room

Very small systems can be supplied from a portable generator, but larger systems will benefit from a purpose stand-by plant. The emergency-plant room should accommodate a diesel generator adequate for the total load of equipment, air-conditioning, and lighting. Typically, the load will be from 3 to 20 KVA (the upper limit being where a large co-located base station is included). This room contains an electrical switchboard and must have an exhaust fan. Measures should be taken to contain any fuel spillage in this room, including surrounding the room with curbing 150 mm high (or making the floor 150 mm lower than the rest of building). Figure 22.1 shows a typical emergency-plant room. To allow easy access and safe passage, raised ramps should be provided for doorway access.

The generator room should be soundproofed because the noise levels can be quite high.

An essential part of the emergency-plant room is to remove the heat generated by the diesel. The installation manual for the generator will show the recommended mounting details. Because the airflow from a large generator is hot and the volume is large, there may be a need to duct the airflow above the level of passersby. Notice that

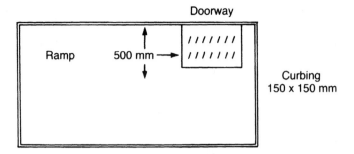

Figure 22.1 The emergency-plant room should be lower than the rest of the building so that fuel spillages will not seep into the main equipment room.

the emergency plant has its cooling fan arranged so that the airflow blows *out* (the opposite of a motorcar).

The room should be designed to facilitate the airflow. An air intake at least 1 m² is needed and an exhaust fan, placed high in the room in the vicinity of the radiator, can be used to assist the airflow.

The engine exhaust pipes and muffler are dangerously hot. If the generator has auto-start, then a sign indicating "Caution: This generator may start at anytime" should be posted in a conspicuous position.

Uncrating Area

The uncrating area, a part of the switch room, is deliberately left vacant to provide space for additional equipment. The area serves as a workplace for installation and maintenance staff and is the last area of the switch room to be occupied. Good access should be available for vehicles transporting equipment, and the area should be fitted with doors 1.5 m wide.

Storeroom

The switch building will probably house its own spare parts and is an ideal site for the main storeroom. The value of spare parts is usually about 5% to 10% of the network equipment value, so good, safe storage is essential. A storeroom 3 × 4 m would be adequate.

Walls

The walls should be made of bricks or cavity blocks. Because cavity blocks cannot be cut to size as bricks can, when cavity blocks are used, room dimensions must be exact multiples of the block length. The internal walls to the power room/storeroom, equipment room, and emergency-plant room should be non-load-bearing and surfaced with a fire-stop plasterboard or other fire-retardant material with a 1-h rating. The internal ceiling height should be 0.6 (minimum) to 0.9 m higher than the switch rack or the minimum height specified by the local planning authority, whichever is the highest. Such considerations ordinarily yield a wall height of around 3.4 m (equipment height 2.5 m), although suppliers have rack heights from 2.2 to 2.9 m. Of course, these measurements do not apply if under-floor cabling is used.

When an existing building is used and the ceiling is significantly higher than the optimum, adding a false ceiling is worthwhile to substantially reduce the air-conditioning load.

Roofing

The building must be totally waterproof because water on the equipment can cause total malfunctioning. The roof should have a minimum pitch of 5 degrees (assuming steel decking is to be used). The appropriate local high-wind code should be applied. Particular attention should be paid to the guttering and downpipes, which should be a "leaf-free" type to decrease the likelihood of blockage. The gutter sections should be designed to prevent overflow of blocked systems into the building, as shown in Figure 22.2.

Insulation Cladding

Table 22.1 shows the minimum insulation in fiberglass batting or equivalent that should be provided. Insulation should be used in walls and ceilings to reduce air-conditioning costs. All materials should be fire-resistant.

Floor Loading and Construction

The floor should be designed to carry live loads of 9.5 kPa throughout the building. Suspended floors should be used only where the site would require excessive fill.

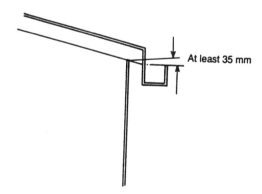

Figure 22.2   Because gutters can become clogged, it is recommended that the outside part of the gutter be lower than the wall side to prevent overflow down the walls and perhaps ultimately to the equipment.

TABLE 22.1 Thermal Insulation for Equipment Shelters

Climate Zone	Insulation Thickness (mm)
Temperate	100
Mediterranean	75
Subtropical	50
Tropical	75

Ideally, the whole building should be on level, consolidated ground. A reinforced concrete raft slab, incorporating edge beams to the perimeter and ground beams under the walls, should be used. Newly filled land requires time to compact and this must be allowed for in planning time-tables if it is an issue.

The raft slab should be placed on a consolidated base, leveled with sand, and covered with a waterproof membrane.

The floor should be elevated sufficiently above ground level to ensure against the entry of water under the worst flood conditions. (A flood-free site should always be selected.)

Ceilings

Ceilings should be of plasterboard or similar material and insulated to the recommended thickness.

Windows

Windows should be of laminated glass. External windows, for security, should be high and covered with a metal security screen that also diffuses sunlight. To reduce air-conditioning costs, double glazing or thick glass (60+ mm) can be used for external windows. Adequate provision should be made for cleaning the windows (a sliding construction with a key lock facilitates cleaning).

Appearance

Because the main switch and control room will probably be located in a residential area, building design is important. The building should not look like a residence, but it should blend in with the area and should not be conspicuous. The visual effect of the large equipment-access doors can be reduced by painting the top of the doors in a dark color.

Tornado/Hurricane Areas

In tornado/hurricane areas, brick or blockwork should be reinforced with galvanized tie rods, from the footings to the roof beams, providing post-stressing.

Figure 22.3 A cable window with a number of unused cable boots which line up with the internal cable tray.

Structural Steelwork

All structural steelwork should be painted to protect against corrosion. Galvanizing is not necessary except for lintels.

Cable Window

A cable window measuring about 500 × 500 mm provides cable access into the equipment room (see Figure 22.3). This window will be positioned so that its bottom edge is rack height above the floor. The window should be made of galvanized, painted mild steel, with one plate on the outside of the wall and one inside. A cable tray (gantry) supports cables from the tower, see Figure 22.4.

Figure 22.4 The gantry supports the cable that runs up from the tower to the building.

INTERNAL FINISHES

A dust-free environment, ease of cleaning, and hard wear are the main factors to consider when choosing internal finishes.

Floors

The toilet and entrance lobby should be floored with ceramic or vinyl tiles. All other areas can be finished with vinyl or similar tile.

Walls

All brick walls should be cement, rendered with a rubber float tool to a fine-sand finish. Partitions to the emergency-plant and battery rooms should have a 1-h fire rating. Steel-studded plasterboard can be used in other areas except the toilets. The surface coating should be a gloss enamel finish to minimize maintenance.

Doors

All internal doors should have a durable gloss enamel finish. Door frames in brickwork should be pressed steel and finished with the same coat as the adjoining walls. The doors to the uncrating area should be 1.5 m wide and able to swing 180 degrees to be flat against the external walls. All external doors should be faced with sheet metal.

EXTERNAL FINISHES

External finishes should be maintenance-free; painting should be avoided. Aluminum frames should be anodized, and steel fixtures should be formed from material with an aluminum/zinc coating and finished with polyester.

All external doors should be fitted with dual heavy-duty mechanical locks vertically separated by 1 m. An electric combination lock for use during normal hours of operation is also recommended. A combination lock allows the staff to move freely but keeps out others.

EXTERNAL SUPPLY

Three-phase, four-wire (or sometimes three-wire, single-phase for small-systems) power is required. Electrical utilities usually offer a range in price; select the option that is most economical.

The main switchboard should be located in the battery room. The power board should be fully enclosed with hinged doors for access to all components except main switches and changeover switches.

ELECTRICAL POWER OUTLETS

All equipment and control rooms should have adequate power outlets. In most locations, double outlets every 2 m is adequate. The outlets should be located at least 150 mm above the floor.

EXTERNAL-EMERGENCY PLANT

Because the emergency-plant room requires adequate access to a diesel generator, the doors to the room should be at least 1 m wide.

Provision should be made to attach an external emergency generator. A manual changeover switch with three positions should be provided: normally on, off, and external emergency plant.

A suitable three-phase external socket should be fitted to the building.

ESSENTIAL POWER

When using the emergency plant, only the essential power is provided. This power supports

- Main switch rectifiers
- Power and battery room ventilation fans
- Control room
- One single-phase future expansion with circuit breaker
- Emergency lights and outlets
- Essential air-conditioning

AIR-CONDITIONING

The whole building should be air-conditioned except for the battery and power room and emergency generator room. Separate air-conditioning for the switch room and control room should be powered from the emergency plant. It is probably not necessary to run the rest of the building from the emergency power.

Temperature should be kept in the range 20 °C to 30 °C, with a relative humidity of 20% to 60% (noncondensing) and dust to a maximum of 60 mg/28 cu m^3 of air by weight (5 μm diameter).

BASE-STATION HOUSING

The base-station housing can take many forms—for example, existing buildings, transportable huts, and specialty built structures (usually to meet town planning requirements). With the exception of very remote sites, all new structures or acquisitions should be adequate in size for ultimate expansion to full-base size. This full-base size depends on the system type and frequencies allocated. Due to intermodulation and frequency plan problems, it is generally not practical to place more than one fully equipped base (which uses the full cell-frequency allocation) at any one site. This problem does not occur in low-density regions.

The housing should be designed for ease of equipment-rack expansion, especially in installations where estimates of channel requirements at particular sites are, at best, guesswork.

Base-station housing can present many challenges for the installation engineer. It is possible to use a standard shelter for new sites; for installations in existing buildings, however, it is necessary to conform to the space and layouts available. This constraint often also applies to building rooftop installations, where sufficient space for an optimum layout is not generally available. It is often necessary to use small spaces, often with awkward shapes, in order to conform with rooftops that were never designed to accommodate a base station.

Dead-load limitations can also restrict the type of installation that can be placed on a rooftop; this is particularly true if a tower structure is also required.

Shelters can be made of concrete, metal, or fiberglass. The cost of a well-designed shelter built with any of these materials is similar, and all materials can be expected to be suitable for about 20 years. Locally available materials and expertise may well determine the most cost-effective choice.

Concrete generally provides the most durable shelter and, where necessary, a virtually bullet-proof enclosure. The concrete must be scaled about every four years but otherwise requires very little maintenance. Concrete hardens with age, however, and can suffer cracking in extreme weather conditions.

The incidence of bullet damage in the United States, although not widely reported, is believed to be on the increase. It is generally considered that high-powered rifles represent the biggest threat to the equipment. A number of operators, particularly in rural areas, are turning to bulletproof structures that range from 4-in. concrete-walled buildings to prefabricated "bulletproofed" shelters.

A consideration when "bulletproofing" is how much reinforcing is necessary. More resistance means more cost. In the United States the Underwriters Laboratories code "Standard for Bullet-Resisting Equipment, UL 752" provides a good definitive basis for the level of bullet resistance.

A concrete shelter, which was not necessarily chosen for its bulletproof qualities, is shown in Figure 22.5.

Steel structures require a little more maintenance but can be expected to last about 20–25 years. Steel and fiberglass panels are easily moved and are most suitable for assembly in awkward places like rooftops. A steel base station in England is shown in Figure 22.6.

Figure 22.5   A concrete base station.

Figure 22.6   A steel transportable base station.

Fiberglass, with careful upkeep, can be expected to last 15–20 years. Care should be taken to ensure that panel joints and scaling (caulking) are permanent and do not require annual maintenance. Fiberglass and steel are both less vandal-proof than concrete.

Transportable huts can be prefabricated and have all hardware installed before being moved onto the site. These huts can also be placed on top of office blocks and other high-rise buildings. When using high-rise buildings that are not the property of the operator, locating equipment in prefabricated huts is recommended because the huts can be moved relatively easily in the event of future leasing disputes. It is often necessary to construct the huts on site because of access problems that make the placement of a complete hut on the rooftop impractical.

Using transportable huts reduces installation costs (because of standardization). It may be possible to assemble the huts and the base-station equipment before transporting them from a central workshop site. Attention should be paid to adequate thermal insulation to reduce air-conditioning costs. The huts should be of a type that is easily dismantled, or they should be designed for transport by helicopter, vehicle, or crane.

It is easier to meet deadlines if prefabricated huts are used, because they can be under construction even before the site is selected. Base stations may be constructed of brick, as shown in Figure 22.7 but it should be remembered that brick will not, by itself, provide adequate insulation.

Figure 22.7   A brick base station.

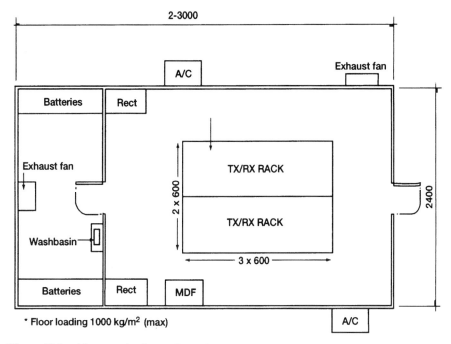

Figure 22.8   A base-station layout for equipment that can be installed back-to-back, for a large site.

In areas where the water supply is undependable and wet cells are used, the roof of the hut should have adequate gutters to collect water. If the hut contains lead-acid batteries, a secure water supply is needed for washing off acid spills. In addition, safety warnings should be posted around the shelter, and signs showing how to treat acid burns and, in particular, damage to the eye, should be prominently displayed. Figure 22.8 shows a typical base-site floor plans.

Floor Loading

The floor loading should be designed for the actual equipment loading, typically 700–800 kg/m^2.

Weight

The choice of construction materials will seriously affect the weight of the structure, which can vary from a few tons for wooden and fiberglass, 5–6 tons for aluminum clad, and around 15 tons for a concrete shelter. The weight can be a significant factor in the cost of foundations and transport. Some roads have weight limitations (which may apply year-round or seasonally due to heavy rain or snow/ice) that can make transport in an assembled state impossible. Disassembly for transport is usually a possibility but not an economical one. Heavy buildings often will require two cranes for site placement.

Lighting

Adequate lighting can be provided by eight (4×2 twin units), 40-W fluorescent lights within the hut. Incandescent lights should not be used because of their low power efficiency and high heating load.

Security

It is usually a good idea to build a human-proof fence around the hut/tower installation to prevent vandalism. The fence should clear the structure by 2 m and should not provide an access to the building roof. The antenna structure should have some means that prevent attempts to climb it. Serious legal liability could result should an intruder fall and be injured. All doors should be fitted with entry alarms that are monitored 24 h a day at some central location.

A number of innovative signs have been attached to unmanned base station sites to deter trespassers, some of which are shown in Figures 22.9, 22.10, and 22.11.

Special attention should be paid to doors, which can be a weak point. Good-quality locks with at least two latching points should be provided.

Figure 22.9   The sign on an unmanned transformer station in the United Kingdom.

Figure 22.10   A warning to Racal-Vodafone cellsite in the U.K.

An independent means of communicating to the main operations control center should be provided. A conventional telephone and landline, or a cellular phone link, is adequate. This is needed if complete failure of the base station, its link, or the switch occurs.

Figure 22.11   A warning sign on a microwave site in Honolulu.

Equipment Mounting

The floor of the building should be constructed to allow the racks to be anchored by suitable bolts. The wall construction should be such that equipment (such as power boards and microwave links) can readily be attached. In the case of metal construction, mounting can use the supporting studs and the steel frames between them. For fiberglass construction, it may be necessary to include a plywood layer in the wall that can be used as a structural support. With fiberglass, it is important not to rupture the outside wall because a rupture can lead to problems with waterproofing.

Insulation

To minimize the air-conditioning load, fiberglass-batting insulation should be placed in the walls and ceilings of the shelters. Table 22.1 gives the proper insulation thickness.

The grade of insulation is a trade-off between cost and energy conservation due to heat losses. A thermal resistance rating in the range R-12 to R-22 covers the range of insulation normally used in such shelters.

Cable Window

It is necessary to bring RF cables for the base station and perhaps microwave antennas into the base-station housing. Because these ordinarily come from the same antenna support structure, a single cable entry of approximately 500 mm × 800 mm suffices. The cable window should be able to support additional cables as the base station expands.

For transportable shelters, the orientation of the shelter cannot always be determined by the operator before construction (the shelter often must be made to fit into available space), so it is sometimes necessary to bring the cables into the shelter through any of the walls. In these cases, the location of the cable window must be flexible.

Electrical

The shelter should be wired for the following:

- Air conditioners
- Rectifiers
- Eight double, general-purpose power outlets
- An exhaust fan (when lead-acid unsealed batteries are used)
- Lights
- Emergency lights that can run off the equipment batteries
- An external power-inlet socket for an emergency portable generator
- General-purpose power outlets, one double unit every 2 m, at least 150 mm above the floor level

Notice that the power consumption of a large base station can be quite high (around 15 kW); if possible, it is a good idea to use a three-phase power supply (some authorities may require it). Approximately 200 W per channel should be allowed for equipment power and a similar amount for air conditioning.

If an external generator is used, it must be housed in a suitable weatherproof shelter. Access for refueling (of diesel, propane or natural gas) is essential. Above ground fuel storage is both cheaper and allows regular inspection of the tank. Below ground is sometimes used because of local regulations or where space is at a premium.

Access

Good access is essential for any base-station site, both for ease of installation and for maintenance. The access road should be a minimum of 4-m wide but preferably 4.5 m. The access gradient should not exceed 15%, and any bends should be at least 20-m radius. Good drainage around the site, tower, and access roads is necessary to prevent water erosion.

Factors that could affect access at particular times—including fires, heavy rain, winds, snow and ice—should be considered.

GROUNDS AND PATHS

For both the switch and the base station, it is important that dirt and dust not get into the equipment rooms. The grounds surrounding the shelters should be paved, and the rest of the ground should be free of dust, mud, or other potential pollutants.

CHAPTER 23

TOWERS AND MASTS

Towers (self-supporting structures) and masts (guyed structures) cost about the same if they are both short (that is less than 10 meters). As the structures get higher, the costs of guyed masts tend to increase linearly (for the same cross section); the cost of self-supporting towers increase exponentially. Because guyed masts require a good deal of land, they are used mainly in rural areas.

Towers and masts require very different amounts of land. Figures 23.1 and 23.2 show the area needed for a mast.

Figure 23.3 shows the amount of land needed for towers of different sizes. Table 23.1 shows the amount of land needed specifically for three-leg and four-leg towers. In this table, T and W are the land dimensions used in Figure 23.3.

No structures of any kind should be built closer to the tower than the edge of the boundaries defined in Table 23.1 because the support of the surrounding soil against turning moments may be diminished. These dimensions are a guide only; the design of a tower or mast depends on such factors as wind loading, local building codes, and local planning-authority regulations.

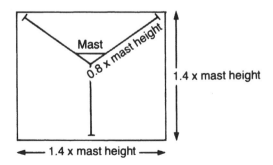

Figure 23.1 Optimal land area for mast.

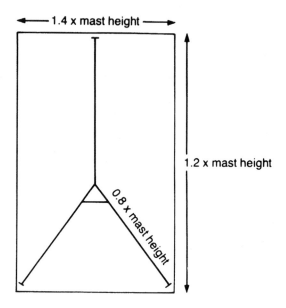

Figure 23.2 Minimal land area for mast.

The choice of monopole, mast, or tower for trunk radio is often made for the operator by the local-government rules or environmental considerations. Sometimes, however, there is a choice, so it is worthwhile exploring the alternatives.

The location of the tower needs some careful consideration. In most countries, it will be sufficient that the tower win local zoning approval (from the local authorities) and does not constitute a hazard to air navigation. Usually, unless the tower is particularly large or the proposed location is zoned residential, there will not be too many objections from the local authorities. However, they may require the planting of trees around the structure (and in some cases even a painting scheme that is more environmentally sympathetic, such as sky-blue or green).

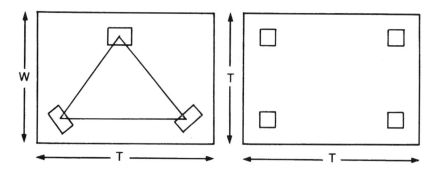

Figure 23.3 Dimensions of land for towers of different sizes.

TABLE 23.1 Land Usage and Weight for Three- and Four-Leg Towers

Three Legs				Four Legs		
Tower Height (meters)	T	W	Approximate Weight (tons)	Tower Height (meters)	T	Approximate Weight (tons)
10	7	7	0.7	20	7	1
20	8	7	1.7	30	9	2.2
30	10.2	9	3	40	10	4
40	11.5	10	6	50	12	8
50	13.8	12	10	60	13	12
60	15.5	14	14	70	14.4	16

Conflicting with the requirements of being inconspicuous will be the requirements of the aviation authorities that the structure be visible to aircraft on VFR (Visual Flight Rules) and that it not be an obstruction to any existing or future flight paths.

Generally, if the structure is smaller than 50 m (in the US special conditions apply above 200 f) and more than 10 km from any airport, it is unlikely to be a problem. In any case it is a good idea to get a ruling from the local aviation authority on both the location and warning markings/beacons that are required.

Information required by the aviation authorities will include

- Accurate coordinates of the tower location
- Structure height
- Structure type (tower/mast/pole)
- Proposed warning beacons and hazard warning paintwork.
- Location of the nearest airport and other airports within 10 km

Also it may be required in some countries (as it is in the United States) that a full inventory of the RF facilities—including the frequency, power, and radiation patterns—be provided. These are sometimes needed to assist in the evaluation of potential interference to IFR (instrument flight rules) navigation equipment.

In the United States it is *compulsory* to get an FAA "determination of no hazard to navigation." This can take about 60 days, or longer if the submission is not complete.

MONOPOLES

In general, a monopole is more aesthetically pleasing, although very few neighbors are likely to welcome any support structure. The monopole, like a building, has a fixed platform height and usually comes in a very limited range of sizes (typically, 15–50 m). It may have an internal ladder and cable tray. Its main structural advantage is small land area required, typically 3–4 m square.

Monoples can be erected in about one day provided that the base has to be poured and cured. Because they are available in a limited range of sizes, they often can be ordered almost off the shelf and have much shorter delivery times than towers.

Monopoles are usually fabricated in tapered sections of about 10 m apiece and fit together simply by stacking the sections (see Figures 23.4).

The support is simply a cage, designed to withstand large turning-moments, embedded in concrete. Shafts typically 8 m deep and tapering from 2 m at the bottom to 3 m at the top form the foundation. Other monopoles can have much wider bases with correspondingly smaller shafts. Bolts 2 m long and 57 mm in diameter, embedded in the concrete, attach to a flange at the bottom of the shaft. Up to 50 bolts can be used to hold the flange.

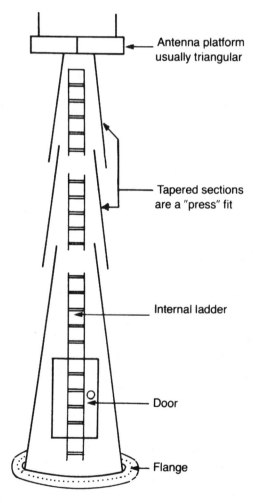

Figure 23.4 The construction of a monopole with an internal ladder as used by Telstra.

These structures can be designed to give the torsional stability required for low-frequency microwave bearers (maximum 1/2-degree twist).

GUYED MASTS

Guyed masts are practical only where land is inexpensive. They often prove to be the cheapest solution in rural environments.

Guyed masts are usually constructed of sections of triangular cross section about 6 m long. The sections are typically 0.5 to 1 m wide per side and are designed to be bolted together. The strength of a mast is essentially in the guying, so proper tensioning of the support cables is vital. Concrete anchors hold the guy wires. For trunked radio applications, the standard cross sections should prove adequate to accommodate the cellular and microwave antennas. Figure 23.5 shows a 50-m mast in Asia.

Figure 23.5 A 50-m mast in Asia. Notice the long grass (a fire hazard) in the foreground and recent evidence of fire in the very near foreground.

Hybrid Structures

Sometimes it is not easy to decide if a structure is a mast or a tower. Figure 23.6 shows a structure that started as a tower and later sprouted a mast on top. This structure is on a rooftop in Manila, the Philippines. The mast may have been built to reduce the total loading on the roof.

Fabrication

The cost of a mast is related more closely to the weight of steel than to the height. Table 23.2 shows that the weight of a mast is almost linearly dependent on its height. It can be seen by comparing Table 23.2 with Table 23.1 that self-supporting masts increase more rapidly in weight than guyed masts, particularly for structures higher than 30 m.

Figure 23.6 Dual platforms significantly improve the rigidity of the antenna assembly.

TABLE 23.2 Mast Height and Weight

Mast Height (meters)	Approximately Weight (tons)
30	1.5 to 3
50	3 to 5
70	4 to 7
100	10 to 12

TOWERS

Towers are self-supporting structures that are most practical when land is expensive. Figure 23.7 shows a tower that supports a number of microwave dishes (the solid dishes) and gridpacks (the wire-formed dishes).

Figure 23.7 Some towers are designed for high-density dish loading. Notice that each of them has a cover for weather protection.

Towers require less land than guyed masts and are capable of supporting a large number of antennas, a factor when you plan to rent tower space to other users. When other microwave facilities are planned (for example, for a wireline carrier), a self-supporting tower is probably the best choice.

A three-sided tower is usually the best value (load-carrying ability per dollar). For the same strength, however, it has a wider base and requires more land than a four-leg tower. A four-sided tower has an extra face and can carry more antennas. For a given strength, it is also smaller.

Towers members can be of various types, including solid, tubular, and channel sections. Tubing is the least expensive material for tower construction; It is available in a large number of sizes and needs little work to make it suitable for towers. Tubing does, however, have a long-term maintenance liability. Moisture can build up inside the tubing and cause corrosion, and in extreme environments, can freeze, thereby splitting the tubing. In coastal areas or in areas near heavy industry, this type of construction could prove to be a liability. Such towers need to be designed with weep holes, and the holes need periodic cleaning and unblocking.

Round-bar members can be made of round solid sections; they do not have the corrosion problem of tubular sections. They do, however, require substantially more steel for the same strength and thus weigh and cost more.

The most common material used in tower structures is the channel section, which can be made from formed-plate or angle sections. Formed plate is cut from rolled plate; it is cut to length with the bolt holes punched while still in the plate form. It is then cold-formed into 60-degree or 90-degree channels along the center axis.

Deformed plate is made from milled 90-degree sections. For 60-degree sections, the plate is bent another 15 degrees on each flange. This plate is less expensive than formed plate, but often it is not precision-formed, which can lead to problems with bowing.

SOIL TESTS

Before a tower, mast, or pole can be erected, it is necessary to conduct a soil test. This involves taking core samples of the ground on which the structure is to be built and then having the samples analyzed. Using the results, the design engineer can determine the load-bearing capacity of the soil and its ability to resist the turning moments of the footings. Only after this test is complete (from one to four weeks) can design of the structure foundations begin.

The cost of foundations, particularly for a large tower, can be very significant. It can even be the major cost in areas of high wind loading. For this reason it is not possible to give a meaningful quote for a tower until soil tests have been done. Particular care should be taken if rock is found to ensure that the sample is representative. Sometimes a soil test will reveal rock when actually what is being seen is only local. A number of samples at different locations will resolve this problem.

Figure 23.8 An "antenna farm" on a rocky outcrop. These antennas are mainly TV antennas mixed with a few links. Notice that the high-gain antennas are often mounted off-vertical, where they won't work well.

OTHER USERS

If the structure is in a particularly prominent position, you should consider, before the tower is designed, the prospect of obtaining additional revenue from leasing tower space to other users. A modest increase in cost at the design stage can significantly improve the load-carrying ability of the structure.

When planning for other users, include them in the overall design by assigning their number, antenna type, and positions on the tower at the design stage. The structural design should also include detailed drawings of the proposed positions of other users so that they can be allocated at a future date without the need for new load calculations.

In general, operators need not fear that including other users will cause interference, provided that they operate outside the trunked radio band and do not transmit very high power, as is the case with UHF TV, for example.

Other users' services sometimes sprout up almost spontaneously in certain prominent areas and are known in the trade as "antenna farms." These "farms" can appear almost anywhere; Figure 23.8 shows one such farm thriving on a rock face in Baguio, in the Philippines.

ANTENNA PLATFORMS

Often it is convenient to provide a platform for the antennas. This platform provides a safe working place and should have handrailing, grated floors, and kickplates (to prevent tools from falling over the edge). The platform will normally be the same shape as the tower cross section (three or four sides) and should have sides of around 3.5 m (whether triangular or rectangular).

Alternatively, extension arms can be used. Extension arms are suitable for small base stations, particularly those with few antennas (for example, omni sites with less than 16 channels).

TOWER DESIGN

The antenna structure must be designed by a structural engineer, but it is worthwhile to consider the design parameters. The structure must account for gravity loads (dead loads) that include structure weight, antennas, and ice, as well as live loads, such as those caused by wind and seismic activity. Invariably, wind, ice, and tower fittings will provide the dominant loads on the tower.

Wind Loads

Until very recently, the dynamic load caused by wind was not fully understood, and towers were designed to withstand a known static load, which was increased by a safety factor (often doubled) to account for dynamic effects. In the light of recent studies, it is clear that early designs tended to overdesign the bases and underdesign the top portions of the structures. Particularly in typhoon and hurricane areas, the top portions of old designs are now being strengthened.

In general, as more collapsed structures are studied and more detailed information of long-term wind peaks becomes available, the minimum requirements of codes for determining wind loads have consistently increased. Old structures should therefore be used only after a thorough survey and inspection.

Wind speeds, recorded by national authorities, are of interest to a tower designer. The designer should know the peak gusts (instantaneous readings) and fastest-minute-wind (the highest velocity sustained for 1 min). These two figures are connected by a ratio of approximately 1.3:1.

Fifty-year peak wind velocities are sometimes interpreted as ones that are expected to occur 50 years apart. This is not an accurate interpretation. A better interpretation is that 50-year peaks are ones that occur with a probability of 2% each year. Therefore, the fact that an old tower is still standing may simply be good luck!

Typical Specifications for a 40-Meter Tower

The tower designer must know a number of things about a tower before beginning the design process. The following lists contain the considerations required for a typical 40-m tower:

- Four-sided (or three-sided).

- 40 meters.
- Designed to EIA RS222D (the U.S. standard), Australian Design Standards, or other preferred standard.
- Stress factor (that is, suburban or rural safety factor).
  For example, in the Australian design code for suburban areas this stress factor is:
  1.7 × factor on steel.
  1.75 × factor on foundations (this factor can be found from the relevant design code).
- Zone specifications and wind loading, depending on location.
- Maximum allowable twist (0.25 degree for 7-GHz microwave or 0.15 degree for 10 GHz).
- Maximum allowable tilt (1% for 7-GHz microwave or 0.5% for 10 GHz).
- Platforms and walkways at the levels where access to microwave dishes will be required.
- A platform of about 3.5 × 3.5 m at the top, with guard rails 1.5 m high and suitable for mounting antennas at the edges. The mounts will be used to attach

Figure 23.9   When towers are to be climbed by staff other than riggers, safety guards should be provided.

antennas with tubular supports up to 70 mm in diameter using three heavy-duty clamps. Up to 10 antennas can be mounted at the top with an equivalent flat-plate area of 0.23 m^2 (See manufacturers' catalogs for particular antenna types.)

- Cable tray will be accessible from the ladder and will be 0.6 m wide.

- Safety guard around the ladder, which will be internal with respect to the tower. (See Fig. 23.9.)

- IAO standard paintwork and an aircraft warning beacon at the top.

- Tower orientation.

- Specify the microwave dishes, type (solid or grid pack) and mounting level. Allow for future expansion (even if expansion is not planned, it will probably be required; a good rule is to estimate the future requirement and then double it).

- Tower footings should be confined within a square plot of land (as specified earlier in this chapter).

SECURITY

Towers are attractive to youngsters, who see them as a challenge to climb. If towers are climbed, untold damage may be caused to the antennas and cables, and even worse the youngsters may suffer serious injury or death as the result of a fall. In order to discourage unauthorized access, a human-proof fence should be installed around the

Figure 23.10 A human-proof fence (with spiked steel posts) around a rural site.

Figure 23.11 Access can be restricted by the use of spikes on each leg and on the access ladder.

Figure 23.12 A warning sign at the base of a tower in England (this is on the same tower as shown in Fig. 23.11).

tower (as shown in Fig. 23.10) or access can be barred by the attachment of spikes around the legs (as seen in Fig. 23.11). In all cases it is advisable to place a notice, similar to the one shown in Figure 23.12, on the base of the tower to deter trespassing.

HOW STRUCTURES FAIL

A free-standing structure such as a tower is most vulnerable in the compression leg (the side away from the direction of the wind). A mast is similarly subject to compression failure but, because of the multiple guying points, has a more complex failure mode. Failure in both instances will probably be due to buckling.

The stress is very sensitive to wind velocity. It varies as the square of the velocity for static loads and as the velocity to the power of approximately 2.5 for dynamic loads. Wind speed varies more or less regularly with height and has an approximately parabolic gradient from ground level to 400 ms.

A less predictable factor is turbulence, although this is probably the major factor in structural failure. Turbulence is poorly correlated along the length of the structure (it is randomly distributed) and varies rapidly with time. In modern studies, the very unpredictable nature of turbulence is taken into account, and it has been found that some turbulence patterns are significantly worse than others.

Topology plays a part and many large towers will be situated on hilltops to gain additional elevation. Hilltops unfortunately produce increased airspeeds over their cresents, and a 10% hill slope can produce a 20% increase in airspeed or a 40% increase in wind loading. This is the reason that windmills and wind generators are usually placed on hilltops.

Stiffness (the ability to resist deflection) is a sought-after characteristic in structures and an important factor for reliable microwave operation. Stiffness is often obtained, however, only by using more metal, which increases the cost and weight. For economic reasons, modern structures are designed to minimize the amount of materials used, so a trade-off occurs. Adding extra dead loads (for example, equipment and antennas) reduces stiffness.

TOWER, MAST, AND MONOPOLE MAINTENANCE

The unscheduled replacement of the antenna support structure can be both costly and disruptive to the installed service and should be avoided if at all possible. The collapse of a tower or mast, particularly in a populated area, can be at best embarrassing and at worst a catastrophe.

Antenna support structures require regular routine maintenance, which is often neglected on the grounds that the structure has been up for years and has not shown signs of fatigue to date.

To appreciate the need for competent inspections, it is necessary to first understand how and why structures fail. These are the major causes of failure:

- Poor design, which inadequately allowed for static or, more frequently, dynamic wind loads
- Overloading of the structures with too many antennas and feeders
- Corrosion, particularly where hollow structural members are used
- Insufficient attention to guy-wire tensions and conditions (corrosion)
- Inattention to the indicators of stress
- Guys corroded or improperly tensioned

As an interesting example of the problems facing masts, the mast shown earlier in Figure 23.5 and again in Figure 23.13 is worthy of a closer look. The long grass in the foreground of Figure 23.5 represents a fire hazard, and evidence in the extreme foreground indicates a recent fire.

Figure 23.13   The dishes with conical radomes (to help water run off) mounted at the base of the tower shown in Figure 23.5. Notice the spalling concrete base.

Figure 23.14   Cracking and spalling of the guy-wire anchor blocks.

Figure 23.15   The buckle connecting the guy wire to the anchor block in Figure 23.14. Notice that the bolts were not tightened and that no washers were fitted. The structure was, however, relatively rust-free.

This mast uses passive reflectors (the large plates at the top) to deflect a microwave link to the ground-mounted receiving dishes illustrated in Figure 23.13. These dishes are protected by conical radomes. The cracking and spalling concrete seen in Figure 23.13 at the base of the mast is a sign of excessive stress.

Masts are held up by guy wires that are anchored into concrete blocks. Signs of stress were evident at all of the anchor points at the site in Figure 23.13. Figure 23.14 illustrates cracking and spalling at these points at this site. All of the anchor points inspected on this structure showed signs of spalling. This mast was well-painted and relatively rust-free, but as Figure 23.15 illustrates, little attention was given to mechanical details. The buckle linking, the guy wire to the anchor point in Figure 23.15, shows that the bolts were not tightened and washers were not used. The large central bolt is about 40 mm in diameter.

Routine inspections should be carried out about once a year for structures located near the coast and every two to three years at sites more than 100 km from the sea, as well as after severe storms or periods of prolonged heavy icing.

INSPECTION

Very few trunk radio companies are large enough to employ a full-time, qualified structures inspector. Those that can will invariably be nationwide operators.

Because of the special nature of support-structure maintenance, the operator will generally find that there are few companies with the necessary expertise and that the availability of those companies is limited. Having found a competent operator, it is therefore a good idea to arrange the maintenance on a contract basis. The company should have a good structural engineer and experienced inspectors who can climb and inspect every portion of the structure. The inspection process should begin with a review of the existing documentation about the structure and its fixtures. It should then proceed step by step, using a checklist like the one provided at the end of this chapter.

If only trunked radio or mobile two-way (PMR) antennas and microwave links are mounted on the tower, the inspection can be carried out without disturbing the operation. The inspector should avoid prolonged periods of exposure (more than 10 mins) within 1m of the antenna. The relevant local RF radiation limits should be observed.

STIFFNESS

A structure that is too flexible is subject to excessive stress and is liable to failure. All structures have resonant modes about which they vibrate. The primary mode for a free-standing tower involves its whole length and results in maximum movement at the top. The tower will sway under wind loads and the period of this sway is a measure of its stiffness. This period is the time to complete one full cycle (that is, from the vertical position through to the maximum deflection and back to the vertical is one half a period). This period can be measured by observation (difficult and inaccurate), by a video camera (better), and by an accelerometer (best).

Accelerometers are usually located at three or more positions along the length of the structure; the results are relayed to the ground for later analysis. Equipment records motion in two directions, as well as torsion. The optimum period is a function of the structure height, strength, design, and mass. For a 180-m tower, a 2-s period is good; a 4-s period would indicate excessive flexibility.

Because early design codes did not fully appreciate the effects of dynamic wind loading, underdesign of the top portions of the structures was common (together with overdesign of the lower portion). As a result, the flexibility of the top portions often, over time, causes high levels of stress. Strengthening the top portions is thus often required at a cost of approximately 10% of the structure cost. Operators will probably encounter this problem only if they use an old, existing tower; design techniques today properly account for the distribution of stress. A good indicator of stress is localized flaking paint and, in some instances, corrosion. Flaking paint is best detected soon after a storm when the recent stress highlights the problem.

REPAIR

Any repair and maintenance indicated by inspection should be undertaken as soon as practical. Finding suitable contractors to do the work may be difficult.

Towers should be painted once every five to seven years, depending on the environment. Painting and touch up for corrosion can be done by many contractors, particularly by those who specialize in heavy industry or bridges.

Replacing bolts, adjusting antennas, and low-stress members can be done by a suitably qualified rigger.

Stress problems are more serious, however, and require the intervention of a structural engineer. The stress problems could be due to weakened members but are more likely design-related. After analysis, the structural engineer can recommend the necessary modifications. The replacement of high-stress members requires the services of a specialist structures contractor.

Stress can be reduced by lowering the wind loading of attachments, but it more often involves adding structural members. The structural engineer usually considers various alternatives to reduce stress and recommends the most cost-effective one.

Welding of strengthening members often destroys galvanizing and other protective coatings, so protective coats will be needed.

TOWER INSPECTION CHECKLIST

A tower inspection should include the following steps:

Tower

1. Check foundations, ground points, and straps.
2. Check for corrosion and condition of painting.

3. Check welds for cracks, using ultrasonic equipment where necessary.
4. Check for signs of stress, particularly flaking of paint or bowing of members.
5. Check all bolts for proper tension and corrosion. (Some may actually be missing).
6. Check guys for proper tension and possible corrosion. In some areas, anticorrosive agents must be applied.
7. Note the position of all fixtures, and when these positions differ from the records, note the details (including photographs).
8. Check for bent or fractured members.
9. Check for tower twist or distortion (sometimes twist can be detected by checking that the tower lines are true).
10. Check the condition of the galvanizing.
11. Check for corrosion in hollow members; this can sometimes be detected by hitting members with a hammer and listening for falling rust. (In some instances and particularly in corrosive environments such as along the coast or in heavy-industrial areas, a low-stress member can be removed and replaced by a new one. This member can then be examined in a laboratory for strength and corrosion.)
12. Keep a permanent log of the inspection.

Grounding

13. Check that all clamps and ground straps are secure and in good condition.
14. Check that bolts are covered by an anticorrosive material.
15. Check that the lightning rod is secure and in an effective position relative to all antennas (higher than any antenna and at least three wavelengths away). (Some cellular installations dispense with lightning rods and use DC ground antennas instead. This is sometimes essential where space on the tower top will not allow for a reasonable separation between the antennas and the rod.)

Antennas

16. Check that all antennas are vertical or at the correct angle of downtilt.
17. Check the physical condition of the antenna; it should be free of cracks, dents, and burns.
18. Check that all bolts and clamps are secure.
19. Check that the antenna grounding is secure.
20. Check that the feeder grounding is secure.
21. Check that the feeder support is adequate and not causing wear or fatigue.
22. Check for any slippage of the feeder.
23. Listen for any audible signs of gas leakage in pressurized systems.

24. Check that the antenna "tail" connector is properly sealed.

Anchorage and Foundations

25. Check that concrete anchors are free of spalling (flaking) or cracks.
26. Check that anchor bolts are tight.
27. Check that grounding is secure.
28. Check that anchor rods are not rusted or corroded.
29. Check for any signs of anchor slippage or creep.

Guy Wires

30. Check for any signs of rust or broken strands.
31. Check that the connectors to guy wires are in good condition.
32. Check that the turnbuckles are in good condition.

Tower Lighting

33. Check that all beacons are in working order.
34. Check that all beacons are in good condition.
35. Check that beacon drain holes are clear.
36. Check that beacon reflectors are in good condition.
37. Check that beacons are free of signs of moisture.
38. Check that beacon lenses are clean.
39. Check that beacon wiring is in good condition.

Ice Shield

40. Check that the ice shield is secure and undamaged.

CHAPTER 24

UNITS AND CONCEPTS OF FIELD STRENGTH

There are a diversity of units of field strength in use in RF engineering today. Mobile engineers tend to prefer the unit dBμV/m, but some still use units like dBm which have their origins in the land-line network. This chapter seeks to clarify the usage of the different units and the concepts behind the measurement of field strength.

In free space, the energy of an electromagnetic wave propagates through space, and, according to an inverse-square law, with distance. The energy is dispersed over the surface of an ever-increasing sphere, the area of which is R (where R is the sphere radius). The total energy is constant because there is no loss in free space. The energy measured in watts/square meter (or any other units) will be constant in total and so the energy per unit area will vary as $1/R^2$. Notice that this equation holds for all frequencies. This concept is illustrated in Figure 24.1.

The total energy within a solid angle is a constant at any radial distance from the origin.

In a mobile-radio environment, the signal is attenuated much more rapidly than in free space and follows approximately an inverse fourth power law with distance. There is a common misconception that this attenuation increases rapidly with frequency. It will be shown later in this chapter that, although the attenuation is an increasing function of distance, it is not a very strongly frequency-dependent function.

However, because the capture area, or aperture, of an antenna decreases directly with frequency, the energy captured by the antenna is directly a function of frequency. Thus, a quarter-wave antenna at 450 MHz is twice as long as a 900-MHz antenna and thus can capture more energy from a field with the same intensity. This difference in capture area or aperture is what mainly accounts for the better long-range performance of lower-frequency systems.

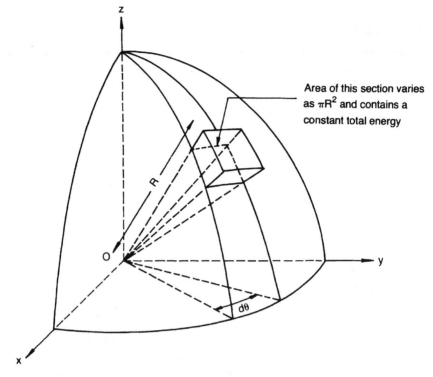

Figure 24.1 The originating energy from the origin, O, is dispersed uniformly over a spherical surface as it propagates in space.

Of course, this discussion must be limited to a frequency band where the propagation mode is similar. If the range 150–1,000 MHz is considered, then the assumptions will hold.

Mathematically, the effective aperture or capture area A of an antenna is

$$A_{\text{eff}} = \frac{\lambda^2 G}{4\pi}$$

where

$$\lambda = \text{wavelength}$$

$$G = \text{antenna gain}$$

$$A_{\text{eff}} = \text{effective aperture}$$

Therefore, if the wavelength is increased by a factor of 2, the effective capture area will increase by a factor of 4. The energy received will also increase by a factor of 4. Thus, the power collected by a 450-MHz dipole antenna compared to a 900-MHz dipole antenna is 10 log 4 = 6 dB higher in a field of the same intensity. The same result can be derived by visualizing the antenna as immersed in an electric field of V volts per meter. A longer antenna is swept by more lines of field strength and so induces a higher voltage.

If a 900-MHz antenna is compared to a 450 MHz antenna of the same type, the 450-MHz antenna is twice as long and will intercept twice the electric field potential. The resulting increase in received field strength is 20 log 2 or 6 dB.

By comparing the aperture we can obtain the relative gain, as shown here:

$$\text{The relative aperture is then } \left[\frac{900}{450}\right]^2$$

$$\text{or a relative gain of } 10 \log \left[\frac{900}{450}\right]^2 = 6 \text{ dB relative gain for the 450-MHz system}$$

Field strength is a measure of power density at any given point. The main units of field strength are dBµV/m, dBµV, µV, and dBµ. The dBµV/m unit measures power density of radio waves. The other units measure received energy levels. Most mobile-radio engineers prefer the dBµV/m unit, while microwave engineers prefer dBm, and bench technicians use µV. This preference is partly historical and partly the practicality of the units in the different fields.

To visualize how these units relate, consider Figure 24.2, which shows a test dipole in the plane at right angles to the direction of propagation. It measures the electric field strength in the direction of the antenna.

The units dBµV, dBm, and µV measure the power or voltage received by a dipole antenna in the field. Voltages in mobile equipment are usually measured at 50 Ω, but any impedance can be used. These units are defined as

$$\text{dBµV} = 20 \log \times \frac{\text{voltage at transformer output (terminated in 50 Ω)}}{1\ \mu V}$$

$$\text{dBm} = 10 \log \times \frac{\text{power measured at the transformer}}{1\ \text{mW}} \text{ (impedance-independent)}$$

µV = voltage at transformer terminal into 50-Ω load in microvolts

The unit dBµV/m is the voltage potential difference over 1 meter of space, measured in the plane at right angles to the direction of propagation and in the direction of the test antenna. It is defined as follows:

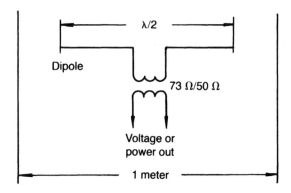

Figure 24.2 The concept of measurement of field strength in μV/m using a dipole.

$$\mathrm{dB\mu V/m} = 20 \log \frac{\text{voltage potential per m}}{1\ \mu V}$$

To see how all these units relate to each other in a real environment, a case study based on the work of Okumura et al. (a classic work on mobile propagation) will help.

Consider a typical base station that has a reference height of 200 meters and a transmitter ERP of 100 W. Table 24.1 lists the received field strength for various transmitter frequencies in the far field; for example, 10 km.

From the table, it should be clear that the actual receiver signal varies enormously with frequency, even though the transmitter power, site, and antenna height remain fixed. Therefore, if the units μV, dBm, or dBμV are selected, the results of a survey of one frequency cannot easily be translated to frequencies that are significantly different. However, if dBμV/m is selected, only a few decibels separate the readings. Furthermore, if additional sites are studied (that is, different heights for the transmitter and different distances), this relationship is retained. Thus, any field strength measured in dBμV/m measures energy density at a given point and is dependent on frequency, only to the extent that atmospheric and clutter attenuation is dependent on

TABLE 24.1 Field Strength at 10 km for a 100-W TX at 20 m in an Urban Environment[a]

Frequency (MHz)	dBμV/M	μV	dBm	dBμV
150	49	71	−70	+37
450	47	17	−82	+25
900	45	8	−89	+18

[a]This table was derived from a paper by Okumura et al. entitled "Field Strength and its Variability i n VHF and UHF Land-Mobile Radio Service," *Review of the Electrical Communication Laboratory*, Vol. 16, Nos. 9 and 10, Sept-Oct 1968, pp. 835–873.

frequency. Therefore, it is possible to use results from a survey done at one frequency to draw conclusions about another if an allowance of 2 dB per octave is used. Some caution should be exercised when using these broad generalizations. However, they can be most useful approximations of coverage.

RELATIONSHIP BETWEEN UNITS OF FIELD STRENGTH AT THE ANTENNA TERMINALS

Assuming a 50-Ω termination, a dipole receiving antenna (unity gain), and a zero-loss feeder, the relationship between units of field strength at the antenna terminals is as follows:

$$E(\mu V/m) = \frac{\mu V}{39.3924} \times \text{frequency (MHz)}$$

Starting with dBm all into 50 Ω, we obtain

$$\mu V = 2.236 \times 10^5 \times 10^{dBm/20}$$

$$dB\mu V/M = 20 \log \times (5.676 \times 10^3 \times \text{frequency (MHz)} \times 10^{\,dBm/20})$$

$$dB\mu V = 20 \log (2.236 \times 10^5 \times 10^{\,dBm/20}) \tag{24.1}$$

Starting with μV all into 50 Ω, we obtain

$$dBm = 20 \log \frac{\mu V}{2.236 \times 10^5}$$

$$dB\mu V = 20 \log \mu V$$

$$dB\mu V/m = 20 \log \times (\mu V \times \text{frequency (MHz)}/39.3924) \tag{24.2}$$

Starting with dBμV/m all into 50 Ω, we obtain

$$dB\mu V = dB\mu V/m - 20 \log \frac{\text{frequency (MHz)}}{39.3924}$$

$$dBm = 20 \log \times \left(\frac{10^{dB\mu V/m/20} \times 1.76168 \times 10^{-4}}{\text{frequency (MHz)}} \right)$$

$$\mu V = \frac{39.3924 \times 10^{dB\mu V/m/20}}{\text{frequency(MHz)}} \tag{24.3}$$

Starting with dBμV all into 50 Ω, we obtain

$$\mu V = 10^{dB\mu V/20}$$

$$dBm = 20 \log \frac{10^{dB\mu V/20}}{2.236 \times 10^5}$$

$$dB\mu V/m = dB\mu V + 20 \log \frac{\text{frequency (MHz)}}{39.3924} \qquad (24.4)$$

CONVERSION TABLES

Because it is very easy to make a mistake when applying the formulas to translate between units, the conversion table in Appendix 8 can be very helpful. The conversions apply for the situation shown in Figure 24.2. If a different antenna impedance is considered (a 300-Ω folded dipole antenna, for example), then the results cannot be used directly. Similarly, in a real-life environment, the signal levels will probably be measured at the receiver input, as shown in Figure 24.3, and the necessary corrections must be applied.

The relationship between the variables must be adjusted by the antenna gain minus the cable loss. At 900 MHz, using a 3-dB antenna, the cable loss is about 3 dB, and thus the correction factor approaches zero. At other frequencies, this approximate relationship will not hold.

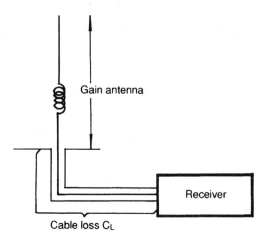

Figure 24.3 Actual field strength measurement.

STATISTICAL MEASUREMENTS OF FIELD STRENGTH

In point-to-point radio, field strength is a one-dimensional variable of time. Most of the time variance is due to log-normal fading, and the nature of the signal variability is well documented. Because it is a simple function of time, field strength in point-to-point radio can be easily understood and measured. The situation is somewhat more complex in the mobile RF environment. Thus the field strength in the point-to-point environment is a simple function of time as shown here:

$$F = f(t)$$

The situation is somewhat more complex in the mobile environment where the field strength also varies with location (space), and so the measured value is a four-dimensional statistical variable:

$$F = f(t, x, y, z)$$

or, if $(x, y, z) = L$ location, then

$$F = f(t, L)$$

The real-life measurement of the field strength of an area is the collective result of a number of measurements made in different points in space and time, as illustrated in Figure 24.4. Thus, if a measure of field strength is required to typify an area, then its statistical nature dictates the measurement is made. Let's assume a measurement of the average field strength along a 500-m section of a road is required. If all samples are taken at the Nyquist rate (in space at locations $L_0, L_1, \ldots, L_m$) and at one instant (T_0), the result would be an average value at a particular time, T_0.

Mathematically, this can be expressed as

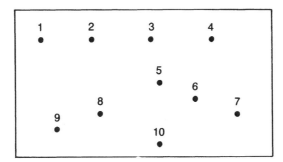

Figure 24.4 The measurement of field strength over an area is made at unique points in space and time. Actual readings vary with time (due to log-normal fading) and space (due to log-normal and Rayleigh fading). Any statement about the field strength in this area is a statement about a variable of space and time.

$$F_0 = \frac{\sum_{i=0}^{m} f(T_0, L_i)}{m+1}$$

($m + 1$ = the number of readings taken at positions $L_0, L_1, \ldots L_m$).

In practice, making simultaneous measurements is difficult and the individual space measurements would be made at different times ($T_0, \ldots, T_n$). The average value in time and space of the field strength in that region is then given by F as shown here:

$$\text{Average field strength} = \frac{\sum_{K=0}^{n} F_K}{n+1}$$

Virtually all real measurements are an average (or some other statistical measure) of sample in different locations in space and time.

Other statistical measures are widely used. For example, all readings can be recorded and separated into two equal groups depending on level. Thus, it is possible to obtain the value above (or below) which 50% of the recorded samples occur. This is known as the median value (not the same as the average value). If all readings were taken at a single instant, the median value F_{50} can be found and the result would be at time T_0

$$F_{50} = f \text{ median } \{f(T_0, L_i)\} \qquad \text{for } i = 0 \text{ to } m$$

where f_{50} is the level below which 50% of the samples 0 to m lie.

Most published data on mobile RF propagation plot the field strength as the median 50%/50% level. This is probably because what was measured was the mean, which for a normally distributed sample will be the same as the median. The mean is the easiest thing to measure, and so the 50%/50% plots predominate.

For a number of distributions the mean and median are equal, although this is not necessarily so for all samples. As an example of how the average and median values can differ, consider the following readings:

$$25, 29, 33, 33, 35, 45$$

The average is

$$\frac{25 + 29 + 33 + 33 + 35 + 55}{6} = 34.99$$

But the median is 33 (because half of the sample is above 33).

In practical terms, because survey readings are taken at different points in space and time for a given region, the average of a set of readings in a region is the true time/space averaged measurement. Because time fluctuations of field strength are usually fairly fast relative to measurement periods, this assumption holds fairly well. Log-normal fades produce field-strength variations with a periodicity of a few seconds, while a measurement is usually taken over 1- to 5-min intervals. Therefore, measurements of field strength taken from a moving vehicle and used to calculate F_{50} will result in field strength for 50% of locations and 50% of the time.

Because the number of samples in space equals the number of samples in time, it is also reasonable to use the same technique to obtain the 70%/70% or 90%/90% values. However, because F_{50} is a function of space and a different one of time, a simple survey technique cannot be used to directly obtain the value for 90% of locations for 50% of the time. To obtain this value it would be necessary to obtain the standard deviation of the time-dependent variation independently (approximately the standard deviation of the log-normal fade).

Interpreting a trunked radio requirement for the field strength to be, for example, 25 dBµV/m for 90% of locations and 90% of the time is a difficult task because the concept has never been adequately defined. In particular, the concept originally applied to small regions of space and its application to service areas is vague. It might mean that within the service area only 10% of all readings will be below 25 dBµV/m and this is how most people think of it. However, the goal of a system designer is not to have 10% of the service area substandard. Probably what is really meant by this is that 10% substandard coverage is a worst-case scenario and is an upper limit rather than a design parameter. One would not expect, for example, a batch of resistors with 10% tolerance to be designed to be 10% out of tolerance. The 10% limit merely means the manufacturer's tolerance or worst case is 10%.

For a 450-MHz system, a 30 dBµV/m field strength is an accepted design level at the boundary. When it is the boundary that is defined to be 30 dBµV/m, it is clear that within the area defined by the boundary the average level will be higher than this. So because a 30 dBµV/m average in practice is a field strength that will ensure a "reasonable" service for 90% of locations and 90% of the times, it follows that the quality of area covered should be significantly better than the minimum in 90% of locations or times.

Some European administrations using NMT systems define the boundary to be the 20 dBµV/m level. This level has some interesting implications. At this level, conversations are only just intelligible, and calls are likely to be dropped. It is below the effective service level.

At 20 dBµV/m the noise is similar to that of the familiar 12 dB SINAD used to measure mobile sensitivity. At this level the processing gains of systems of different deviation will be equalized.

CONCLUSION

It is much simpler and less ambiguous to define the field strength of the service area boundaries, using the following as an example: 30 dBµV/m average (approximately

= 30 dBμV/m median) or 25 dBμV/m in 90% of locations and times. Any specification wherein the time and space variables are not equal is impossible to realize in practice (that is, 90% of places, 95% of the time is physically meaningless unless what is meant by each parameter and how it can be measured are clearly defined).

CHAPTER 25

PRIVACY

Privacy is not an inherent feature of analog trunk systems. Radio scanners can be used to eavesdrop on analog trunk radio calls.

Radio scanners are wideband radio receivers that, under microprocessor control, can scan through a preset group of channels and stop when a carrier (an active channel) is encountered. The receiver stops on that channel indefinitely (unless programmed otherwise) while there is a carrier present. Once the carrier has gone (in the trunked radio instance, the call terminates), the scanner moves on through its list of programmed channels until a new carrier is detected, at which time it stops and proceeds as before.

For a few hundred dollars, anyone can purchase a scanner. There are a remarkable number of people who have not only done so, but who are content to spend hours every day listening to an interminable number of conversations on the off-chance of picking up some scrap of information, perhaps of an industrial, political, or even personal nature. The more organized "snoops" may have tape recorders attached to the scanners, and some may even have commercial outlets for various kinds of information. For this reason, it is wise to warn all analog system users to avoid talking about secret business transactions, or at least to do so in a circumspect manner.

This may not be a problem for some people, as in the case of one rather well-known Australian politician who, when asked if the lack of privacy on his car phone was a problem, replied, "Heavens no, I never say anything that makes any sense anyhow."

While it is true that it is difficult to lock onto a particular mobile, it should be remembered a snoop needs only to have a receiver at a high trunked radio location to receive most base stations. Because some modern scanners are bus-controllable and the signaling protocols are public knowledge, it would not be too difficult to build a scanner programmed to "track" a particular mobile.

There are basically two types of scanners on the market. Currently, the least expensive type is usually limited to 512 MHz and therefore is unable to receive the 800- and 900-MHz systems.

Other, more expensive scanners cover the spectrum up to about 1 GHz and beyond and do cover all bands. These receivers are usually not particularly sensitive and are therefore limited to relatively nearby base stations. However, there are still usually plenty of channels within range (10–15 km) to keep snoops, both amateur and professional, quite contented.

The way in which conversations are received warrants some attention. All conversations between a mobile and the base station are full-duplex (that is, from the mobile, the transmission occurs on one frequency and the return signal, to the mobile, occurs on a different frequency, often 45 MHz away). This fact has led many people to conclude that a scanner can therefore hear only half of a conversation, a conclusion that is, unfortunately, wrong. In fact, as shown in Figure 25.1, scanners hear both sides of the conversation.

As seen in Figure 25.1, although the two conversation paths appear to be separate, the path that transmits to the vehicle F_2 has a very wide range AGC (automatic gain control) amplifier, which has the function of ensuring that constant audio levels are transmitted. This system works so well that side tone (or leakage from transmission to reception) is amplified to the level of normal audio.

Side tone is deliberately introduced into telephones (including cellular telephones) so that some response to the user's speech is heard in the receiver of the speaker. When this tone is not provided, the earpiece is silent during speech, and this silence is interpreted (usually) as a transmission failure (the telephone sounds "dead"). The leakage may or may not be related to the generation of side tone, but it is invariably present; the mobile transmission as well as reception is transmitted by the mobile base at approximately equal levels. The clarity of the leaked path as received on a scanner varies from poor to very good, depending on the system, but is usually intelligible.

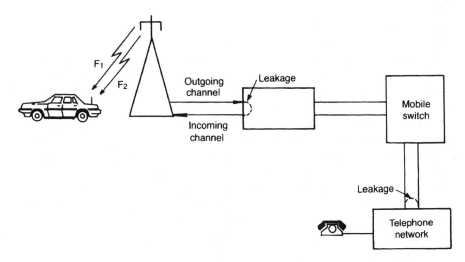

Figure 25.1 Leakage and internal automatic level controls enable scanners to hear both ends of a conversation even though it takes place over different duplex channels.

A number of encryption companies that can provide varying degrees of security have emerged recently. The market originally was aimed at the security sector, but this market is just too small. Besides, most secret services are so paranoid that they will not buy encryption from third parties for fear of leakage of the codes or keys from the vendors.

Criminals, "crazies," and underground organizations are in the market for encryption. This part of the market taints the whole industry and has perhaps led to suspicion on the part of potential corporate users.

Commercially available devices that encrypt two-way radio communications have been available since the 1970s, but the market has been very thin.

There is, however, a need to protect corporate information, and it is this area that is currently being targeted.

DIGITAL ENCRYPTION

By nature, digital trunked radio transmissions are sent to air encrypted. The degree of encryption is sufficient to eliminate the amateur with a scanner, but in general it will not be enough to ensure complete security of the voice path.

The voice security offered by digital systems will be high compared to its analog counterpart, but it should not be assumed that the code is unbreakable. A plus for digital is that encryption can be done at the software level.

ANALOG PRIVACY BY FREQUENCY INVERSION

Privacy has rarely been added as a standard feature to analog systems for one simple reason—cost. To see how this cost arises, the simplest method of privacy or scrambling (frequency inversion) is examined here. Figure 25.2 shows how frequency inversion causes the speech spectrum to be inverted.

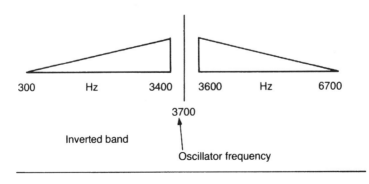

Figure 25.2   A simple frequency inverter.

In Figure 25.2, the normal speech band of 300–3400 Hz has been inverted by a simple modulation technique so that the content of the high frequencies (3,400 Hz) now appears as the low-frequency (300 Hz) part of the inverted band. Similarly, the 300-Hz component now appears as the high-frequency 3400-Hz component. This renders the speech virtually unintelligible.

The inversion can be achieved by a simple mixer and filter combination, as shown in Figure 25.3. The mixer produces the sum and difference of the input signal frequency and the 3700 Hz oscillator. If the mixer is followed by a low-pass filter that cuts off at around 3,700 Hz, then all the sum components are filtered out and only the difference frequency is passed. Thus, if the input signal is 300 Hz and the oscillator is 3,700 Hz, the output will be 3,400 Hz (3,700–300 Hz). Similarly, if the input is 3,400 Hz, the output will be 300 Hz. These frequencies are now said to be inverted. Demodulation consists simply of "reinverting" and uses the same circuitry as shown in Figure 25.3.

This method has two disadvantages. First, it is simple and can therefore easily be decoded (unscrambled). This disadvantage can be overcome by splitting the audio band into a number of components and inverting them separately; this is commonly done. More sophisticated scramblers use this method, but go even further. They divide the speech band into many smaller segments, which are not only inverted separately but sometimes transposed according to a user settable pattern. Clip-on, acoustic-coupled devices with thousands of combinations are commercially available.

The second disadvantage—the inherent increase in signal-to-noise performance—is more fundamental and has no easy solution. In its simplest form, noise necessarily introduced by the scrambling and unscrambling of this signal degrades the signal path by approximately 9 dB in S/N performance. This degradation reduces the useful range of a base station to about 60% of a base without scrambling. More elaborate decoders exist with filters that eliminate most of the noise caused by the inversion process. With such devices, the S/N degradation is reduced to about 1 dB and so the range reduction is not so serious. When time-dependent inversion is used, then the filters must be switchable and follow the pattern of the encoder.

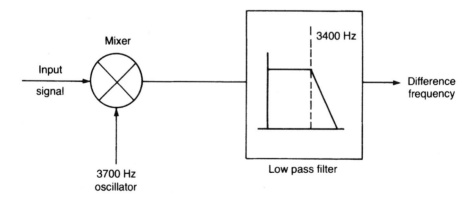

Figure 25.3 A double sideband AM modulator and filter produces frequency inversion.

Unless scrambling capability is provided as a feature of the trunk switch, it should be noted that any "add-on" scrambler requires a decoder/encoder at each end (whether mobile or not) where scrambling and descrambling are to occur.

Multiple splitting of the voice band is achieved as shown in Figure 25.4. The band is divided into two halves and then each half is separately inverted around its own center frequency. A more sophisticated version of this has the center frequencies and the split moving in real-time with up to 32 split points being commonly available. This encryption is known as variable split band (VSB) with rolling code.

Analog scrambling is effective and difficult to decode in real-time. However, it can be analyzed and the inversion points and frequencies determined from regularities in the waveform and so be decoded. More complex analog scrambling systems inherently introduce more noise and also degrade the speech quality.

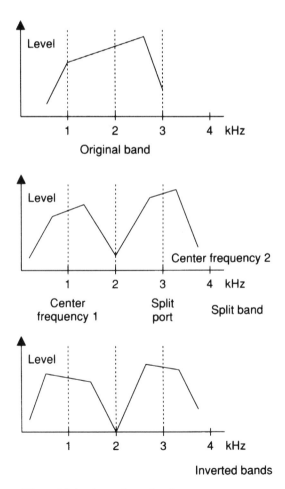

Figure 25.4 A two-inversion frequency encryptor.

DIGITAL ENCRYPTION (OF ANALOG SIGNALS)

The alternative is to use digital scrambling. Digital scrambling starts by sampling the analog signal and then processing it. Within some bounds the degree of "processing" will not be related to the voice quality, provided the reconstruction does not introduce errors. Transpositions in the time domain must be limited to a few milliseconds so that delays are not noticed by the receiver. In fact, nearly all digital systems use scrambling techniques to process the signals in a way that ensures that long strings of "1's" or "0's" do not occur, even when encryption is not the intention.

Other methods to discourage eavesdropping are in place. In most countries, it is either illegal to listen to cellular and other telecommunications, or it is illegal to use any information gathered by such eavesdropping. These laws are largely "paper tigers," however, because they are virtually unenforceable. There is little value in legislating that listening to the trunk radio bands is illegal unless that legislation can be enforced. Some administrations have gone a little further and prohibited import or manufacture of scanners that cover certain bands. However, usually only the cellular bands are blocked. This measure results in the marketing of scanners with discontinuous coverage. This method is also not particularly effective, because there are usually plenty of radio shops willing to "modify" these scanners for a relatively small fee so that full coverage is once again available.

One interesting way of discouraging eavesdroppers takes into account the way that the scanners actually work and turns scanner use into a disincentive in its own right. As noted before, scanners skip over unused channels and stop only on those that transmit a carrier. If all channels always transmitted, then the scanner could function only manually. By having all channels always transmitting, and by having them transmit loud and unpleasant sounds, the eavesdropper soon finds that the hours spent at the scanner are not particularly pleasant. This discourages a number of potential snoops.

On the negative side, however, there are quite a few costs. All channels must be turned on at all times. This significantly increases the power consumption of the base station and decreases the life of the power amplifier stage of the base-station channels. Unless some interlocking device stops this mode of operation during power failures, the time that the batteries can support a base station is significantly reduced.

For this method, base-station antennas must operate at all times at maximum power; this contributes to corrosion and antenna faults. Finally, because all channels are always transmitting, co-channel and adjacent-channel interference always operates at its maximum level and contributes to a net decrease in the traffic capacity and availability of the system.

In general, one can see that where attempts are made to add privacy to analog systems, the costs are high. This, of course, does not apply to digital systems, where very high levels of encryption can be applied at little or no cost. Digital systems can thus be expected to provide very high levels of privacy.

CHAPTER 26

ANALOG MODULATION/DEMODULATION METHODS

Modulation is simply the process by which information is impressed on a carrier; demodulation is the opposite, involving extracting the information from the carrier. The earliest and simplest modulation was on–off keying (OOK). The transmitter either was sending or it was not. A detector could be very simple because all it had to do was detect the presence of a carrier.

To transmit voice a more sophisticated method was needed, and amplitude modulation (AM) fitted the bill. It simply involved making the RF signal proportional in level to the voice level. Demodulation was simple (an antenna, tuned circuit, and crystal were all that was needed), and this form of modulation was dominant for decades. In the 1930s, frequency modulation (FM) was developed. Although it was easy to modulate, it was not so easy to demodulate. In theory, FM uses an infinite bandwidth, but in practice most of the energy is contained in a relatively narrow bandwidth. FM had the advantage over AM that high-quality signals could be sent and received. This was due to the processing gain, obtained by spreading the modulation over a bandwidth wider than the base band (or original signal modulation bandwidth).

It is the processing gain that is at the heart of modern modulation methods whereby attempts are made to use a method that gives enhanced signal-to-noise performance from a given carrier power level.

Most analog systems use FM for speech and FSK (frequency shift keying) for data. Other modulation systems could be used, but this combination gives the best performance when signal-to-noise (S/N) and simpler modulation methods are the main considerations.

Because of the threshold effect (discussed later in this chapter), the mathematics of noise performance above and below the threshold are very different.

RECEIVER PROCESSING GAIN

The baseband reference gain or processing gain (G_B) is defined as the *S/N* obtainable at the detector output compared to the *S/N* at the receiver input if the noise is considered to have the bandwidth of the base band only. Thus, if:

The received power	$= P_r$
Noise (Hz)	$= N_d$
Modulating signal bandwidth (Hz)	$= B_m$

TABLE 26.1 Processing Gain of Different Modulation Systems

System	$G_B = \dfrac{SNR_A}{SNR_R}$
SSB-SC	1
DSB-SC	1
DSB	$\dfrac{m^2}{1 + m^2}$
AM	$\dfrac{m^2}{2 + m^2}$
PM	$(A\phi)^2$
FM	$\dfrac{3}{2}\beta^2$
Additional gain with FM preemphasis	$\dfrac{\pi}{6}\left[\dfrac{W}{f_1}\right]$
Compander	G_B is a function of level and the compression ratio being maximum at low levels of modulation

SSB = single sideband
DSB = double sideband
SC = suppressed carrier
AM = amplitude modulation
PM = phase modulation
FM = frequency modulation
m = modulation index $0 \le m \le 1$
$A\phi$ = maximum phase deviation for PM
β = modulation index $= \dfrac{\text{deviation}}{\text{audio bandwidth}}$
W = baseband bandwidth of modulating signal
f_1 = 3-dB point for preemphasis and deemphasis

then

$$SNR_R = \frac{P_r}{N_d \times B_m}$$

(26.1)

where SNR_R is the reference signal-to-noise ratio.

The signal-to-noise improvement factor is then

$$G_B = \frac{SNR_A}{SNR_R}$$

(26.2)

where SNR_A is the actual signal-to-noise ratio. This ratio becomes more meaningful when the results for various systems are tabulated, as shown in Table 26.1.

For SSB – SC and DSB – SC, $G_B = 1$, then, as a relative measure of noise performance, the S/N performance of any modulation mode compared to SSB –SC or DSB – SC can be used. The values of G_B for 12-kHz FM trunk systems are shown in Table 26.2, comparing results with other modulation systems.

Clearly, FM is superior in all cases, with the difference between the various systems being a function of their peak deviations. For this reason, FM was chosen for the speech channels.

The data channels generally use FSK, which can yield excellent S/N performance particularly at low data rates. FSK is used frequently for signaling in noisy mobile environments. At the time most of the analog trunk systems were developed, techniques for good digital performance over reasonable bandwidths for speech channels had not emerged, and so only analog systems were considered for speech. Today, techniques are available that transmit good-quality speech over bandwidths less than the base bandwidth.

TABLE 26.2   The Relative Noise Performance or Processing Gains of Different Modulation Systems as Signal-to-Noise Ratio

System [a] G_B	Other Modes	Trunk Systems $D = 4.7$ kHz
SSB	1×	
DSB	1×	
AM ($m = 1$)	$\frac{1}{3}$×	
PM[b]		2.4× (3.8 dB)
FM		3.6× (5.6 dB)

[b] PM using same bandwidth.
[a] All cellular systems are assumed to have 3-kHz baseband.

THRESHOLD EFFECT IN FM SYSTEMS

All angle-modulation techniques exhibit a threshold effect; the result of this effect is that a received signal moves from acceptable quality to unacceptable very rapidly as the signal level drops below a certain critical value. FM systems have a processing gain in S/N performance that is a function of their modulation index, which explains why FM is also chosen as the modulation method for high-quality commercial broadcasts. FM stations use 75-kHz deviation, which ensures a very high processing gain. The consequent lower threshold, however, requires a high input signal. A very noticeable improvement in commercial FM-received S/N can be noted with increased input signal level, especially if the receiver is operating in a poor reception area.

At some level, L_{it}, the baseband reference gain, drops off sharply, and within a few decibles of input level can drop to less than unity. L_{it} can be shown (by somewhat arduous theory or by practical measurement) to be in the range of 10–20 dB for practical FM systems. This effect is not present in SSB or DSB systems, but it is present in AM detectors that envelope detectors. But because AM systems have processing gains less than unity, this effect is not so important. SSB systems that use synchronous product detectors do not exhibit this threshold phenomenon, so this mode can be used as a reference.

Figure 26.1 illustrates the threshold effect. Figure 26.2 shows a synchronous product detector.

All types of linear modulation (DSB, AM, SSB, VSB) can be detected by a synchronous product demodulator. For low-level data detection in AM systems, using a synchronous product detector can avoid the threshold effect. There is, however, little practical value in doing this for nondata circuits because the advantages are only realized at very low S/N.

The $f(w_c)$ for an AM signal can be represented as

$$K \cos w_m t \cdot \cos w_c t$$

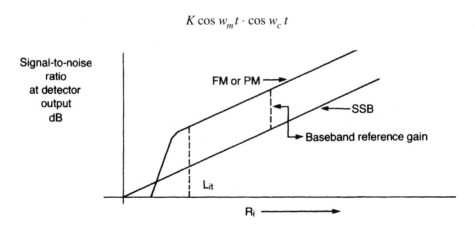

R_i = Receive input signal-to-noise ratio (dB)

Figure 26.1 Process gain of PM or FM over a linear system such as SSB.

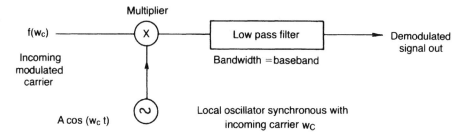

Figure 26.2 Synchronous product detector.

where w_m is the modulation frequency and w_c is the carrier frequency. If the detector product,

$$P = K \cos w_m t \cdot \cos w_c t \times A \cos w_c t = AK \cos^2 (w_c t) \cdot \cos w_m t$$

is taken, we can expand:

$$\left[\text{using } \cos^2 (w_c t) = \frac{\cos(2w_c t) + 1}{2} \right]$$

$$P = \frac{AK}{2} (\cos 2w_c t + 1) \times \cos w_m t$$

$$P = \frac{AK}{2} (\cos 2w_c t \cdot \cos w_m t + \cos w_m t) \tag{26.3}$$

Because the term $\cos 2w_c t$ equals the second harmonic of the carrier frequency, a low-pass filter can easily remove this product, leaving only

$$\frac{AK}{2} \cos w_m t$$

(the original modulation).

For a reasonably good-quality signal (acceptable *S/N* performance), the receiver must operate at an output *S/N* level of about 35 dB (30-dB *S/N* is considered the lowest level at which the noise is not obviously intrusive). This is well above threshold.

However, trunk radio systems often operate at *S/N* levels of around 25 dB.

BANDWIDTH

The bandwidth required for an FM system is given by Carson's rule, which states that 98% of the power in the sidebands is transmitted if the bandwidth of the system is such that

$$B_T = 2 (\Delta F + f_m) \tag{26.4}$$

where B_T is the bandwidth, ΔF is the maximum deviation, and f_m = maximum modulation frequency.

PREEMPHASIS AND DEEMPHASIS

The noise at the output of an FM detector has the following density function:

$$S_{(f)} = \frac{N_i \times f^2}{2P_r} \tag{26.5}$$

where $S_{(f)}$ is the noise density function, N_i is the noise power density at the receiver input, f is the frequency, and P_r is the power received.

From this equation, it can be seen that the noise power is inversely proportional to the input power, and that this is responsible for the FM quieting effect (that is, the noise level decreases as the input carrier level increases). Figure 26.3 shows the noise power output of FM as a function of deviation. Also, the noise power has a parabolic spectrum as it is proportional to f^2.

The detected S/N to input S/N can be shown as

$$SNR_{\text{het}} = \frac{3}{2} \times D^2 \times \left(\frac{B}{W}\right) \tag{26.6}$$

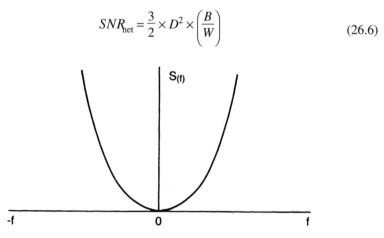

Figure 26.3 Noise power output of an FM detector as a function of instantaneous deviation.

where D is the peak deviation ratio, B is the bandwidth, and W is the base bandwidth.

Thus, the noise power increases rapidly as the bandwidth increases. Fortunately, this can be partially compensated for by preemphasis. This involves boosting the transmitted signals in proportion to their frequency.

This is usually achieved using the transfer function

$$P_E = K\left(1 + j\frac{f}{f_1}\right) \tag{26.7}$$

where P_E is preemphasis, f is the baseband frequency, f_1 is the cutoff frequency, and $j = \sqrt{-1}$.

The f_1 is the point at which the modulating signal is boosted by 3 dB. The slope of this curve approaches 6 dB per octave, which is the inverse of the noise spectral density function $S_{(f)}$. In practice, a 6-dB emphasis has been shown to be both readily realizable and capable of producing good results. Increasing the boost has not been found worthwhile. Naturally, deemphasis must be applied at the receive end, which has an inverse form:

$$D_E = \frac{S_0}{(1 + jf/f_1)} \tag{26.8}$$

where D_E is the deemphasis transfer function, f is the frequency, f_1 is the 3-dB point, and S_0 is the input signal deemphasis circuit.

It can be shown that the improvement in S/N performance due to preemphasis and deemphasis is given approximately by this equation:

$$\text{Improvement} \approx \frac{\pi}{6} \times \frac{W}{f_1} \tag{26.9}$$

where W is the base bandwidth and f_1 is the 3-dB point.

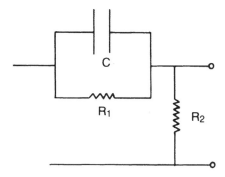

Figure 26.4   A simple RC preemphasis network for FM.

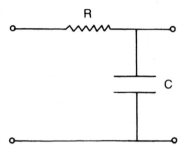

Figure 26.5 Deemphasis circuit.

Nearly all FM systems employ some form of emphasis and deemphasis, including commercial FM and cellular and trunked radio systems. Thus, a preemphasis network can be a simple RC network, as shown in Figure 26.4.

It is normal that $R_1 \gg R_2$ so that the time constant of this network is $R_1 \times C$ for the low-frequency cutoff point and $R_2 \times C$ determines the high-frequency cutoff.

Figure 26.5 shows the deemphasis circuit. This circuit is sometimes described by its time constant $R_1 \times C$ with values of 50, 75, and 100 μs being common.

SIGNAL-TO-NOISE IMPROVEMENTS WITH A PHASE-LOCKED LOOP

The ultimate performance of an FM system in high-noise conditions is determined by the threshold level as can be seen in Figure 26.1. For some applications a threshold extension can improve overall performance and range.

These techniques have particular application in satellite and deep space communications.

A phase-locked loop (PLL), which can be used in FM receivers, can be designed to improve *S/N* by 2.5 to 3 dB in the region below the threshold by incorporating a loop response that has spike suppression. Above the threshold, the processing gain of a PLL is the same as a conventional discriminator.

Figure 26.6 shows a simple PLL demodulator. Known as the FM feedback (FMFB) technique, it uses superheterodyne detection. Usually the detection will occur at the IF (intermediate frequency) stage. The difference is that the local oscillator is replaced by a PLL, the instantaneous frequency of which is controlled by the frequency of the modulated signal.

The net effect is to reduce the effective bandwidth of the IF stage and thereby reduce the noise power at the output. For the equilibrium of this loop, it is required that

$$\frac{d\phi}{dt} = \frac{d}{dt}\left(G \int_{-\infty}^{t} V(\lambda) \cdot dt \right)$$

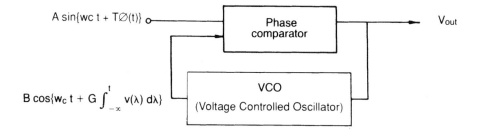

A sin{wc t + T∅(t)} o————[Phase comparator]———→ V_out

B cos{w_c t + G ∫_{-∞}^{t} v(λ) dλ}   [VCO (Voltage Controlled Oscillator)]

where w_c = carrier frequency
            t = time
            ϕ (t) = modulation frequency
            G = constant
            V = output voltage

Figure 26.6   Simple phase-locked-loop detector.

and if

$$w = \frac{d\phi}{dt}$$

then

$$w = G \cdot V(t) \quad \text{or} \quad V(t) = \frac{W}{G} \tag{26.10}$$

where w_c is the carrier frequency, t is the time, ϕ (t) is the modulation frequency, G is the constant, and V is the output voltage. Thus the output voltage is directly proportional to the modulation.

This demodulator is a first-order (or single-pole) device; a further improvement can be obtained by using a second-order transfer function.

Such a PLL can yield an additional 2.5- to 3-dB threshold improvement. More elaborate PLL detectors are available (with second- or third-order transfer functions).

COMPANDING

Companding is the process of compressing the signal before transmission and expanding it at the receiving end. Figure 26.7 illustrates this process.

A compander, as illustrated in Figure 26.7, reduces the distortion generated at the transmit end by reducing the voltage excursions of the modulating signal. It also improves the *S/N* of an analog linear system by boosting the level of the low-level

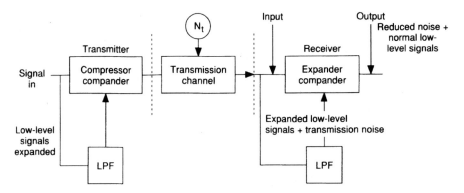

Figure 26.7 A compander system. The terms "compressor" and "expander" are somewhat misleading because the compressor compander compresses signals above the mean level and expands those below it, whereas the expander compander does the opposite.

signals. The transmission noise (N_t) is added (logarithmically) to the signal on the transmission channel but is reduced by the compander ratio at the receiver. In FM systems, a compander also improves the S/N performance by increasing the deviation of the low-level signal components. Figure 26.8 shows a typical transfer characteristic of a compander.

This same compression technique is used in some hi-fi tape recorders to reduce the relative S/N and is known commercially as HX noise reduction in some tape recorder systems.

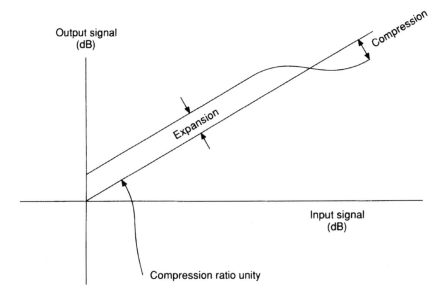

Figure 26.8 Typical transfer characteristic of a compander.

CHAPTER 27

DIGITAL MODULATION

Digital modulation can be divided into three basic types: amplitude shift keying (ASK), in which the carrier is either turned on or off; frequency shift keying (FSK), where the carrier-center frequency is changed from one value to another; and phase shift keying (PSK), where the phase of the carrier is switched. All of these forms of modulation are used in telecommunications, but PSK is becoming the most extensive.

Although in a random-noise environment ASK and FSK have the same performance, ASK does suffer from additional errors due to the amplitude variation that is characteristic of radio signal fading. PSK can be shown to have a 3-dB advantage over ASK in a random-noise environment, and the information rate can be doubled, in the same bandwidth, by using 4-level (or QPSK) phase shifting.

Digital modulation came into vogue in telecommunications in the 1960s. At this time the use of the technique was severely limited by the bandwidth needed to transmit intelligible voice. The standard voice channel of 64 kbits was derived from the Nyquist consideration that a signal needs to be sampled at twice the highest frequency in the sample (nominally 4 kHz) and that an 8-bit word containing 256 levels was required to give a subjectively acceptable dynamic range. Hence 64 kbits is equal to $4000 \times 2 \times 8 = 64,000$ bits/second or 64 kbits/s. This form of encoding is covered by CCITT standard G711.

The general method of encoding as an N-digit binary number is known as pulse-code modulation, or PCM. The basis of PCM is to represent the analog level of the signal by a pulse-coded number. For voice using basic PCM the 64-kHz bandwidth required can be seen to make rather inefficient use of the spectrum compared to the 3.4-kHz needed to transmit the same data as an analog signal.

The main advantage of digital transmission over analog is its noise performance. With analog systems each repeater, each amplifier, and in fact every component of the network will add some degradation to the S/N. Digital systems offer the possibility of noiseless regeneration, and with the use of an appropriate error-correction code a repeater can "clean up" an incoming signal, making it possible to communicate at any arbitrarily small bit-error-rate desired (provided sufficient bandwidth is available).

Digital systems also offer the following advantages over analog:

- Integration of voice and data services
- Store-and-forward capability
- Bandwidth efficiency if CODECs are used
- Security
- Immunity to interference

In order to make digital transmission more spectrally efficient, it is necessary to incorporate a considerable amount of memory and processing power. With the advent of mass-produced digital ICs and ASICs in the 1970s, commercial digital transmission became economically practical for point-to-point applications, such as microwave and fiber-optic applications.

DIGITAL ENCODING

Digital PCM signals are encoded as an N-binary level and decoded by passing the reconstructed signal through a low-pass filter. The output power of the low-pass filter is directly related to the sample duration, and it is in order to make the filter effective that a sample rate of 8 kbits/s (rather than 2×3.4 or 6.8 kbits) was initially chosen.

Because each sample must be rounded off to a discrete binary level, there will be quantizing errors that will produce noise and distortion. Where the quantizing results in errors, distortion arises. This distortion is particularly severe for low signal levels. In practice, to minimize this source of noise, companding is used to increase the level of low level signals and to decrease the level of the highs as seen in Figure 27.1.

It can be shown that if the slope of the companding curve is proportional to the input level, then the quantizing distortion will be independent of the input level. That is, $dy/dx = kx$, where k is a constant. Such a relationship is satisfied by $y = k \log x$. It should be noted, however, that as x approaches 0, so $\log x$ tends to minus infinity, and so the relationship needs to be modified for small input levels.

The two widely used log compression laws in digital transmission are the A law and the mu law.

The A Law

The A law is defined as

$$y = Ax/\{1 + \ln A\} \quad \text{for } 0 < x < 1/A$$

$$y = \{1 + \ln (Ax)\}/\{1 + \ln A\} \quad \text{for } 1/A < x < 1$$

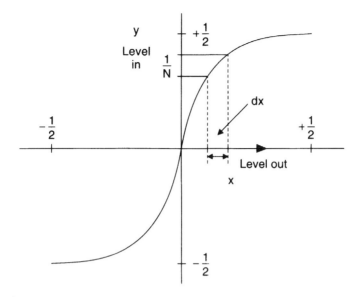

Figure 27.1   A digital compander transfer characteristic.

The MU Law

The mu law is defined as

$$y = \{\ln (1 + \mu x)\}/\{\ln (1 + \mu)\} \qquad \text{for } x \text{ positive.}$$

Practical values for μ and A can be obtained by using the rule of thumb that each quantum step should be at least 16 dB below the RMS level of the speech and that the quantum step should not exceed 1.4 dB. Both these laws can only approximate the uniform distortion criteria because of the need to avoid a singularity at $x = 0$.

TABLE 27.1   The Spectral Efficiency of a Number of N-ary Systems

		Spectral Efficiency bits/s/Hz (max)
BPSK	64 kbits/s	1
QPSK	32 kbits/s	2
16 QPSK	16 kbits/s	4
64 QPSK	64 kbits/s	6
256 QPSK	256 kbits/s	8
1024 QPSK	1,024 kbits/s	10

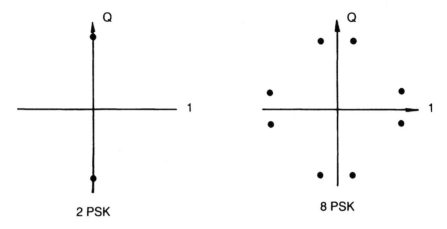

Figure 27.2 Phase shift keying (phasor diagram).

A half-rate encoding known as adaptive PCM or ADPCM using 32 kbits/s sampling rate is widely used.

Simple binary encoding, which generally took the form of a binary phase-modulated signal (BPSK), is not spectrally efficient. Soon higher levels of phase modulation were used, and the hierarchy shown in Table 27.1 was established.

Note that these modulation types are also known as 4-ary, 16-ary, and so on, with a general system with m phases being known as m-ary. Higher -ary systems have lower costs per channel, but also have higher bit-error rates for the same signal levels.

This modulation is illustrated in the Figures 27.2 and 27.3.

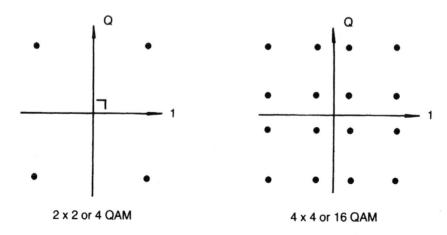

Figure 27.3 QAM modulation constellations.

MODULATION SYSTEMS

The main modulation techniques used in digital links are

- QAM (quadrature amplitude modulation)
- PSK (phase shift keying)
- FSK (frequency shift keying)
- CP-FSK (continuous phase frequency shift keying)
- FFSK (fast frequency shift keying, which is really a special case of CP-FSK)

These techniques are best illustrated by signal constellation diagrams that illustrate the phase and level relationships between the system states. Figure 27.2 illustrates PSK. Figure 27.3 illustrates QAM.

Digital systems are inherently secure because a scrambler is needed to prevent adjacent channel interference due to spectral peaks which may occur during quiet passages, synchronization bursts, or other times when a repetitive pattern may be sent. A typical digital encoder is shown in Figure 27.4.

In the mobile radio environment it is very desirable to have an encoding system that is both spectrally and power efficient. It is also desirable, as well as a regulatory requirement, to avoid interference to the adjacent channels. Economically priced high-power, high-efficiency power amplifiers, of the type that can be expected to be found in a mobile phone, are nonlinear. This means that they will have a gain that rolls off with increasing output level. The gain of a typical PA is shown in Figure 27.5. The nonlinearity of the PA stage can be reduced by "backing off" the peak output power by 3 or 4 dB, as shown in Figure 27.5.

For many applications, particularly in the field of satellites, QPSK is the most common choice. Because of this QPSK is often the standard by which the relative merits of other modulation systems are judged.

The unfiltered spectrum of a QPSK signal is shown in Figure 27.6. Notice that the spectrum is not uniform within the base-bandwidth, and that the side lobes are only

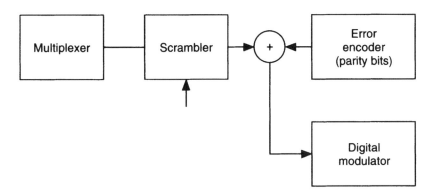

Figure 27.4   A digital encoder.

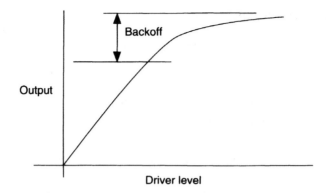

Figure 27.5   The output stage (PA) of a power-efficient amplifier will be nonlinear.

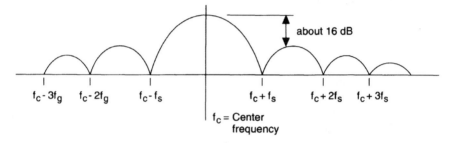

Figure 27.6   The spectrum of a QPSK signal, includes significant levels of sidebands.

about 16 dB down from the center frequency level. The nonlinearity of the PA can boost the sidebands to almost the level of the center frequency, and so a bandpass filter must be inserted to attenuate these frequencies.

Usually a raised-cosine filter will be used as a bandpass filter, and it will have an idealized response as shown in Figure 27.7. To a considerable extent the shape of this

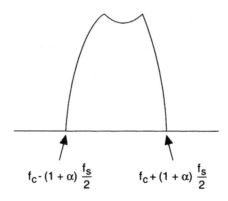

Figure 27.7   A raised-cosine filter characteristic.

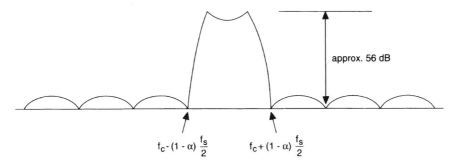

approx. 56 dB

$f_c - (1 - \alpha) \dfrac{f_s}{2}$ $f_c + (1 - \alpha) \dfrac{f_s}{2}$

Figure 27.8   The spectrum of a QPSK signal, which has been passed through an idealized raised-cosine filter.

filter will determine the spectral efficiency (defined as the bit rate to the bandwidth) of the system. In practice, efficiencies of around 1.7 bits/sec/Hz can be achieved with $\alpha = 0.2$.

In turn, the spectrum of a QPSK signal filtered by a raised-cosine function will have an idealized waveform as depicted in Figure 27.8.

The filtered spectrum does result in an attenuation of the side lobes to 40 dB below the center frequency level, but it needs to be kept in mind that they are still there and can appear at the output at much higher levels due to PA nonlinearities if precautions are not taken.

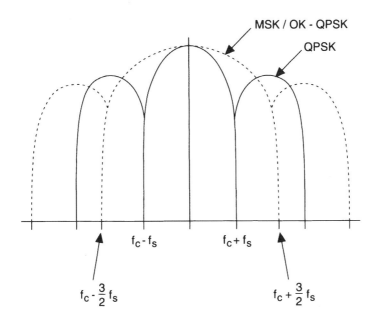

MSK / OK - QPSK

QPSK

$f_c - f_s$ $f_c + f_s$

$f_c - \dfrac{3}{2} f_s$ $f_c + \dfrac{3}{2} f_s$

Figure 27.9   Spectral power density for MSK/OK-QPSK modulator compared to QPSK.

SPECTRALLY EFFICIENT ENCODING

It has been seen that the envelope variation in QSPK becomes a spectral problem at the PA stage. Spectrally efficient modulation can be achieved by employing techniques to smooth the envelope. QPSK allows phase transitions of ±90 and ±180 degrees, which when band-limited will cause very large envelope variations. By limiting the phase transitions (OK-QPSK) or smoothing them (MSK) it is possible to obtain an envelope as shown in Figure 27.9. The modulation used by GSM is derived from MSK modulation.

An MSK modulator differs from a QPSK modulator in that the rectangular-shaped pulses are converted into half-sinusoidal pulses by a pulse-shaping filter. MSK is a special case of continuous-phase frequency shift keying.

DEMODULATION

Demodulation can be *coherent*, meaning that the receiver uses the transmitted frequency and phase information, or *incoherent*, meaning that it does not. A coherent system has superior noise performance but is more complex and costly to implement.

Π/4 QPSK

This modulation method has spectral advantages over 8 QPSK and yet is simple to implement, it is not patented and in particular it is very simple to demodulate. In fact a standard FM discriminator can be used as a demodulator. As can be seen in Figure 27.10, the π/4 QPSK constellation looks like a pair of four QPSK diagrams

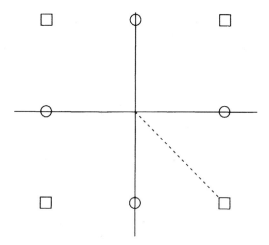

Figure 27.10 π/4 QPSK can be visualized as two superimposed 4 QPSK constellations. Transitions must alternate between the phases.

superimposed. The phase excursions are limited by requiring that changes of phase must alternate from one constellation to the other. Thus instead of up to 180-degree changes being permitted in 8 QPSK, changes are limited to 135 degrees. This can be verified by considering Figure 27.10; and starting from either the constellation with the circles, from any starting point it can be seen that a transition to the "square constellation" can only involve a 45- or 135-degree phase shift. The penalty that is paid is that the information content is reduced from 3 bits/phase in eight QPSK to 2 bits/phase.

SPREAD SPECTRUM OR CDMA

The definition of a spread spectrum signal is one that

- has a transmitted signal, which has a much wider bandwidth than the baseband (message bandwidth)
- has an occupied bandwidth, which is independent of the baseband signal
- has a baseband spreading generated by a code sequence which is independent of the baseband

The three basic methods of generating a spread spectrum are

- Encoding based on encryption with a reproducible pseudorandom digital sequence (which is the essence of CDMA)
- Frequency-hopping where a large number of frequencies can be selected in a pseudorandom manner
- Chirp, where the carrier frequency is swept with each pulse

The interest of the military in CDMA was initially based on the relative security that the techniques provided, because it is practical to transmit useful spread-spectrum intelligence at levels that would be below the noise threshold of conventional receivers. Equipped with the right receivers, however, it is not as difficult as might at first be suspected to intercept spread-spectrum messages. This is particularly true of frequency-hopping where the sweeping can be used to detect it.

Another military application of spread spectrum is in radar ranging, which can be done with precision without the use of high-powered, short-duration pulses.

Spread-spectrum signals can coexist with narrow-band emissions and these emissions will be seen as just background noise. Other spread-spectrum signals that coexist with it will be seen as noise also, provided that they are not correlated to the wanted signal. In practice, there will always be some correlation; and as the number of interferers increases, so the chance of correlations increases.

Spread-spectrum techniques have long been used by the military because of their high immunity to interference and their high security. It is the interference immunity that particularly appeals to the designers of future cellular systems.

One of the things most difficult to grasp about spread-spectrum techniques is the ability of the system to work at very poor signal-to-noise levels. Shannon's theorem can be used to illustrate how this can be done. Consider the relationship

$$C = W \log_2 [1 + S/N] \qquad (27.1)$$

where C is the maximum bit rate, W is the bandwidth in hertz, and S/N is the signal-to-noise ratio.

This tells us that any bit rate can be achieved provided that the correct mix of S/N and bandwidth are adhered to. If it is desired to operate at a very poor S/N, for example sending 2,400 baud data over a channel with $S/N = 0.01$, Equation 27.1 gives, after transformation,

$$W = C/\{\log_2[1 + S/N]\}$$

$$= 2400/\{\log_2 (1.01)\}$$

$$= 167,185 \text{ Hz} \qquad (27.2)$$

Thus it appears that it is possible to communicate over such a noisy channel if 167 kHz of bandwidth is used. This expression can be simplified by noting that

$$C/W = \log_2 \{1 + (S/N)\}$$

and changing bases of the logarithmic expression using the identity

$$\log_a F = \log_b F/\log_b a$$

then

$$C/W = \log_2 e \times \log_e \{1 + (S/N)\}$$

or

$$C/W = 1.44 \times \log_e \{1 + (S/N)\}$$

Note that $\log_e \{1 + (S/N)\}$ expands to

$$S/N = (1/2)[S/N]^2 + (1/3)[S/N]^3 - (1/4)[S/N]^4 + (1/5)[S/N]^5 \ldots \text{etc.}$$

For very noisy signals, S/N will be small (less than 0.1), and thus to a good approximation the higher powers of S/N can be ignored. Thus the relationship reduces to

$$C/W = 1.44 \times (S/N) \qquad (27.3)$$

or

$$W = (S/N)^{-1} \times C/1.44$$

Using Equation 27.3 for the conditions in the previous example, we obtain

$W = 0.01^{-1} \times 2400/1.44 = 166,666$ kHz, which is a useful approximation to the required bandwidth.

CDMA

In essence, the CDMA principle is simple. As shown in Figure 27.11 the transmitted digital signal is multiplied by a pseudorandom sequence. The frequency of this sequence is such that it is significantly higher than the information signal and so the resultant modulated waveform "smears" the modulation out over a wide spectral bandwidth. A conventional wide-band receiver would interpret the signal as noise.

In order to decode the signal, the receiver must use the same random sequence as the sender. By multiplying the received code with the decode key, the intelligence is removed.

However, if other signals, either wide-band or narrow-band, are received, they will not correlate and so will be received as noise. With the use of robust error-correction techniques, much of this noise can be filtered out.

Because the pseudorandom sequence is different for each call, many calls can be set up on the same bandwidth and transmitted simultaneously. In this form, spread spectrum is an example of CDMA.

CDMA is now used widely in cellular microwave link applications and has been used in experimental trunked radio systems.

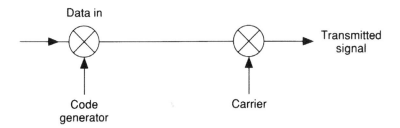

Figure 27.11   A spread-spectrum modulator.

MODULATION

The usual modulation technique for spread spectrum is phase modulation using either 180 degrees phase-shift or binary phase-shift keying (BPSK); FSK is sometimes used, although it does have inferior S/N performance.

When the technique of encoding is a pseudorandom sequence multiplied by the data and then phase-modulated, it is known as "direct-sequence" spread spectrum (DSSS). An important parameter of direct-sequence modulation is the "chip" or the duration of the smallest bit in the code sequence. This determines not only the spectral bandwidth of the modulated signal but also the processing gain of the receiver, both of which increase as the chip duration decreases. The modulating pseudorandom sequence is called the *chip sequence*.

The bulk of the information energy is contained in a bandwidth of $F_0 - F_c$ to $F_0 + F_c$, where F_0 is the center frequency of the transmission and $F_c = 1/ (t_c)$, where $t_c =$ chip duration.

The modulator can be a simple exclusive OR gate as depicted in Figure 27.12.

The chip sequence and data stream need not be synchronous; but if they are synchronous, the clock recovery is simplified.

Other techniques include frequency-hopping (as employed in GSM) and time-hopping where the time bursts have a pseudorandom duration.

Another technique known as *chirp modulation*, which consists of linearly sweeping over a wide frequency range, is also sometimes classified as spread spectrum, although the lack of a pseudorandom generator means that this is not true spread spectrum.

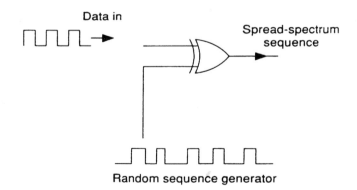

Figure 27.12 Exclusive OR gate.

DEMODULATION

This proceeds in the reverse order of modulation (i.e., just multiply the received signal by a synchronized copy of the chip sequence). The additional complication that the frames must be synchronized by using an autocorrelation pseudorandom code (one that will still correlate even if a phase shift has occurred) and by allowing the received and the regenerated pseudorandom code to slip in phase (by adjusting their clock frequencies) until the best correlation is found.

Because direct-sequence demodulation is mainly done in the digital domain, the decoders are simple and cheap compared to frequency-hopping demodulators.

The demodulation can be done at the RF, IF of baseband levels. The demodulator is sometimes known as a "despreader" and indeed the baseband demodulators consist simply of a chip-sequence multiplier followed by a low-pass filter, the purpose of which is to filter out the noise that is still spread.

FREQUENCY HOPPING

Frequency hopping involves sending bursts of code called *chips*, on a number of different frequencies, determined by a pseudorandom sequence generator which controls a frequency synthesizer. The processing gain (improvement in signal-to-noise ratio) of a receiver is equal to the number of available channels or to the total RF bandwidth divided by the IF bandwidth.

GSM uses this form of spread spectrum.

CHIRP SPREAD SPECTRUM

This type of modulation is largely confined to radar techniques that improve the range and resolution for a given power.

TESTING

The FCC has permitted limited testing of spread-spectrum systems under Part 15 of the Federal Communications rules governing the operation of radio systems without an individual license. The frequencies permitted are 902–928 MHz, 2.4–2.4835 GHz, and 5.725–5.850 GHz using direct-sequence or frequency-hopping techniques. Transmit power is limited to 1W, although no limit is placed on the antenna gain.

MULTIPATH IMMUNITY

FM systems operating at frequencies around 900 MHz have distinct nulls within the local interference patterns, which have wavelengths around 15 cm. Because DSSS operates over a wide bandwidth, it becomes unlikely that all of the signal will be in a null. The wider the operational bandwidth, the greater the immunity.

PROCESSING GAINS

The processing gain of a DSSS system is simply B_{rf}/B_s, where B_{rf} is the RF bandwidth and B_s is the signal or base-band width.

CODECS

CODECs, which are digital compression devices, are the key component leading to the spectral efficiency of digital systems. It has long been known that speech contains a high degree of redundancy. This can easily be demonstrated in a noisy room full of people talking all at once (such as at restaurants). Despite the background of babble, most people will not have too much trouble following the conversation of the person next to them. Even though a lot of the actual sound will be missed, the listener can reconstruct the original sound from the bits received. So it was recognized quite early in the development of electronics that clever encoding could reduce the bandwidth needed for speech.

Bandwidth compression of speech in the analog domain is possible, and the first practical example of this was in 1936 when "Dudley's Vocoder" was demonstrated. It consisted of ten 300-Hz bandpass filters, which were used to extract the speech envelopes and then to pass these envelopes through a 25-Hz bandpass filter. The original 3,000 Hz was thus reduced to 250 Hz (plus guard bands).

These ten "representative" levels were then transmitted to a receiver, which reconstructed the original signal. The speech was said to be intelligible.

Truly controlled and efficient bandwidth compression of speech of a commercial standard had to await the development of economical digital processing. At the heart of current efforts in improvements in spectral efficiency is the CODEC, which compresses the bandwidth of digitally encoded speech. The redundancy of speech is such that the theoretical maximum compression is about 1.5 kbits/s. This limit is rapidly being approached; although CODECs are improving, the attempts to-date to gain the advantages of further compression have been disappointing, in practical systems.

Modern CODECs fall into three main families: waveform, source, or hybrid coding. The first of these, waveform coding, is not usually considered to be a compression technique, but on deeper analysis it is.

Waveform Coding

The early work on waveform coding was done by Bell Systems in the 1930s and 1940s. Commonly known today as pulse code modulation (PCM), it compresses voice into 64 kbits. As is well known, this encoding samples at the 8-kbit rate and uses 8-bit bytes. The 8 bits allow for only 64 levels, and it was the recognition that voice levels are perceived by the ear not linearly but logarithmically that permitted effective use of 8-bit coding. Use was made of companding to expand low signal levels and compress higher levels, as shown in Figure 27.13, so that more bits are used to represent the lower-level signals.

Use was also made of the, by then well-known, fact that telephone systems transmitted good intelligible speech over bandwidths of around 3,400 Hz. So the input audio signal was first passed through a bandpass analog filter. The Nyquist theorem says that the sample rate should be at least twice the bandwidth of the signal being sampled, and therefore 8-kbits/s sampling would be adequate.

To illustrate how this represents compression, consider the example of a compact disk that samples at 44,000 Hz and uses a 16-bit digital to analog conversion process for an overall sampling rate of 704 kbits/s. Compared to this, PCM represents an 11:1 compression.

Because of this early work, PCM became a network standard, and most modern CODEC systems today are designed to multiplex channels into 64-kbits/s streams (or into higher multiples of 64 kbits/s).

The next step was a 2:1 compression technique to derive two 32-kbits/s channels from one 64-kbits/s channel. The method used is called adaptive differential PCM (ADPCM). Instead of sending the actual signal, ADPCM sends the difference between successive samples. The subjective quality for voice is equal to PCM, and the technique has been given international recognition by the CCITT under recommendation G. 721.

Another form of waveform encoding is continuously variable slope delta (CVSD) modulation, which has been used for voice at rates from 12 to 16 kbits/s with reasonable success but with high background noise levels.

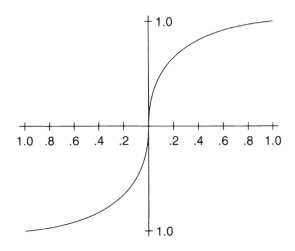

Figure 27.13   A companding curve.

Source Coding

Source coding involves modeling speech in a way that imitates the way the sounds were formed. The sounds of the human voice lie in the 100- to 10,000-Hz band and have an average power of a mere 10–20 μW. Compare the voice power level to that of a drum, which can produce about 25 W peak power, or an orchestra, which may generate 100 W. The vocal cords are responsible for producing the raw sound energy. They resonate at frequencies of 200–1,000 Hz for women and 100–500 Hz for men. The sounds produced are rich in harmonics, and it is the filtering of the generated sounds by the cavities in the mouth, throat, and nose that produces the characteristic sounds of vowels. The lips, tongue, and teeth are used to produce consonant sounds.

The ear is a very selective receiver, with the perceived sound level being both frequency and level dependent. All frequencies have a threshold level below which they will not be heard; if the level is high enough, sounds will not only be heard but they will be felt. The ear has a response to sounds varying from around 25 Hz to 18 kHz, but the range is very age-dependent. From about 20 years of age the frequency range of the average person begins to deteriorate in frequency, linearity, and dynamic range until about 60 years of age, when the average hearing bandwidth is only about 5 kHz (which accounts in a large part for the frequent complaint of older people that the youngsters "don't speak clearly anymore"). People subjected to high noise levels over long periods of time, such as in heavy industry, will have premature hearing impairment.

It is well established that for people with average hearing, good-quality intelligible speech can be transmitted within a bandwidth limited from 300 to 3,400 Hz. This bandwidth is now the internationally accepted "toll quality" range.

The earliest of the source coding techniques is known as linear predictive coding (LPC). The system models voice as two types of sounds, vowel-like (called voiced sounds) and consonant-like (called unvoiced sounds) plus a time-varying digital filter.

An intelligent source coding CODEC breaks down the components of speech into its original parts by modeling the vocal tract. The vocal chords are seen as the source and the vocal tract a filter. The speech components of English can further be broken down into

Vowel sounds: A E I O U
Nasal sounds: M & N
Fricatives: S Z TH
Plosives: K P T

The compressed voice is sent in frames (timeslots) that transmit the following:

- The digital filter model
- The sound type (voiced or unvoiced)
- Amplitude

- Pitch information

Commercial CODECs that can achieve around 4 kbits/s and use LPC are available but they still leave something to be desired, particularly with female voices. LPC relies on the fact that speech is highly predictable in that no rapid changes in level or frequency are expected and the frequency range is quite limited. It is thus possible to sample speech, predict the next sample, and transmit only the difference between the prediction and the expectation. LPC CODECs, although spectrally efficient, are only useful for modeling voice sounds and are not suitable for transmission of other sounds. Even multiple speakers talking simultaneously can confuse the algorithm. Subjectively, the voice quality is only fair.

As the degree of LPC encoding increases so the CODECs will become more specific to voice and maybe even to particular characteristics of voice, so that a CODEC that sounds fine when used by a native speaker of, for example, English may not sound so good when the speaker uses Chinese. The GSM CODEC, for example, was tested with the five main European languages (which have commonality—English and German or French and Spanish) and it may or may not be suitable for other languages. It is interesting to speculate on how other sound sources such as music or background noises like traffic noise will sound after they have been processed by a CODEC built for speech!

Also with increasing efficiency the CODEC algorithms become more complex, and so require more complex processing and more complex coding. This all translates to a higher chip cost.

Hybrid Coding

Hybrid CODECs use both waveform coding and source coding. By the mid-1980s there were three main hybrid techniques in common use: residual excited linear prediction (RELP), adaptive predictive coding (APC), and multipulse coding. Each of these are time-domain coders, which perform processing on the signal in real time. Like source coders, the hybrids use source models, but they use it differently. The model is used to identify and remove the redundancies.

RELP uses linear prediction to model the vocal tract, which is then used to inverse-filter the speech. It is the remaining (residual—from the first letter of the name RELP) signal that is sent. This signal can then be put through a low-pass filter (of around 800 Hz) and transmitted. It can be reconstructed by a reverse process at the receiver end. This method does have problems with the high frequencies found in female voices.

APC also uses a linear predictive residual, performing a number of time-domain processing operations to minimize the information that needs to be sent to transmit this information.

Multipulse compression is similar to APC, but additionally the residual is sent in pulses that are spaced in time and length depending on the information content. A multipulse code known as regular pulse excitation (RPE) is used by INMARSAT for its aeronautical mobile satellite terminals. GSM also uses a version of RPE.

A more advanced coding, adaptive transform coding, makes use of the different spectral content of speech (as shown in Fig. 27.14). Voiced sounds are at lower frequencies than unvoiced sounds. The ATC algorithm allocates most transmission bandwidth to the spectral components that are most strongly represented after the LPC process and reverse-filtering. This produces good fidelity for both high- and low-pitched tones, and unlike most of the other techniques can transmit fax and modem tones. Because it processes a frame at a time, it does not need significant amounts of memory.

Code Excited Linear Prediction (CELP)

Although ATC produced good-quality voice at rates as low as 9.6 kbits/s, its limitations below this rate limit its application. CELP has become the technique showing the most promise for lower rates. It is the method chosen by the TIA for the US digital cellular specification. This technique is the most complex of the hybrid methods.

CELP does not use a vocal tract model, but rather it compares the sample with a library of individual generators and then transmits the one that is nearest to the sample. At the receive end the same sample (called a "code") is accessed and then processed to regenerate synthetic speech, as can be seen in Figure 27.15.

Time-Domain Harmonic Scaling (TDHS)

Time-domain harmonic scaling (TDHS) is a hybrid procedure that does not fall clearly into any of the coding schemes mentioned so far. It relies on the periodic nature of

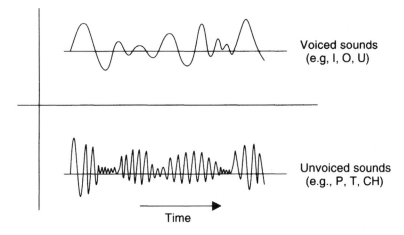

Figure 27.14 The spectrum of voiced and unvoiced sounds.

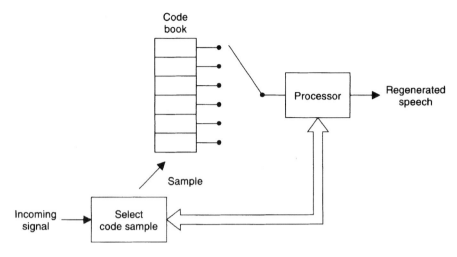

Figure 27.15 A CELP receiver.

voiced sounds and the fact that adjacent samples have similar waveforms. It compresses by sending the average pitch for a number of surrounding samples. It has problems with short sounds and at transition points between voiced and unvoiced sounds.

A new encoding technique known as multiband excitation (MBE) has been developed, which offers enhanced performance over the LPC technique. Commercial units operating at 4.2 kbits are available, and laboratory units are operating at 3 kbits.

DIGITAL SYSTEM PERFORMANCE

While analog system performance is easily quantified by signal-to-noise (S/N) measurements, the situation is a little more complex for digital systems. Because of the increased complexity, there are more ways of performance impairment of a digital radio system. The contributing factors are

- Additive white noise
- Intersymbol interference
- The transmission-medium bandpass characteristic
- Co-channel interference
- Adjacent-channel interference
- Timing errors
- Nonlinearities of phase and amplitude
- Fading

Intersymbol interference is caused by the inevitable spreading of the signal spectrum when it is passed through the filters that are part of the system. The spreading effect is illustrated in Figure 27.16.

Fading can be either flat (wide-band), as is the case in rain attenuation, or selective (due to multipath), where only some frequencies are affected.

The effect of these factors can be measured by the following parameters:

- Bit-error rate
- Error-free-seconds
- Transmitted power spectrum

The easiest thing to measure is bit-error rate, and in its simplest form the measurement can be done using only an exclusive OR gate and a counter as shown in Figure 27.17.

Acceptable bit-error rates in real systems are listed below:

System	BER
Mobile radio	10^{-2}
Good voice	10^{-4}
Microwave	10^{-6}
Quality data	10^{-8}

For a given transmitted power the bit-error rate will increase as the data rate or total bandwidth increases. In order to compare the performance of digital systems a measure that is independent of bandwidth and pulse rate is needed. Such a measure can be defined by dividing the signal energy at the receiver by the noise power spectral density.

The bit energy at the receiver can be defined as

$$E_b = CT_b$$

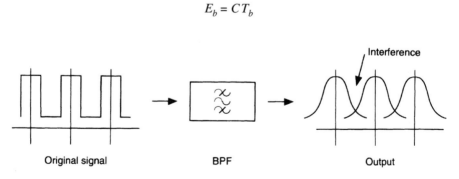

Figure 27.16 Intersymbol interference is caused by spectrum spreading associated with the bandpass filters.

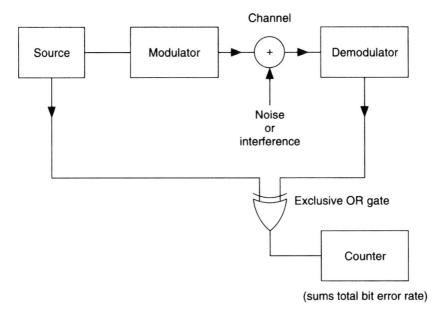

Figure 27.17   A simple bit-error-rate counter.

where E_b is the bit energy at the receiver input, C is the received carrier power, and T_b is the bit duration.

Furthermore, if the noise power spectral density N is defined as

$$N = \{\text{Noise power}\}/\{\text{Bandwidth}\}$$

then E_b/N is a performance measure that is independent of both bandwidth and pulse rate.

CHAPTER 28

NOISE AND NOISE PERFORMANCE

All communications systems are noise-limited in the strict mathematical sense that noise determines the maximum data rate that can be transmitted over any fixed bandwidth and with any arbitrary error rate. The bit rate that can be communicated at an arbitrarily low error rate in a Gaussian noise environment was calculated by Claude Shannon in the early 1940s to be

$$C = W \log_2 (1 + S/N) \qquad (28.1)$$

where C is the baud (or bit) rate, W is the bandwidth in hertz, and S/N is the signal-to-noise ratio. This equation is true for all signal-to-noise (S/N) ratios (including those < 1), provided that a suitable encoding (modulation) system is used.

All analog speech-encoding techniques are very inefficient because speech itself uses very high redundancy and makes poor use of the channel bandwidth.

From Equation 28.1, it can be seen that the baud (or information) rate is improved with any increase in bandwidth and, for reasonably high S/N ratios, the improvement is almost directly proportional to the log of the S/N ratio. FM was an early and successful attempt to exploit this relationship. In FM broadcasting, a wide bandwidth is used to gain an improved signal quality. Commercial FM broadcasting uses a bandwidth five times broader than the baseband (the modulation frequency).

For HF (high frequency, 3–30 MHz) the limiting noise is mainly manmade (for example, electrical and ignition systems), but also includes storms, solar flares, and, more importantly, other transmissions on the same frequency (if they are considered to be noise). Thus, with HF, the improvements in receiver technology cannot achieve much improvement in transmission over any single channel. Frequency-hopping techniques, however, can achieve a marked improvement.

Noise can have many origins, including galactic and extragalactic, thermal, manmade, and even quantum mechanical. Modern radio receivers in the VHF/UHF bands operate close to the theoretical limits of sensitivity imposed by these noise sources.

At VHF and UHF frequencies (30–3,000 MHz), and particularly above about 300 MHz (where most mobile radio systems operate), the background noise is relatively low, and the limits of performance are set by the equipment. Galactic noise can be significant in the region of 40–250 MHz.

At these frequencies, the predominant source of noise is thermal. Modern receivers operate at very low-noise factors (levels of introduced noise) and, short of using very expensive technology (such as masers), few improvements in the basic sensitivity of these receivers are possible.

GALACTIC AND EXTRAGALACTIC BACKGROUND NOISE

Galactic background noise (which originates within the galaxy) and extragalactic background noise (which includes the microwave emissions left over from the Big Bang some 16 billion years ago) are significant radio noise sources.

Galactic noise has its origin in stars, supernovas, neutron stars, black holes, quasars, and other noise sources that are scattered throughout the galaxy. Particular noise sources include the sun, Cygnes A (thought to be a black hole that emits high levels of X-rays and radio but only low-intensity light radiation), and Cassiopeia A (a particularly noisy star at radio frequencies). In addition, because the universe is composed of some 10^{12} galaxies, it is not surprising that some of the noise emanates from outside our local galaxy.

Antenna noise temperatures are used to define the background noise levels and have nothing to do with the actual temperature of the antenna itself. The noise power is directly proportional to the bandwidth of the receiver. It is important to note that the noise power is independent of antenna gain.

The radio noise is very much a function of frequency, because the Earth's atmosphere acts as an attenuator at high frequencies, while the ionosphere attenuates the lower frequencies. The quietest area is from 1 to 10 GHz, where galactic noise is at a minimum. This region between 1 and 10 GHz is most amenable to application of special, low-noise antennas. Trunked radio systems usually operate at antenna noise temperatures of about 2.5 K to 100 K. The large range of antenna temperatures is due to the fact that the antenna temperature depends on how the antenna is oriented at the time of measurement. At lower frequencies (below 1 GHz), the maximum temperature occurs when the antenna is pointed at the galactic poles. At higher frequencies, the maximum temperature occurs when the antenna is pointed at the horizon.

THERMAL NOISE

Because of random molecular movement caused by thermal energy, all passive and active components at any temperature above absolute zero generate a certain amount of wide-band energy. In electronic devices, this energy manifests itself as system noise and imposes fundamental limits on usable sensitivity of receiving and detecting systems.

Thermal noise (also known as resistor, Johnson, Nyquist, and circuit noise) is proportional to the absolute temperature of the conducting device. J. B. Johnson of Bell Laboratories first described this noise in 1927, and H. Nyquist first described it theoretically in 1928. It is best visualized as the random motion of electrons induced in a resistor by thermal energy. The energy is spread uniformly across the frequency band. This type of noise is known as white noise (i.e., all frequencies are equally represented, as opposed to pink noise in which there is a bias toward the lower frequencies).

In any conductor the available noise power can be determined by this relationship:

$$P = KTB \text{ watts} \tag{28.2}$$

where P is the available noise power, T is the temperature in degrees absolute (Kelvin), K is Boltzmann's constant $= 1.38 \times 10^{-23}$ joules/kelvin, and B is the bandwidth in hertz.

Notice that the noise power depends only on temperature and bandwidth and is not dependent on resistance.

Substituting into Equation 28.2, $B = 1$ Hz and $T = 290$ K (standard room temperature), the noise power is 4×10^{-21} W/Hz or -174 dBm/Hz.

Equation 28.2 seems to imply that infinite power is available in the noise source, as the bandwidth is unlimited. This, of course, is not the case; and as the bandwidth approaches optical frequencies at room temperature, quantum corrections come into play that band-limit the independence on frequency so that the total power is constrained.

For a conductor with a resistance R (see Fig. 28.1), it can easily be shown that

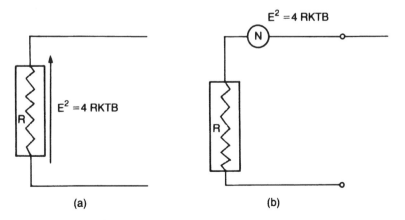

Figure 28.1   Equivalent voltage of a resistive noise source, where (a) is the voltage across a resistor and (b) is the equivalent voltage-generating circuit. R is the equivalent resistance and N is the equivalent noise voltage generator.

$$E^2 = 4\,RKTB \qquad (28.3)$$

where R is the equivalent resistance and E is the equivalent noise voltage generator.

Notice that the voltage is represented as E^2. The square of the actual value of voltage (that is, a value proportional to energy) is used in lieu of E, the actual voltage, because the average voltage is zero (that is, noise with negative-going pulses is just as likely to occur as noise with positive-going pulses).

The RMS value of voltage is

$$E_{RMS} = \sqrt{4\,RKTB} \qquad (28.4)$$

It is important to note that whereas the energy is directly related to the absolute temperature, the voltage is proportional to the square root of the resistance. In the case where a device has an input resistance of 1 MΩ at room temperature, and a bandwidth of 1 MHz it can be seen that the voltage across the input resistance is

$$E_{RMS} = \sqrt{(4 \times 10^6 \times 1.38 \times 10^{-23} \times 290 \times 10^6)}$$

$$= 126\ \mu V$$

This means that all signals below 126 μV would be lost in noise.

ATMOSPHERIC NOISE

Atmospheric noise is largely due to lightning discharges and is consequently very seasonal. It predominates in the frequency range up to about 20 MHz. Atmospheric noise is not generally a factor at the higher trunked frequencies except in abnormal circumstances.

MANMADE NOISE

Due mainly to low-frequency devices such as motors, neon signs, power lines, and ignition, manmade noise sources tend to decrease rapidly in intensity with increasing frequency. Typically, suburban areas are about 15 dB quieter than city centers, and rural areas are about 15 dB quieter than suburban areas. This noise source is significant up to about 1 GHz, but it is generally not a serious problem above 500 MHz.

STATIC NOISE

Static noise is caused by ionospheric storms, which cause fluctuations in Earth's magnetic field. It is also caused by sunspot activity and can be a serious problem in long-distance communications.

SHOT NOISE

Shot noise is generated within active circuitry and is caused by the movement of electrons under applied voltages, such as would be encountered within a transistor. Low-noise devices, which are used in critical parts of the circuitry such as RF stage, have been designed to minimize shot noise.

PARTITION NOISE

Partition noise is similar to shot noise and occurs when a current meets a junction within an active device, which causes the current to divide.

CROSS-TALK

Cross-talk is familiar to all telephone users and is caused by signal leakage from one circuit to another. It occurs when a signal cross-talk limits the frequency reuse of trunked systems. In some large trunked networks, frequencies are allocated to the operator and it is the operator who takes responsibility for frequency management (not the regulator). Frequency reuse is often vital for capacity, but can impair service quality from a distant user operating on the same channel interferes with the channel in use.

Cross-talk also refers to signal leakage between cable pairs in conventional wireline links.

SUBJECTIVE EVALUATION OF NOISE

An established way of measuring the effect of noise on a system is to measure its level relative to the wanted signal. The most common form of expressing this is as the logarithm of the ratio:

$$\text{signal-to-noise ratio} = 10 \times \log \{(\text{signal} + \text{noise power})/\text{noise power}\}$$

When the S/N is expressed as a ratio of voltage or current levels, the expression becomes

$$S/N = 20 \times \log \{(\text{signal} + \text{noise})/\text{noise}\}$$

The ultimate determination of S/N performance is the perception of the user. In the audio environment, it is possible to classify S/N in terms of quality. Table 28.1 shows some categories of everyday experience of S/N.

Another method that was developed by the mobile-radio and amateur-radio community is to evaluate S/N on a scale of 1 to 5. Table 28.2 shows this method and its approximate S/N equivalents.

TABLE 28.1   Some Common *S/N* Levels in Everyday Systems

Type of Signal	*S/N* Ratio (dB)
Limit of operation of five-tone sequential pager	0–3
Barely readable two-way radio	5–10
Telephone voice quality	25–40
Hi-fi analog recording	55–65
Compact disc	80+

TABLE 28.2   Signal Quality as a Function of *S/N* Ratio

Signal Quality	Approximate *S/N* (dB)	Signal Number
Broken and unreadable	5	1
Broken and just readable	10	2
Readable with some difficulty	15	3
Readable with noise	20	4
Clearly readable	25+	5

The fact that this table ends with "clearly readable" is indicative of mobile two-way standards, where a high-quality signal is not generally sought. An *S/N* of 20 dB would be a low limit of acceptability for trunk radio subscribers and would be acceptable in fringe areas only.

NOISE FACTOR

In order to look more closely at the noise performance of receivers, it is necessary to introduce the concept of noise factor. It is important to understand that the noise factor is defined only for amplifiers or devices for which the input is a pure random noise source. Thus a 3-dB noise factor amplifier connected to a purely resistive load will increase the output noise level by the amplifier gain plus the 3 dB. Where the input contains additional noise (for example, from a noisy signal), the nonthermal noise will be increased only by the amplifier gain.

This manifests itself in such a way that the noise contribution of a series of cascaded amplifiers is largely determined by the noise factor of the first amplifier in the chain, as will be shown later.

Noise factor can be defined as

$$F = \frac{\text{available } S/N \text{ power ratio at input}}{\text{available } S/N \text{ power ratio at output}}$$

(28.5)

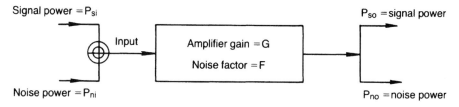

Figure 28.2   A single-stage amplifier, where the noise contribution of the noise factor F results in the amplifier adding to the output noise.

where

$$F = \text{noise factor}$$

Figure 28.2 shows the noise factor.
    From Equation 28.5, you can see that

$$F = \frac{P_{si}}{P_{ni}} \times \frac{P_{no}}{P_{so}} \qquad (28.6)$$

For linear amplifiers, F is always greater than 1 (that is, noise will be added).

$$G = \frac{P_{so}}{P_{si}}$$

where G is the gain of the amplifier. Similarly, from Equation 28.6:

$$F = \frac{P_{no}}{GP_{ni}} \qquad (28.7)$$

THE AMPLIFIER'S CONTRIBUTION TO NOISE (REFERRED TO THE INPUT LEVEL)

It is often useful to determine the contribution of the amplifier to the overall noise of a system. The output noise referred to the input is P_{no}/G. Thus, the noise contributed by the amplifier (referred to input level) is

$$\text{AMP noise} = \frac{P_{no}}{G} - P_{ni}$$

$$= FP_{ni} - P_{ni}$$

$$= P_{ni} (F - 1) \qquad (28.8)$$

CASCADED AMPLIFIERS

Amplifiers connected in cascade (series) result in an overall noise factor that includes the contributions of each stage. Figure 28.3 shows cascade amplifiers.

From Figure 28.3, it can be seen that

Equivalent noise at the input of amplifier 2
= noise input to amplifier by amplifier 1
 plus the contribution to the total noise
 by amplifier 2 referred to the input

$$= G_1 \times P_{ni1} \times F_1$$

$$= P_{ni2} \times (F_2 - 1)$$

(from Equation 28.7 and Equation 28.8 Hence

Noise output power of amplifier 2 = $G_2 \times$ the total noise input power

Therefore, the noise output power is

Noise output of stage 2, $P_{no2} = G_2 G_1 \times P_{ni1} \, F_1 + P_{ni2} \, (F_2 - 1) \, G_2 \qquad (28.9)$

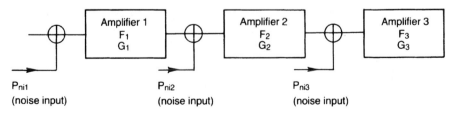

Figure 28.3 Amplifiers connected in cascade can be regarded as having inputs that are the sum of the previous stage outputs, regarded as NF = 1 stage and the net noise contribution.

If the amplifier input and output impedances are matched, and amplifier 2 is at the same temperature as amplifier 1, then

$$P_{ni1} = P_{ni2}$$

$$= KTB \text{ watts}$$

But, from Equation 28.7 we obtain

$$F = \frac{P_{no}}{GP_{ni}}$$

which results in the overall gain for series amplifiers 1 and 2 being

$$G = G_1 \times G_2$$

From Equation 28.9:

$$P_{no2} = G_2 G_1 \times P_{ni1} F_1 + P_{ni2} (F_2 - 1) G_2$$

Then, from Equation 28.7:

$$F_0 = \frac{G_2 G_1 \times P_{ni1} F_1 + P_{ni2} (F_2 - 1) G_2}{G_1 G_2 P_{ni1}}$$

But $P_{ni1} = P_{ni2}$ for impedance matched amplifier stages, so the cascaded noise factor F_0 is

$$F_0 = F_1 + \frac{F_2 - 1}{G_1} \tag{28.10}$$

It can similarly be shown that for additional cascaded amplifiers we obtain

$$F_0 = F_1 + \frac{F_2 - 1}{G_1} + \frac{F_3 - 1}{G_2 G_1} + \cdots \tag{28.11}$$

Two important conclusions are now drawn from these equations:

1. In the mobile environment, the first RF stage virtually determines the noise performance of the receiver (because in general $G_n \gg 1$). A typical mobile RF stage has the following specification:

$$\text{Gain} = 15 \text{ dB (ratio} = 31.62)$$

$$F = 1.5 \text{ dB (ratio} = 1.41)$$

The next stage may have a gain of 30 dB (1,000) and a noise factor of 10 dB (ratio 10). So, the overall performance, denoted F_c is

$$F_c = 1.41 + \frac{10 - 1}{31.62}$$

$$= 1.41 + 0.284$$

$$= 1.694$$

The overall noise factor in dB is 2.28. In other words, the noise factor has not been significantly increased by the addition of a noisy (10 dB) second stage.

It can be further shown that additional amplifiers can be of progressively lower quality (with a higher noise factor and therefore less expensive) without significant degradation in the overall system performance. This fact is most fortunate because the mixer stage in a superheterodyne is very noisy indeed and so, in all high-performance receivers, the mixer is preceded by a low-noise amplifier.

2. The noise factor of a typical UHF receiver can be calculated from *S/N* ratios using Equation 28.4, once measurement details are known. Assume the following:

- *S/N* ratio = 12 dB
- Channel bandwidth = 30 kHz
- Modulation is 1 kHz at 1 kHz deviation
- Measured sensitivity = 0.2 μV
- Processing gain at these conditions = 5 dB
  Then

$$\text{Actual input noise} = \frac{1}{2} \sqrt{4\, RKTB} \text{ volts RMS}$$

$$= \frac{1}{2} \sqrt{4 \times 50 \times 1.38 \times 10^{-23} \times 290 \times 30,000} \text{ volts}$$

$$= 0.0774 \text{ μV}$$

where $T = 290$ K or $17°C$ (an accepted standard room temperature for noise calculations; In practice, this temperature may be somewhat high or low, depending on location).

Hence

$$S/N \text{ at the input} = 20 \log \left(\frac{0.2}{0.0774}\right) + \text{process gain}$$

$$= 8.24 + 5.53$$

The lesson is that S/N ratios mean very little unless the conditions of measurement are clearly stated.

Furthermore, when the measurements are made at very low S/N levels (less than 15 dB at the receiver input), threshold effects can mask the true nature of the S/N performance.

NOISE FACTOR OF AN ATTENUATOR

The mathematics of the noise contribution of an attenuator, such as a transmission line or coaxial cable, are a little complex and involve thermodynamic considerations. Only the result is listed here. Figure 28.4 shows the noise factor contributed by cable loss.

The noise factor of an attenuator F is

$$F = 1 + (L - 1)\frac{T_c}{T_0} \tag{28.12}$$

where

$$L = \text{the attenuation factor}$$

so that

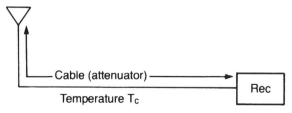

Figure 28.4 Noise factor contributed by cable loss. The feeder cable to an amplifier is equivalent to a series attenuator.

$$L = \frac{\text{power out of attenuator}}{\text{power into attenuator}}$$

$$T_0 = 290 \text{ K}$$

$$T_c = \text{cable temperature}$$

When $T_c \approx T_0$ then $F \approx L$ (the attenuator). Hence, the noise factor introduced by the feeder cable will be similar to the attenuation of the cable itself.

As an example, these calculations can be used to determine whether a low-noise mast-head amplifier would be of value in the mobile environment, given the need to operate in a fringe area. Assume the following:

- The mobile receiver has a noise factor of 6 dB.
- The cable loss is 3 dB.
- A mast-head amplifier of 15 dB gain and 4 dB noise factor is available. (This is a low-grade wide-band amplifier.)
- Receiver gain is 70 dB.

In the case of no mast-head amplifier the overall noise factor is shown in Figure 28.5. For $T_c = T_0$:

$$F_{\text{overall}} = F_c + \frac{F_R - 1}{G_c} \quad \text{(from Equation 28.11)}$$

$$= 1.99 + \frac{3.98 - 1}{0.502}$$

$$= 7.92, \text{ or } 9 \text{ dB}$$

Now, consider the use of a mast-head amplifier, as shown in Figure 28.6:

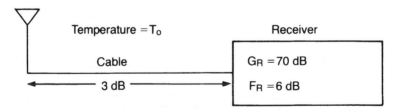

Figure 28.5   Noise factor of a receiver and feeder cable. The values shown are for a practical receiver with a 3-dB cable and a 70-dB receiver gain with a noise factor of 6 dB. $G_c = -3$ dB $= 0.502$, $F_c = 3$ dB $= 1.99$, $F_R = 6$ dB $= 3.98$.

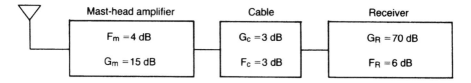

Figure 28.6 A typical mast-head amplifier is placed as close as possible to the receiving antenna. $F_m = 4$ dB = 2.51, $G_m = 15$ dB = 31.6, $G_C = -3$ dB = 0.52, $F_C = 3$ dB = 1.99, $G_R = 70$ dB.

$$F = 2.51 + \frac{1.99 - 1}{31.6} + \frac{3.98 - 1}{31.6 \times 0.502}$$

$$= 2.51 + 0.031 + 0.187$$

$$= 2.72, \text{ or } 4.35 \text{ dB}$$

Thus an improvement of 4.65 dB in *S/N* performance would result in this example and, in marginal areas, could well be worth the effort.

It should be noticed here that the "mast-head" amplifier used is of better quality than the receiver, resulting in an improvement that exceeds the cable loss. When the amplifiers are of similar quality (that is, about a 4-dB noise figure on the first stage of the receiver), the improvement will be similar to the cable loss, because the contributions to the overall noise factor (F), after the first term, tend to be small. This system is not necessarily what would result in practice because mast-head amplifiers with noise factors in the range 0.7–2 dB are available.

This suggests that mast-head amplifiers are useful for survey purposes, particularly when low-powered transmitters are used as a source and measurements are done near the limits of the noise performance of the receiver. Mast-head amplifiers may be difficult to install and are prone to lightning strikes. In a mobile environment, because they are usually wideband devices, they are somewhat prone to intermodulation, which can limit their utility. They cannot be generally used on antennas that are connected to a duplexer.

PROCESSING GAIN AND NOISE

The processing gain, shown in Table 28.1 for an FM system, is

$$G_B = \frac{3}{2} \beta^2 \tag{28.13}$$

where

$$\beta = \Delta f_d / \Delta f_m = \frac{\text{deviation frequency}}{\text{modulation frequency}}$$

However, this formula is valid only at high-input S/N levels, where the noise levels are not high enough to cause the noise spikes familiar in FM systems operating in the threshold region. At very high noise levels (at the receiver-input port), noise spikes are generated by the noise components, which are of sufficient level to cause a phase reversal in the incident wave form. Figure 28.7 illustrates the locus of the vector sum of the carrier and noise signals.

The signal out of the receiver, S_o, is such that

$$S_o \propto d\theta/dt$$

where θ is the phase of the resultant of the incident signal and noise, and t is time.

From Figure 28.7, it can be seen that a phase reversal produces a pulse. If a plot is made of $\beta(t)$, $\theta(t)$, and $d\theta/d(t)$, the form of S_o can be determined. Here, we assume a steady carrier and a noise signal of approximately constant amplitude (but larger than C) that is rotating uniformly with respect to C.

Figure 28.8 shows a typical FM discriminator output filter.

Figure 28.9 shows that this phase reversal produces a pulse. Similarly, it can be shown that when the locus of the resultant does not encircle the point P (as shown in Figure 28.7), a different pulse form arises (as seen in Figure 28.10).

The noise pulse in Figure 28.10 can be shown to be less energetic than the noise pulse in Figure 28.9. Thus, the effect of noise increases rapidly with deterioration of the incident signal. The actual S/N under these conditions can be shown to be

$$(S/N)_o = \frac{(3/2)\beta^2 (S/N)_i}{1 + (12\,\beta/\pi)(S/N)_i \exp\left[-\frac{1}{2}\frac{(S/N)_i}{\beta+1}\right]} \tag{28.13}$$

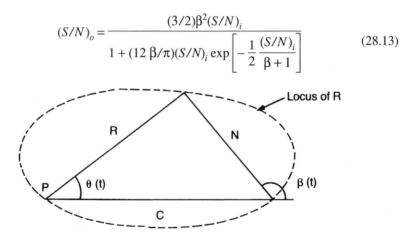

Figure 28.7 The locus of an FM noise pulse. The received signal is the vector sum of the carrier level (C) and noise level (N). This figure shows the locus of the resultant (received) vector (R), which causes the phase reversal associated with FM "picket fencing."

Input = R, where R = C + N

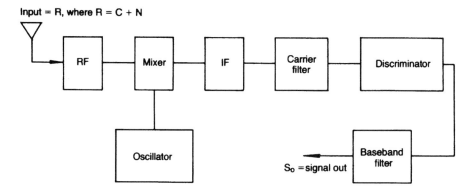

Figure 28.8 FM receiver block diagram.

Note that for large S/N (as encountered in service) the exponential tends to zero and $(S/N)_o \to 3/2\beta_2(S/N)_i$, as before.

The term in the denominator of Equation 28.13 can be seen to tend to 1 when $(S/N)_i$ is large. To determine the levels at which this term becomes effective, it is traditional to define the threshold as the point where the effect of these terms is to reduce the $(S/N)_o$ by 1 dB, as shown in the following equation:

$$(S/N)_o = \frac{1}{1 + (12\ \beta/\pi)(S/N)_i \exp\left[-1/2\ \dfrac{(S/N)_i}{\beta + 1}\right]}$$

$$= 10^{-0.1}$$

$$= 0.7943$$

This function can now be plotted for the three major systems: 900 MHz, 12.5 kHz FM; AMPS (30 kHz); and commercial FM. Table 28.3 shows the S/N performance of various systems.

Using Equation 28.13, the S/N, as measured at the discriminator output of systems with various modulation indexes, can be determined. This function was plotted from AMPS, for a 900-MHz, 12.5-kHz bandwidth FM system, and for commercial FM in Figure 28.11. *Note*: The S/N used in the equation is a power ratio *not* the decibel level; so for 20 dB, for example, $S/N = 10^{20/10} = 100$.

Notice that the AMPS system has a significant relative processing gain up to the threshold (18 dB $(S/N)_i$), but this advantage decreases rapidly as the $(S/N)_i$ decreases below the threshold. Above the threshold, all the systems have significant gains over a zero processing-gain system such as SSB. However, at very low levels of $(S/N)_i$ (for example, below 12 dB), the SSB system will outperform the FM systems.

Notice that the threshold for the trunked system is 14 dB or 4 dB below that of the AMPS system.

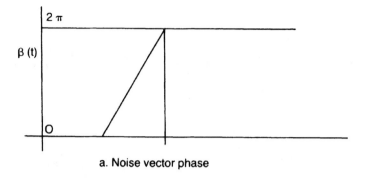

a. Noise vector phase

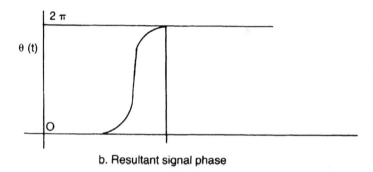

b. Resultant signal phase

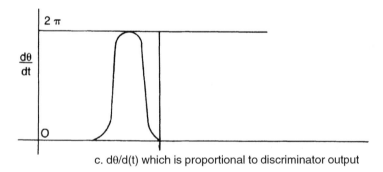

c. dθ/d(t) which is proportional to discriminator output

Figure 28.9   Noise output pulse generated by a noise pulse that causes the resultant locus to pass through 2π radians.

The various systems have a processing gain that tends to unity in the range 12–13 dB *S/N* output. For this reason, it is usual to measure *S/N* at 12 dB SINAD, because this measure reflects the quality of the hardware (noise figure) and will be the same for equal-quality systems regardless of their processing gains.

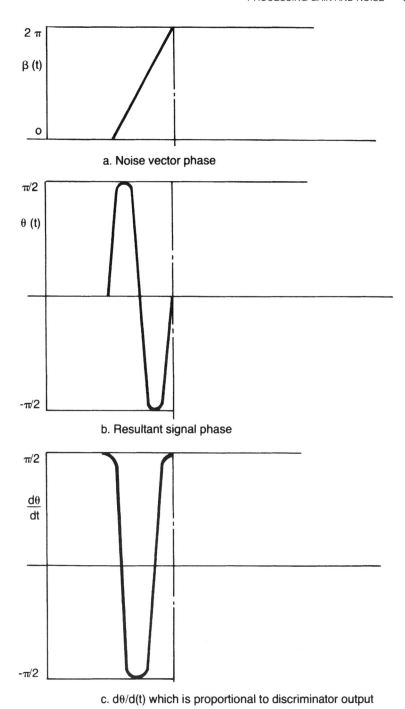

a. Noise vector phase

b. Resultant signal phase

c. dθ/d(t) which is proportional to discriminator output

Figure 28.10   Noise output pulse from a noise vector N, which is smaller than the carrier and rotates through 2π radians.

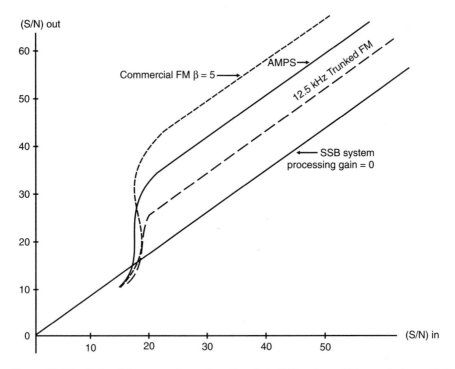

Figure 28.11 Plots of the processing gains of various FM systems. This graph shows *S/N* performance versus carrier signal to noise of AMPS, commercial FM, and a 900-MHz mobile FM system with a 12.5-KHz bandwidth, all from Equation 28.13.

TABLE 28.3 *S/N* **Performance of Various Systems Illustrating the Relative Processing Gains**

$(S/N)_i$	Commercial FM $\beta = 5$	AMPS	900 MHz, 12.5 kHz Trunked FM
50	66	63	55
40	56	53	45
30	45	43	35
25	41	38	30
20	34	33	25
19	30	31	24
18	25	28	23
17	20	24	21
16	17.3	20	20.3
15	14.3	16.3	18.7
14	12	13.3	16.3
13	10	10.8	13.4
12	8.6	8.9	10.6
11	7.4	7.4	8
10	6.5	6.1	5.9

In all real receivers (as opposed to the theoretical receivers considered previously), the processing gain does not improve the *S/N* indefinitely, and a threshold is reached where increases in *S/N* input do not result in increased *S/N* output.

S/N PERFORMANCE IN PRACTICE

Handhelds are the weak link in the chain and are usually assumed to be operated from stationary positions inside buildings. Thus, handhelds do not ordinarily operate in high multipath environments. In general, the full benefits of the processing gain of the FM systems is usable in modern trunked systems.

The situation is similar, but a little different, if only mobile units are considered. With mobile units, there is some significant multipath, and because of occasional excursions below the threshold (contributing relatively high levels of noise), there is some reduction in the relative processing gain. The other requirement of good *S/N* performance, however, minimizes this effect.

A very different situation occurs in mobile communications systems where voice quality is secondary to range (such as CB and amateur transactions). In this instance, the systems are often operated below the FM threshold, and the performance depends only marginally on deviation. Such systems are usually designed to have low deviations and narrow channel bandwidths.

ABSOLUTE QUANTUM NOISE LIMITS

Without going into extensive theory, quantum mechanics demand that radio noise will exist in a vacuum even at 0 K, even though the vacuum is completely shielded from outside radio influences. In essence, the theory states that a condition of "absolute nothingness," free of any noise, is not achievable, even theoretically, in a perfect vacuum.

Some very sensitive measurements (like those used to measure gravity waves) are now approaching the limits of accuracy permitted by quantum mechanics. These noise effects are different from thermal effects (which are predictable by classical physics), and even though much smaller, they may one day limit the speed of future high-technology data communications by placing a fundamental limit on error rates. Such limits are now being approached in some experimental fiber optic applications. New techniques to reduce quantum noise are being explored.

CHAPTER 29

CODING, FORMATS, AND ERROR CORRECTION

DIGITAL SIGNALING

Digital systems can be constructed so that each repeater, or link, not only amplifies but can regenerate the whole code. Provided that sufficient error correction is used, the reconstructed code can have an arbitrarily small error rate while achieving data rates of 2,400 bits/s or more within the conventional 3-kHz speech bandwidth.

For digital data, fast-frequency shift keying (FFSK) commonly is used. This consists simply of transmitting a 1,200-Hz tone to represent a logical "1" and a 1,800-Hz tone for "0". Detection can be completed in one and a half cycles.

The mobile environment is a particularly harsh one for the transmission of high-quality signals; and thus, particularly at high bit rates, it is inevitable that some bits of data will be lost or distorted to the extent that erroneous messages will be received. Because of this, any workable mobile digital code will either have to be very slow or incorporate a robust error correction technique.

In two-way communications, automatic repeat requests (ARQ) are commonly used. Error correction is complete if only error detection is used because a detected error can prompt a resend. Although the basic techniques used for error correction and error detection are similar, error correction is much more complex.

WORDS

Words are groups of code containing the data to be transmitted. A word can be constructed of any number of bits of data. The distance between two code words is defined to be the number of position-bits within the words that differ. Of particular importance is the minimum distance between any two words, which defines how closely the words can match. Where the number of errors exceeds the minimum distance, an erroneous word may be misread as a different but yet valid word.

Digital data are generally structured into words, which are in turn grouped into blocks. Where error correction code is included in the blocks, it is known as block code.

PARITY

Parity is the simplest form of error detection and correction. In its most basic form, a digital word of, for example, 8 bits might contain one parity bit. This bit is inserted to ensure that the total number of "1's" in the word is either odd or even (depending on the parity selected). In this example the information part of the word is 8 bits. To illustrate how this works, assume that the digital information

$$1011001$$

is to be sent.

The total number of "1's" is 4. To construct the word for even parity the last bit would be a 0; that is, the 8-bit word is

$$10110010$$

while for odd parity the last bit would be 1 and so the word is

$$10110011$$

In a simple ARQ system, this is all that is needed because once an error is detected, then a request to repeat the word can be sent. Notice that this code will identify a single error only and can incorrectly accept a word with two errors if they cause the parity to remain correct. Despite these limitations, this method is sufficient for many applications.

Extending this concept, it is possible, by using two parity bits per word, to correct actual errors even in the forward error correction (FEC) mode, which corrects efforts received rather than relying on a resend.

To simplify, consider a word of 4 bits total. If each word contains 1 parity bit and each block has one parity word, then the actual bit in error can be detected. For even parity the word sequence may be

	Information	Parity
Word 1	101	0
Word 2	111	1
Word 3	001	1
Word 4	010	1
Block parity	001	1

Note that even parity has been used for both the word and block check bits. It would be equally valid to use odd parity for the word check bits and even for the block (or vice versa).

Assume now that there was an error in the second bit of word two (that is, the X is a zero instead of one). We now have

	Information	Parity
Word 1	101	0
Word 2	1×1	1
	Error in parity	
Word 3	001	1
Word 4	010	1
Block parity	001	1
	$\vdots$	
	error in parity	

and it can be seen that the error can be detected at the intersection of the error check column and check row.

The block code may consist of any number of message bits, k, together with $n - k$ error detection bits (where n is the total number of bits in the block). Such code is known as (n, k) code. The dimensionless ratio $r = k/n$ ($0 < r < 1$) is known as the code rate.

CONVOLUTIONAL CODES

Convolutional codes differ from block codes by viewing the data as a continuous message sequence and will generate parity bits continuously with the data flow. This type of encoding is distinguished from block codes by the use of memory, so employed that the encoding is dependent on code previously transmitted.

Convolution codes are relatively low efficiency codes (typically $r = 0.5$), which, because they can correct a continuous string of errors, are ideal for correction when error bursts, due to impulse noise, are likely to be encountered.

HAMMING CODES

Because mobile receivers work in a high noise environment, where an error rate of more than 1 in 16 is common, more sophisticated techniques are needed. One commonly used technique is the Hamming Code.

The Hamming code, devised by Richard Hamming in 1950, is capable of correcting multiple errors. Like most early error correction techniques, the Hamming code is largely a product of trial and error rather than a systematic and rigorous mathematical approach. The number of redundant bits is determined by the formula

$$2^n \geq k + n + 1$$

where k is the number of bits in the data word and n is the number of redundant bits.

The code rate for this technique is $r = k/n = 1 - 1/n \times \log_2 (n + 1)$, and it is a reasonably efficient code for long code words.

The redundant bits may be placed anywhere in the word, but depending on the transmission mode there may be positions that will improve the noise immunity.

The Hamming bits are determined by the position of each bit and then determine the resulting exclusive-or value (XORed) with a value for each information bit with a value of "1." At the receiver the Hamming bits are extracted and XORed with all "1's." The result gives the position of single bit errors.

MODULO-2 ARITHMETIC

At the heart of most modern error correction codes is modulo-2 arithmetic. In essence it is simply described by the following rules for binary numbers:

a. All digits must be 1 or 0.
b. $1 + 1 = 0$
c. $1 + 0 = 0 + 1 = 1$
d. $0 + 0 = 0$
e. Addition of binary numbers is accomplished with the above rules plus the rule that there is no carry function.

Addition

For example, add 111001 to 100011, as below:

$$\begin{array}{r} 111001 \\ + \ 100011 \\ \hline 011010 \end{array}$$

The rules a to d above are just those performed by an XOR gate.

Subtraction

Subtraction is identical to addition.

Division

Division is carried out in the same manner as ordinary division. For example, divide 111 into 11001:

$$
\begin{array}{r}
101 \\
111\overline{)11001} \\
111 \\
\hline
101 \\
111 \\
\hline
010
\end{array}
$$

Multiplication

Multiplication is performed the same as conventional multiplication except that the addition is modulo-2. For example, multiply 111 by 110:

$$
\begin{array}{r}
111 \\
\times\ 110 \\
\hline
000 \\
111 \\
111 \\
\hline
10010
\end{array}
$$

The operation of modulo-2 arithmetic will not produce results that agree with conventional arithmetic, but, importantly, it will produce consistent results that can be reproduced in easily constructed hardware. The most commonly used codes, which are subsets of the cyclic block codes (to be discussed later), were derived in the mid-1950s, when the first mathematically derived codes began to appear. Using modulo-2 arithmetic, the ease of realization is not so much a coincidence but at the time may have been a necessity. Of course, today with microprocessor technology, there are few restrictions on the mathematical complexity of the calculations.

In many texts you will find that modulo-2 operations are described using encircled conventional arithmetic signs, as shown in Figure 29.1.

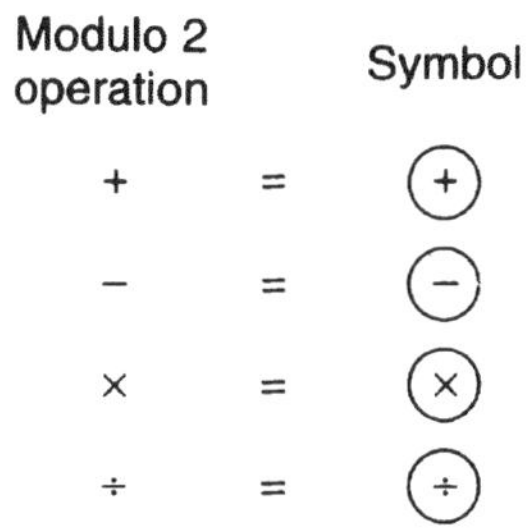

Figure 29.1 Modulo-2 arithmetic operators.

CYCLIC BLOCK CODES

The most commonly used codes today are derivatives of the cyclic block codes. A cyclic code is one in which any code word can be end-about shifted to form another code word. Consider the code formed by the message bits $m1, m2, m3, m4$ and the code check bits $c1, c2, c3$ to form the code word

$$m_1\ m_2\ m_3\ m_4\ c_1\ c_2\ c_3$$
$$1\ \ 0\ \ 0\ \ 1\ \ \ 0\ \ 1\ \ 1$$

Shifting one place to the left results in the word

$$0\ 0\ 1\ 0\ 1\ 1\ 1$$

where 0010 is a valid message word and 111 the corresponding check sequence. The prominence of cyclic codes is due mainly to the fact that they can be easily implemented by straightforward modulo-2 hardware circuits. The word and its corresponding check bits are generated by dividing the message bits (padded with zeros to the total word length, n) by a binary number of length $(n - m + 1)$. This binary number is often referred to as the generator polynomial. The important part of this division is the remainder, which forms the check bits. A characteristic of the word so generated is that it is a perfect multiple of the divisor. Thus a simple check for error by the receiver is to divide the received word by the divisor and any remainder other than zero indicates an error. With a careful choice of divisor, error correction is also possible.

To see how this works, it is instructive to consider the analogous situation in conventional arithmetic. If the data bits were equal to 13 and the divisor was 5, then by performing the division 13/5 we find a remainder of 3. To make the code word perfectly divisible by the divisor (5) it would be necessary to *subtract* the remainder, making the code word 10. However, recalling that in modulo-2 arithmetic addition and subtraction are identical, then *adding* the remainder will produce a number perfectly divisible by the divisor.

The next task is to select a suitable divisor. There are no simple rules to select a divisor, and the codes have high redundancy. The codes are divided into classes based on the choice of divisors.

BCH CODES

BCH code (BCH stands for the inventors, Bose, Chaudhuri, and Hocquenghem, who derived the code in the 1950s) are the most commonly used codes today. The code is specified by its total and message length as (n, m).

BCH codes are such that they can be devised to correct any given number of random errors per code word. For block lengths of up to a few hundred bits they are

TABLE 29.1 BCH Codes Up to Word Length 31

			Divisor								
n	m	t									
7	4	1								1	011
15	11	1								10	011
15	7	2							111	010	001
15	5	3						10	100	110	111
31	26	1						11	101	101	001
31	21	2				1	000	111	110	101	111
31	16	3		101	100	010	011	011	010	101
31	6	7	11	001	011	011	110	101	000	100	111

n = total word length

m = message bits per word

t = maximum number of detectable errors

among the most efficient in terms of total block length and code rate. Table 29.1 gives the BCH divisors for word lengths up to 31.

POLYNOMIAL CODES

The most general class of cyclic codes are known as polynomial codes. A general code word can be described by a polynomial as

$$f(x) = 1 + x + x^2 + x^3 + \cdots + x^{n-1}$$

where x can only take a value of 1 and the power of x signifies its position in the word. It is important to realize that this polynomial is not a conventional one, and for the purposes of this text it can be regarded as a shorthand descriptor of the code bits. The factor x^k, where $x = 0$, is implied for all values of $k < n - 1$ where the kth power is not in the equation.

For example, the CCITT V41 (256, 240) code uses the 17-digit divisor

$$10001000000100001$$

which could be described by the polynomial;

$$g(x) = 1 + x^4 + x^{11} + x^{16}$$

The generator polynomial can get much more complex as, for example, the generator used for the CDMA paging, sync, and access channels, which is

$$g(x) = x^3 + x^{29} + x^{21} + x^{20} + x^{15} + x^{13} + x^{12}$$

$$+ x^{11} + x^8 + x^7 + x^6 + x^2 + x + 1$$

For the forward CDMA traffic channel, the generators are

a. 9600 bps

$$g(x) = x^{15} + x^{11} + x^{10} + x^9 + x^8 + x^4 + x + 1$$

with a frame length of 192 bits in 20 ms

b. 4800 bps

$$g(x) = x^8 + x^7 + x^4 + x^3 + x + 1$$

with a frame length of 96 bits in 20 ms
The reverse channel generator is

$$g(x) = x^{16} + x^{12} + x^5 + 1$$

GOLAY CODE

The Golay code is generated by the divisor, or generating polynomials:

$$g(x) = 1 + x^2 + x^4 + x^5 + x^6 + x^{10} + x^{11}$$

or

$$g(x) = 1 + x + x^5 + x^6 + x^7 + x^9 + x^{11}$$

The Golay codes can detect any combination of three errors in a 23-bit block.

POCSAG

Pocsag, the international code format for paging, is a (31, 21) BCH code, with the generating polynomial:

$$g(x) = x^{10} + x^9 + x^8 + x^6 + x^5 + x^3 + 1$$

To this 31-bit code word is added one extra even parity check bit for the whole code word.

INTERLEAVING

Depending on the nature of the transmission path, an improved performance measured in bit-error rate can sometimes be obtained by interleaving bits of words so that in small duration fades the corruption of any one word can be kept to within the capabilities of the error correction code. Interleaving can thus improve performance without adding redundancy. Because paging generally is used for only relatively short messages, with little opportunity to interleave, this technique is more commonly applied in other mobile data systems.

CHAPTER 30

SAFETY

ELECTROMAGNETIC RADIATION (EMR)

There has been a lot of concern expressed recently about the safety of people exposed to RF radiation (EMR). Much of this is prompted by the ubiquitous mobile phone. There have been literally hundreds of studies that have attempted to uncover any link between RF and health, and to date there is no conclusive evidence of any problem.

RF radiation is known as *nonionizing* to distinguish it from the ionizing radiation caused by nuclear decay products. The essential difference is that radioactive decay products are far more energetic and will displace electrons from atoms and thus leave them with a residual charge (known as *ions*). RF radiation produces much less energetic particles called *photons*, which are absorbed by the electron cloud around the atom but are not displaced by it. The net effect of RF radiation on an atom is to raise the average electron energy, which in turn raises the temperature of the atoms. Microwave ovens exploit this principle.

Analog systems produce a relatively steady RF field, while TDMA systems add a new twist to the controversy. For some time now, there have been studies linking power lines with some forms of cancer. The studies are far from conclusive and often conflicting. Some studies find a link, and some do not. Surprisingly, of those finding a link, not all are negative: Some find less cancer among the radiated than among the control groups.

TDMA systems generate pulses at around 200 Hz (the actual frequency depends on the frame rate). This pulsing can often be heard as a buzz in other equipment being used around TDMA sources. Because the low-frequency frame buzz can be equated to the 50/60-Hz power line, emissions TDMA systems come under extra suspicion. Most digital trunked systems are TDMA.

The radiation from power lines falls off in intensity as the inverse cube of the distance from it. This is because the radiation is from a dipole (all magnetic fields are generated by dipoles); and in the case of power lines, the electric field that comes from two nearby oppositely charged wires is also a dipole. RF radiation on the other hand falls only as the inverse square of distance, and thus its effects are felt at greater distances than it would be from power lines.

The most prominent effect, and the one most easily measured, of RF on the human body is heating. Considerable concern has been expressed that handheld mobiles, generating something from 0.5 to 5 W held near the brain, are worrisome to some people. The brain actually consumes quite an amount of power, and dissipates about 20 W. Estimates of the energy that is absorbed by the body put the figure at around 20% for a handheld. So a 5-W handheld will add about 1 W (or 5%) to the heating of the brain. PTT operation will reduce the heat absorbed to about 50% of that figure. Because intense brain activity and physical exertion can increase dissipation by much more than this, there seems to be little reason to be concerned about the heating effects.

Probably the most extensive review of RF effects on health has just been undertaken by the Cellular Telephone Industry Association (CTIA) in the United States. Their charter was to do the following:

- Conduct a literature review of previous studies on health effects.
- Conduct experiments on dummy heads.
- Become involved in ongoing epidemiological studies.
- Conduct experiments with live subjects.

Although the charter is wide and the study extensive, there is room for cynicism, because the CTIA could hardly be considered an impartial party, there being enormous repercussions to their members should they return a negative finding. The findings suggest nothing new.

There can be no doubt that over the last few generations the level of RF exposure of the average person has risen by huge factors, and yet there is no evidence that the age-corrected incidence of cancer has risen at all. By early 1999 the consensus in the scientific community was that there is little or no evidence of adverse effects from exposure to EMR of the magnitude that would be encountered in the mobile radio industry, and that too much effort and resources are being devoted to these studies. There are studies however that link EMR with biological changes at the cell level, but none of these have yet definitively indicated a cause for concern.

USE OF MOBILES WHILE DRIVING

A number of studies have linked increased accident rates with the use of mobiles while driving. In most countries it is illegal to use a radio device while driving, unless its operation is hands-free. It would appear that accidents are around 50% more likely to happen to drivers who simultaneously use a mobile. This increased accident rate is equated to being marginally over the 0.05% blood alcohol limit (the legal limit in most countries).

A factor that is often overlooked is the danger of dialing while driving. This often involves the distraction first of finding the number, and then the need to take the eyes off the road to dial. Voice-operated dialing would of course counter that problem.

INTRINSICALLY SAFE MOBILES

In most countries today there are restrictions on using mobile radios near petrol bowsers and in blasting areas. There is no evidence that the RF can ignite fuels, but the concern is that the switching (e.g., the PTT operation) could produce small sparks, which in turn could cause vapors to ignite. Intrinsically safe mobiles are designed not to produce any sparking in operation, and are required around refineries and other hazardous areas.

Blasting operations can be compromised by RF, particularly where electrically fired detonators are used. It is possible (but unlikely) that voltages could be induced into the firing leads and set of charges prematurely.

AC VOLTAGES

Nearly all trunked radio systems run from mains power sources. Serious injury or death can result from quite small currents passing through the human body. As little as 10 mA can cause muscular paralysis, which in itself may be distressing, but more importantly it can cause the victim to be unable to release their grip. No live equipment should be worked on, and even when the equipment has been turned off and checked, it is wise to test it first by touching it first with the *back* of the hand because currents induced this way will cause the hand to move away from the source, as an automatic reflex.

At 30 mA, respiratory paralysis can occur and breathing may stop. Usually breathing will start again once the current has been removed, but resuscitation may be necessary. The most dangerous currents are between 75 and 250 mA. At this level, once the exposure exceeds a few seconds, ventricle fibrillation can be induced, causing the heart to beat in an uncoordinated manner, which renders it unable to circulate the blood. The fibrillation is likely to continue even after the source of the current has been removed, and the victim may well die unless first aid is rendered immediately.

Curiously, at currents higher than about 300 mA, the heart is clamped into a state that probably will set it beating normally once the current is removed. Serious injury or death can still result from prolonged exposure to these currents or in the case of very high currents from burns. This level of current is unlikely to be encountered at mains distribution voltages.

When working on mains voltage equipment, there should always be at least two people present, and they should have been trained in cardiopulmonary resuscitation (CPR). This training needs to be repeated regularly to ensure currency. Because irreversible brain damage will occur within 4–6 min of heart or lung failure, time is of the essence. Brain death occurs in about 10 min.

APPENDIX 1

PROPAGATION AND FADING

Light (and thus radio energy) does *not* travel in straight lines! Over short distances in a uniform *medium*, it will *mostly* travel in straight lines, and from this comes the incorrect assumption that it always travels in straight lines.

Radio energy and light are propagated as photons, which are small packets of energy with energy levels proportional to the photon frequency. Photons are *not* constrained to travel in any particular path, but they do show a preference for "the path of least action." In the case of photons the path of least action is also the path of least time. Often this will be a straight line (or very nearly) but more often it will not be.

Consider a jet plane traveling between two distant cities. The pilot will probably fly it to achieve the least cost for the trip, but the flight may also be planned to take the least amount of time. In either case, after take-off the plane will climb to the optimum height for cruise and descend only in the landing phase. In the thinner air at around 37,000 ft, the plane can travel faster relative to the ground than it can in the denser air below. In principle the higher the plane flies, the less the air resistance and the faster it can fly but there will be limits set by the design of the plane which set optimum heights for the engine operational efficiency, wing lift, and control surface response. If thinning of the air were not a restriction to the engines and controls, the optimum height for a long flight would involve a low orbit. For short flights the pilot may choose a level well below the above optimum because the flight is too short to justify a climb to the maximum altitude (this still gives a least-time flight).

Photons also travel faster in less dense air; and as the air density decreases with height above sea level, the photons can travel faster in the upper atmosphere. From a quantum mechanical viewpoint, photons do not propagate as a field, but they instead travel between electrons. In the case of a radio link, the electrons of interest to us are the ones on the surface of the transmit and receive antennas. The *preferred* path between the two antennas will be the one that has the least time lapse for the photons, and in general in the atmosphere this will be one that curves upward much like the path of the airplane discussed above (except there being no "optimum" cruise height). In a

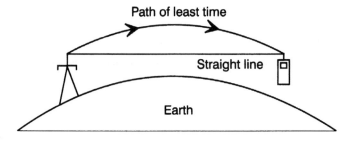

Figure A1.1   The propagation path of least time.

uniform atmosphere the photons will travel in a smooth curve as shown in Figure A1.1. Generally, this upwardly curved path will be such as to allow the signal to travel further than the horizon as seen in Figure A1.2.

The rule that light propagates along the path of least time was first suggested by Fermat in about 1650, and it remains the simplest rule one can follow to understand propagation.

It is well known that temperature inversions can cause anomalous propagation. Ordinarily the temperature of the air will decrease steadily with altitude at about $10°$ C per 1,000 m of altitude. In a temperature inversion, this gradient is reversed and there is a layer of air where temperature increases with height. Mostly, but not always, this will occur at night.

Sometimes anomalous propagation can have profound effects. A temperature inversion can lead to ducting, in which a layer of air forms that has just the right refractive index to confine an RF signal in much the same way that a fiber-optic cable does. Pilots operating ground-to-air radios at around 120 MHz often find that they can clearly hear ground movements (taxiing reports) from airports a thousand or more kilometers away, over radios that ordinarily have a range of around 70 km.

A similar thing happens off the north east coast of Australia: For a few months a year, propagation at 80 MHz from French-speaking islands thousands of kilometers out to sea will regularly cause interference into PMR systems operating along the

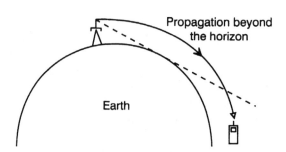

Figure A1.2   Propagation beyond the horizon.

coast line with the language clearly indicating that the source is not local. This interference is seasonal and sometimes persists for months.

To some extent the propagation of any radio signal, and more particularly distant ones, will be subject to path vagaries, and even in real time, variations of 4–6 dB are not uncommon. Sometimes even relatively close line-of-sight systems can suffer total fade-out, due to spurious propagation.

APPENDIX 2

NOISE AND DISTORTION

In a perfect world where all devices are linear, there would be no distortion. A perfectly linear device, like an ideal amplifier, might have a response characteristic like that shown in Figure A2.1. It is a perfect straight line. Now most real-world devices are less than perfect, and so they may have a response more like that shown by the dashed line. To see what effect that may have, it is necessary to recall that any curve can be approximated to any arbitrary degree of accuracy by a polynomial. Now a linear response curve takes the form

$$\text{Output} = \text{Constant} \times \text{input} \quad \text{or} \quad Y = K \cdot X$$

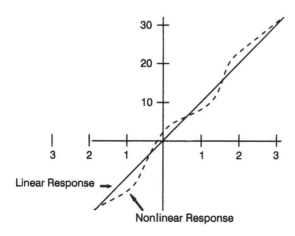

Figure A2.1 A linear and nonlinear response curve.

where Y is the output, K is a constant, and X is the input.

A nonlinear response can be characterized as the polynomial

$$Y = K \cdot (X + e \cdot X^2 + f \cdot X^3 + g \cdot X^4 + \cdots)$$

where, because the distortion is small, the terms e, f, g, and so on, will also be small.

Consider now the effect of the first distortion term only and let the input be sinusoidal, that is,

$$X = K \cdot \cos(2\pi f t)$$

where t is time and f is frequency. Then the equation

$$Y = K(X + e \cdot X^2)$$

becomes

$$Y = K(\cos(2\pi f t) + e \cdot \cos^2(2\pi f t))$$

recalling that

$$\cos^2(A) = \frac{1}{2}(1 - \cos(2A))$$

Then

$$Y = K\left(\cos(2\pi f t) + \frac{e}{2} + \frac{-e \cdot \cos(4\pi f t)}{2} \right)$$

So now we have a term with an amplitude equal to $K \cdot e/2$ which is twice the frequency of the fundamental; that is the second harmonic.

Proceeding in the same way, the third and higher harmonics can be derived from the higher powers of X.

It is worthwhile noting that the *magnitude* of the second and higher harmonics will be a function of the actual nonlinearity of the device, and in general it will be very hard to quantify (except when caused by very predicable devices). However, one thing is certain: There will be *some* higher harmonics in any real-world device.

There is one other term in the equation that is worth a second look at. The term $K \cdot e/2$, in the expression for Y represents a DC component. Thus, in addition to producing harmonics, a nonlinear device will also rectify (that was how the old-fashioned crystal set worked). When the carrier signal has some AM modulation such as an AM radio station or the frame buzz in GSM, K becomes itself a low-frequency variable that the nonlinear device will detect. For this reason an unshielded audio amplifier (being a very nonlinear device especially at broadcast frequencies) in the vicinity of a powerful AM station may demodulate that signal and render it audible. GSM frame buzz (at around 217 Hz) is likewise demodulated by wireline telephone handsets, audio circuits, hearing aids, and, according to some reports, even some dental fillings.

CASCADED AMPLIFIERS

The overall noise performance of an amplifier (or a receiver) can be measured by its noise factor: The lower it is, the better. Amplifiers can be characterized by their gain (usually expressed in decibels) and their noise factor (again usually expressed in decibels). When a number of amplifiers are connected in series as in Figure A2.2, the overall noise factor is given by

$$F_0 = F_1 + \frac{F_2 - 1}{G_1} + \frac{F_3 - 1}{G_1 \times G_2} + \frac{F_4 - 1}{G_1 \times G_2 \times G_3} + \cdots$$

where values of F (the noise factor) and G (the amplifier gain or loss) are expressed as ratios (not the more usual logarithmic form). F_0 is the overall noise factor.

To see why the LNA (low-noise amplifier) is important, we will look at the case where an attenuator is connected in series with a good-quality LNA. The noise factor will be dependent on whether the attenuator is connected before or after the LNA.

Consider the arrangement in Figure A2.3.

Let's assume that the attenuator has a loss of 3 dB. This translates to a gain of 0.5 ($10^{3/10}$) and a noise factor of 2. We assume that the LNA has a gain of 10 dB ($10\times$) and a noise factor of 0.6 dB ($1.148\times$).

Substituting these values in the above equation, we obtain

$$F_0 = 2 + \frac{1.148 - 1}{0.5} = 2.295$$

so the noise factor in decibels is $10 \times \log 2.295 = 3.6$ dB

Now reverse the connections and put the LNA first as in Figure A2.4 and substitute

$$F_0 = 1.148 + (2 - 1)/10 = 1.248$$

or in logarithmic form the overall noise factor is $10 \times \log 1.248 = 0.962$ dB.

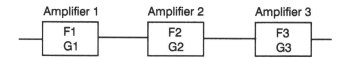

Figure A2.2   Cascaded amplifiers.

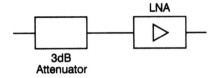

Figure A2.3   An LNA preceded by an attenuator.

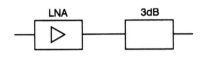

Figure A2.4   An LNA followed by an attenuator.

The important thing to notice is that the contribution of the loss and high-noise factor of the attenuator after the LNA was reduced in significance by a factor proportional to the gain of the LNA.

Because of this, the closer the LNA is to the antenna, the lower will be the overall noise factor of the system. Mounting the LNA on top of the tower before the feeder losses will improve the noise factor by the amount of the feeder losses.

APPENDIX 3

ISO MODEL

A standard structure for the implementation of data/communications networks has been put forward by the International Standards Organization (ISO), which allows a structured definition of any system.

The ISO model divides the functions of a system into layers, which can be independently logically structured. The ISO system was meant to apply to computer and data networks, but can be applied to digital communications, which increasingly are becoming the same thing.

The ISO framework is shown in Table A3.1, with the functionality for cellular radio shown.

Layer 1 comprises the actual physical and mechanical components of the circuits; although layers 2 and 3 are mainly implemented in hardware, they can involve considerable software. The higher layers are essentially software, except for the user part.

TABLE A3.1   The ISO Structure Applied to Cellular Radio

Layer Number	Definition	Function
7	Application	End user
6	Presentation	End user
5	Session	Call set up/clear down
4	Transport	The user-to-user link
3	Network	Radio resource management
2	Data link	Structuring the layer 1 information into meaningful messages
1	Physical	Coding
		CODECs
		Modulation types
		Frequencies

APPENDIX 4

AMPLIFIER CLASSES

A lot of discussion in trunk radio centers around the final PA stage and the class of amplifier. This is particularly true of digital systems, which will work only if the final amplifier is of a very high quality and low distortion. There are six amplifier classes: A, AB, B, C, D, and E. These classes originated in the early days of radio and were used to differentiate between the various valve amplifier configurations.

CLASS A

Class A is the simplest of all the amplifiers and consists of an active device, such as a transistor or FET, which is always conducting. The active device must be biased to ensure that it operates in its linear region. By the nature of its operation, it will draw a significant current even when there is no input to it. This means that it is not power efficient. For many applications in which good linearity is essential (as in most digital applications), class A is still the choice (see Fig. A4.1).

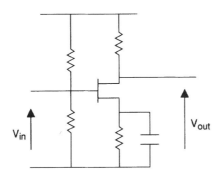

Figure A4.1 A typical class A amplifier.

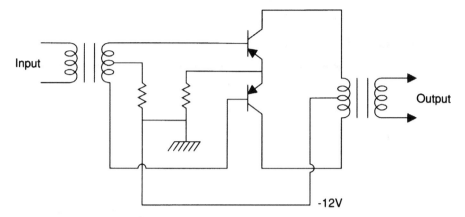

Figure A4.2   The transfer function (output versus input) of a typical class A active device.

Most active devices will have a transfer function that looks something like the curve shown in Figure A4.2. A high-quality class A amplifier will be biased to operate in the linear region. In this region the average power into the output load will be a constant regardless of the input signal level (provided the input signal can be regarded as random). In fact, early textbooks on class A amplifiers recommend checking class A amplifiers for distortion by putting a DC ammeter in series with the output load and then observing the meter reading as a signal is applied. No fluctuations mean "no distortion," but variations in the ammeter with applied signal level indicate nonlinearity.

When driven hard, the device will tend to saturate; and when driven too lightly, the device will again go nonlinear. It is the lightly driven nonlinearity that causes problems with other more power-efficient amplifier types.

The significance of all of this for mobile is that class A amplifiers are widely used in the PA stages of digital transmitters (because of the need for low distortion). As a consequence of the inefficiency (powerwise) of this type of amplifier, in dual-mode mobiles it will be usual to have a second, more power efficient amplifier for the analog part.

CLASS B

Class B amplifiers overcome the efficiency problem by using two active devices, each switched for one-half of the cycle (180 degrees). When a class B amplifier has no input, both devices are off and so it draws no current. It is therefore efficient, but because it requires the active devices to operate from full on to full off, it necessarily causes them to operate in regions where their linearity is poor. The most common example of a class B amplifier is the push–pull amplifier, which was widely used in audio circuits for decades from the 1930s when it first appeared. An example of a transistor push–pull amplifier is shown in Figure A4.3.

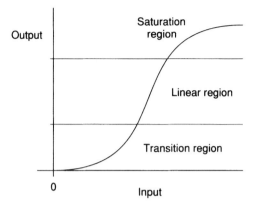

Figure A4.3   A typical class B push–pull amplifier.

CLASS AB

This is essentially a class B amplifier with some forward bias so that the active devices draw some current even when they have no input. This keeps them operating more linearly than a straight class B amplifier, but they are still a compromise compared to class A for linearity. Typically the quiescent current is 2% to 10% of the maximum working current.

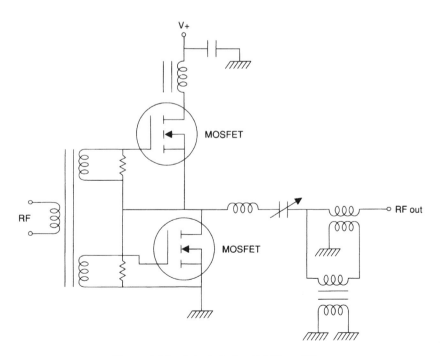

Figure A4.4   A typical class D RF amplifier.

CLASS C

Class C amplifiers are similar to class B but they are self-biased. This means that the active device does not begin to conduct until the input reaches a high enough level to bias the device on. These amplifiers are very nonlinear, but they are highly efficient and tend to be used in applications where linearity doesn't matter much, such as in FM and PM transmitters.

CLASS D

Class D is a switching mode amplifier, which by its nature is very efficient (typically around 80% and wide band). It is used widely in AM broadcasting transmitters. A typical AM class D amplifier is shown in Figure A4.4.

CLASS E

Class E amplifiers are a subset of class D with tuned circuit outputs, which improve the efficiency. The trade-off is that this amplifier will be a narrow-band device.

APPENDIX 5

SMART ANTENNAS

Increasingly, cellular and satellite systems are coming equipped with "smart" antennas. Surprisingly, these smart antennas not only have more gain than fixed ones, but can be expected to cost less (in the near future) and to include interference canceling. These antenna systems are so good that it is hard to see how conventional antennas can survive in the long term except for the most basic of applications. To date, none seem to be available for trunked radio systems, but it must come.

To understand how a smart antenna works, it is necessary first to look at the phased array, around which a smart antenna is built. All multi-element antennas are phased arrays (which include the common collinear dipole). Consisting of a number of radiating elements connected by a phasing harness, the desired pattern is achieved by adding the received signals from each of the dipoles in a way to produce the desired antenna lobe pattern.

To get a feel for how a phase array works, consider an array of two antennas placed nearby. Using the software provided, let's look at some array designs. Select "Mobile RF Calculations" from the main menu, and then select "Phased Antenna Arrays" from the submenu. Look first at a single antenna by entering 1 (wavelength) for the antenna separation, 1 for the phase delay, and 1 for the number of antennas. As you might expect, you will get a uniform circular gain of unity, as seen in Figure A5.1.

Next run the same program for the following figures:

Antenna separation = 0.25
Phase delay = 0.25
Number of antennas = 2

This will give you the football shaped pattern shown in Figure A5.2. What you are seeing is reinforcement in some directions and cancellation in others.

Now try the following:

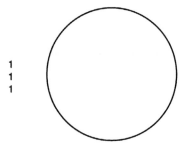

Figure A5.1   The uniform pattern of a single omnidirectional antenna.

Antenna separation = 0.5
Phase delay = 0
Number of antennas = 2

Now you have a coverage that is decidedly directional, with peaks toward the top and bottom of the page and nulls at right angle to the peak direction, shown in Figure A5.3.

Now, let's assume that this is just the kind of pattern we a looking for, but we would like to increase the directivity. This is easily done simply by increasing the number of antennas in the array. Increase the number to 9, and you see the result is similar to Figure A5.4. You will find that, in general, increasing the number of antennas while keeping the phase difference constant tightens the pattern.

Now that you have the software running, play with the configurations to see the possibilities. You can use any numbers for the calculation parameters, but note that the separation and phase delay are measured in wavelengths so that a separation (or phase delay) of 0.75 is the same thing as 20.75 or 100.75, because an integer number of wavelengths brings you back to the starting point.

For many applications a directional array like the one you have just looked at may be all that is needed. However, sometimes something more sophisticated is needed.

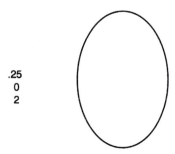

Figure A5.2   Two antennas separated by 0.25 wavelengths.

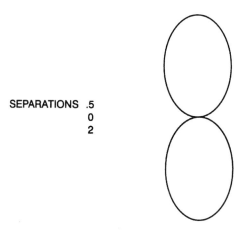

SEPARATIONS .5
0
2

Figure A5.3   Two antennas seperated by 0.5 wavelengths.

Another variant that could be used is a two-dimensional array of antennas, to give even more patterns, and this is what is done in practice.

You will have noticed that there are two ways to vary the pattern: You can either change the antenna spacing or you can change the phasing between the antennas. Changing the phasing is as simple as changing the length of the cables connecting the two antennas as seen in Figure A5.5.

Supposing that it was required to use a fixed array of antennas to provide a number of different patterns—for example, to direct broadcasts in different directions at different times. This could be done by having a number of phasing harnesses and switching in the required harness using positive intrinsic negative (PIN) diodes that can be biased to be open circuit or low impedance short circuit. There is no real limit

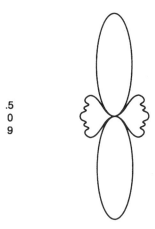

.5
0
9

Figure A5.4   Nine antennas separated by 0.5 wavelengths.

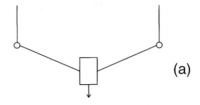

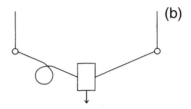

Figure A5.5   The phasing between the two antennas can be changed by lengthening one of the connecting cables.

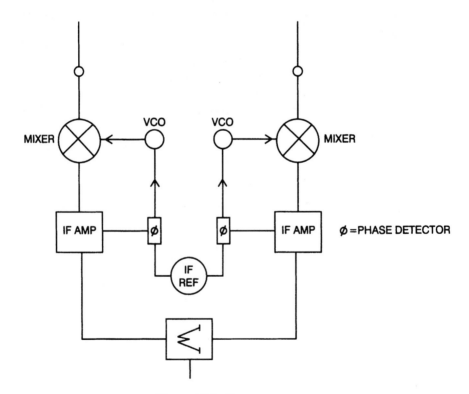

Figure A5.6   Two-antenna array.

to the number of patterns that can be generated in this way, but a large number of harnesses can get messy.

THE ADAPTIVE ARRAY

Adaptive arrays are the heart of the smart antenna. What is needed is a way to vary the relative phase of the antennas continuously. While there are a number of mechanical devices that could be used to vary the phase there is a neat way to do it within a phase-locked loop.

Consider the two-antenna array in Figure A5.6. Here the IF reference oscillator determines the angle at which the signals incident on the two antennas are phase-locked. We now have the basis of an antenna array that can be steered continuously by phasing the wanted signals. When the reference oscillator is controlled by a processor, automatic adaptation to maximize a signal or to minimize interference becomes possible. Even if there is more than one interfering signal, more nulls can be generated to counteract the problem. Additionally, the performance of the systems can be enhanced by simulated patterns, which are generated as part of the signal processing.

APPENDIX 6

THE SOFTWARE:
Program Documentation for
TRUNK (Version 4.0)

TRUNK is a set of powerful DOS-based software routines for mobile-radio engineers. Originally for distribution by CD, the software has instead been made available online as a compressed file at the Wiley ftp site. The address for the site is ftp://ftp.wiley.com/public/sci_tech_med/trunked_radio

The TRUNK software's compressed file should be downloaded and opened in a new directory on your hard drive. For a quick overview of the capabilities of TRUNK, type TRUNKDEM at the DOS prompt or double-click on the TRUNKDEMO.EXE file to load the program from within a Windows file folder. The main program is included in the TRUNK.EXE file.

INTRODUCTION

Trunked and Mobiles Engineer Version 4.0 (TRUNK) is a set of five main programs, containing a total of 53 routines, designed for use by planning and system design engineers, for trunked radio, conventional PMR, paging, and cellular systems. This version is much more than a collection of useful calculations. The programs are designed to be used not simply for calculations, but to enable the designer to answer a lot of "what if"-type questions. Because of this the routines allow the user to look at the same problem from many different angles.

For example, the traffic routines will calculate the circuit requirements for any given traffic. This does the job of the Erlang tables.

But the reverse question can also be asked. Given a number of circuits and a grade of service, how much traffic can be carried? Even this function can be done with judicious use of Erlang tables.

Perhaps you need to know more. Given a known number of circuits, let us say 30, which is carrying a MEASURED 26 Erlangs (and therefore obviously overloaded), what is the actual grade of service and how much traffic is being blocked on this route? This question can be answered directly by the software but cannot be extracted from the Erlang tables. This is just an example of the structure of the programs. They are designed to answer questions as well as to solve one-off problems. The traffic engineering program for circuit dimensioning has 11 subprograms, including complex queuing calculations.

A batch dimensioning program is also available which can dimension, and scale whole networks up or down, in a single run.

The third program is an RF engineering tool to predict mobile RF propagation and path loss for frequencies from 200 MHz to 1,500 MHz. In turn, these programs are divided into a number of routines, each of which performs a particular function. In all, there are 10 different routines. They calculate RF propagation in the mobile, free space, and cable environments.

Then there is a set of 13 handy calculation routines for trunked systems which includes unit conversions, binary and hex to base 10 conversions (and vice versa), frequencies from channel numbers, and decibel and VSWR calculations.

The next program consists of 11 routines for power engineering calculations, for both AC and DC situations.

The routines are as follows:

TRAFFIC ENGINEERING

Nonqueuing

- Calculate circuits from cost factor
- Offered traffic from measured traffic
- Grade of service calculations
- Overflow parameters from offered traffic
- Offered traffic for a given number of circuits and GOS
- Cost factor calculator

Note that either normal (smooth) traffic or rough traffic (variance greater than mean) can be handled. All switches are assumed to be full availability.

Queuing Traffic

- Calculate traffic carried by a given number of circuits, for a given queue length.

- Determine the number of circuits required for a given queue length and offered traffic.
- Find the queue length for a given traffic and number of circuits.

Dual-Mode Base-Station Dimensioning

- The most effective number of second mode (e.g., wide-area channels/local area) for a given traffic.
- The best split of an existing base station with a given number of channels for wide area:
  a. Most cost effective
  b. Highest traffic for the site
- The traffic capacity of a given number of dual-mode circuits.

Mobile RF Calculations

Mobile coverage for a given ERP, frequency and antenna height, and terrain factors.

- Coverage for a given far-field strength
- Range for a given path loss
- Path loss or field strength for a given range
- Microwave path calculations
- Phased antenna array patterns
- Cable loss in decibels for a given length of the more common cables, or any other cable of your choice
- Free-space path loss including:
  1. Far-field free-space loss
  2. Near-field loss for halfwave dipole
  3. Near-field loss received power for a half-wave dipole
  4. Near-field loss for any length dipole
  5. Near-field loss received power for any length dipole

Miscellaneous Calculations

- Convert any of the following units to each other for any frequency and any impedance:
  dBm
  μV
  dBuV/m
  μV/m
  dBuV

W

- Subscribers from traffic and calling rate
- Binary and HEX from base 10 numbers (and vice versa)
- Frequencies from channel number (and vice versa) for AMPS/TACS
- dB calculations which include
  1. Power ratios to dB
  2. Voltage ratios to dB
  3. Current ratios to dB
  4. dB-to-power ratio
  5. dB-to-voltage/current ratios
- VSWR calculations including:
  1. Loss calculated from VSWR into a short or open circuit
  2. Return power loss (watts) to VSWR
  3. Return power loss (dB) to VSWR
  4. VSWR correction for feeder loss

Power Calculations

- Change metric/AWG gauges
- Cable selection
- Cable resistance calculator
- AC power calculator for single- or three-phase including:
  1. Current from power
  2. Current from volt-amps
  3. Power factor correction
- DC power calculator for:
  1. Current from power
  2. Voltage from power
  3. Current from power and resistance
  4. Voltage from power and resistance
  5. Power from current and resistance
- Batch program for dimensioning circuits for cellular or trunked radio networks

GETTING STARTED

- From Windows, double-click on the TRUNK.EXE file in your program folder. If you are using the DOS system or have moved to the DOS prompt from Windows, access the program directory, type TRUNK, and press ENTER. The following menu will appear.

```
TRUNKED RADIO ENGINEER
MAIN MENU

TRAFFIC ENGINEERING
BATCH TRAFFIC CALCULATIONS
MOBILE RF CALCULATIONS
MISCELLANEOUS CALCULATIONS
POWER CALCULATIONS
* EXIT PROGRAM

USE ARROW KEYS TO SELECT
```

In this menu and all others throughout Trunked Engineer, the user selects the desired routine, or subroutine, by highlighting it using the up/down arrow (or cursor) keys and then pressing ENTER.

MIS-KEYING

In most cases when you mis-key, this results in an input that is not consistent with a realistic input; there will be a single beep and the cursor will return to the original entry point where you can overwrite the faulty input.

ESCAPE

The Escape key (Esc) will take you to the next highest level in the menu from virtually any point in the program.

PART 1: TRAFFIC ENGINEERING

Traffic Engineering Routines

This traffic engineering software has been produced with the trunked and mobile radio operator in mind. The package is meant to be used as a real-time calculating tool and not for batch processing. Full availability conditions only are allowed for. This should not be a limitation because today most switches are full availability. The software is in two parts. The first part is a menu-driven set of traffic engineering routines. It can dimension grade of service (GOS) routes and alternate routes, calculate offered traffic from measured, determine carried traffic when a known traffic is offered to a group of circuits, and also calculate the carried traffic when a given traffic is offered to a group of circuits. The ability to deal with rough traffic (which is a consequence of alternate routing) has been built in.

Queuing theory, essential for trunk radio design, is covered in a related set of programs.

   Optimization of dual-mode base stations, taking into account that traffic from the dual-mode mobiles that cannot be carried directly by the dual-mode channels will be carried on the default channels, can also be undertaken.

   The second part is a traffic forecasting routine that enables the user to project future traffic from known traffic today. This can be used for planning purposes and, when used in conjunction with the dimensioning routines, will forecast future trunk requirements.

Getting Started

Use the ARROW keys to highlight TRAFFIC ENGINEERING from the main menu. Press ENTER and the following menu will appear:

```
TRAFFIC CALCULATIONS WITH
AND WITHOUT QUEUING

DUAL MODE BASE STATION CIRCUITS
QUEUING TRAFFIC

NON-QUEUING TRAFFIC

*EXIT
```

(Use cursor to select menu option)

If you select nonqueuing, the following menu will appear:

```
TRAFFIC ENGINEERING CALCULATIONS

CALCULATE CIRCUITS FROM COST FACTOR
CALCULATE OFFERED TRAFFIC FROM MEASURED
  GOS CALCULATIONS
OVERFLOW PARAMETERS FROM OFFERED TRAFFIC
OFFERED TRAFFIC FOR GIVEN CIRCUITS AND GOS
  COST FACTOR CALCULATIONS
*EXIT
```

(Use arrow keys to select)

To select any mode, simply highlight the calculation required using the up and down ARROW keys and press ENTER. Each of these selections is referred to in the rest of the text as a routine.

Calculate Circuits from Cost Factor

Where alternate routing is provided, there is an optimum configuration of circuits for the direct route.

As an example calculation, let's assume that the traffic is smooth, the cost factor of a route is 0.6, and the direct route is offered 5 Erlangs. The first screen will ask if traffic is smooth or rough:

```
ECONOMIC CIRCUIT PROVISIONING
ROUGH TRAFFIC
*SMOOTH TRAFFIC
```

Press ENTER to indicate smooth traffic. If the traffic is not smooth, highlight ROUGH and press ENTER.

The next screen will then ask for variance. (The variance equals the mean for smooth traffic and is therefore redundant for smooth traffic.)

When you press ENTER you will see:

```
THIS PROGRAM CALCULATES THE ECONOMIC CIRCUIT PROVISIONING FOR
A GIVEN TRAFFIC AND A GIVEN COST FACTOR

    traffic? 5
    cost factor? 0.6
    the mean of the overflow is 2.6
    the overflow variance is 3.7

Cost effective number of circuits 4

MAIN MENU
SUB MENU
*MORE CALCULATIONS

Use the ARROW keys to choose MAIN MENU or SUB
MENU (the default is MORE CALCULATIONS) and press ENTER.
```

Notice that when the cost factor is low (around 0.2 or less) the number of economic direct circuits can become so high that the route may be provisioned to Grade of Service standards. To alert you to this, should the GOS be 0.1 or less, then its actual value is given. For example, if the traffic is held at 5 Erlangs and the cost factor is equal to 0.1, then a message indicating that the resultant GOS is 0.07 will appear. At this low GOS, it would not be necessary to consider an overflow path for the traffic not carried. This routine proceeds by iteration; and for cases where the number of circuits is very large, it may take some time to converge. In this case you will see the current iteration in the bottom corner of the screen. This message will disappear once the calculation is complete.

Calculating the Cost Factor

If the cost factors of the routes are not known, they can be readily calculated. Consider the case of 10 Erlangs offered to the direct route with traffic, so that first all the traffic is offered to one circuit only and the rest is overflowed. You find that 0.91 Erlangs will be carried and the rest overflowed. If the process is continued for a large number of circuits, you will find that each additional circuit on the direct route carries less traffic than the one before. By calling the cost of a circuit on the direct route C, for any number of circuits N, the traffic carried by the addition of one more circuit H, then the cost of carrying 1 Erlang of traffic on $N + 1$ circuits is C_1/H. The traffic carried on the backbone route is assumed to be large, so that the addition of overflowed traffic from the direct route will have only a marginal effect on the circuit's traffic capacity. If the circuit from switch 1 to switch 2 has an average traffic of B_2 Erlangs per circuit and an average cost of C_2 per circuit, then the cost of carrying 1 Erlang on this part of the backbone is C_2/B_2. Similarly the cost of carrying traffic on the route between switch 2 and switch 3 is C_3/B_3. As it was noted before, the marginal occupancy H of the direct route decreases as N increases and so there will come a time when the cost of the traffic carried on the direct route just equals the cost of the traffic carried on the backbone. At this point, $C_1/H = C_2/B_2 + C_3/B_3$. This can now be solved for H to give

$$H = C_1/(C_2/B_2 + C_3/B_3)$$

Cost Factor Calculations

Use the ARROW keys to select COST FACTOR CALCULATIONS and press ENTER. The following screen will appear. (*Note*: the full screen develops as you make entries.)

```
THIS CALCULATES COST FACTOR
Per cct cost of the direct route (C1)1
Per cct cost of the overflow     (C2)1
Per cct cost of the overflow     (C3)1
Marginal capacity of route 2     (B2)0.8
Marginal capacity of route 3     (B3)0.8
COST FACTOR IS 0.4
          MAIN MENU
          SUB MENU
          MORE CALCULATIONS
```

Note: cct = circuit.

Calculate Offered Traffic from Measured

Highlight this selection with the ARROW keys and press ENTER. The traffic on a circuit group as measured by occupancy or busy circuits will nearly always be less than the traffic that was offered. This is because the number of circuits on any route will be finite; and since the traffic is normally distributed, then there is a finite possibility that at some time the number of callers wishing to use the circuits will exceed the circuit provisioning. TRUNK has a routine for calculating the traffic that was offered given the traffic that was carried. The routine is iterative, and it proceeds by first guessing the offered traffic and then calculating the carried traffic. For small traffic values the iteration proceeds rapidly as a very efficient search method is used, but for large values (that is hundreds of Erlangs) it may be a little slow. The current state of the iteration process and the current error in the carried traffic will appear on the bottom corner of the screen during the calculation. This information will disappear once the iteration process is complete. For example, assume that you had measured 6 Erlangs on seven circuits and wished to know the actual offered traffic. The screen would show the following:

```
Iterative method for calculating the offered traffic

    what is the carried traffic? 6
    what is the number of circuits? 7
        offered traffic is 10.67
            effective GOS 0.437588
                            MAIN MENU
                            SUB MENU
                    *MORE CALCULATIONS
```

This tells you that the circuits in question were in fact offered 10.67 Erlangs. The effective GOS of 0.437588 tells you that either this is a high occupancy route or it is a heavily overloaded final route. Because this routine is iterative, it may take a while for some of the more difficult calculations. Such calculations are those where the carried traffic approaches the number of circuits provided. This is because if the circuits carry as many Erlangs as the number of circuits, then the offered traffic is infinite. In practice, if the offered traffic as calculated is more than 50% greater than the carried traffic, then you should treat the results with some caution because small errors in measurement of the offered traffic can result in large errors in the subsequently calculated carried traffic. Should you enter such conditions, then in the lower corner of the screen you will see the current state of the iteration process and the currently estimated error in offered traffic.

GOS Calculations

Highlight GOS CALCULATIONS on the TRAFFIC ENGINEERING CALCULATIONS menu and press ENTER.

Grade of service (GOS) calculations are used to determine how many circuits are needed for a route that has no overflow. If the circuit is a backbone, it will probably carry rough traffic. This will be no problem because this routine can allow for that. When you begin, there is a request for setup parameters. The screen will appear as below:

```
THIS PROGRAM RETURNS THE CIRCUITS FOR ANY GOS FOR ROUGH OR
NORMAL TRAFFIC
```

```
SETUP PARAMETERS

      ROUGH TRAFFIC
   *SMOOTH TRAFFIC

      VARIABLE GOS
   *FIXED GOS
```

The program first asks if the traffic is smooth (normal) or rough. Take the example of 4 Erlangs of smooth traffic at a GOS of 0.01. In this case you press ENTER to the first question. Next you are asked if you want to fix the GOS. This function is handy if you intend to do a number of calculations all at the same GOS. If you press ENTER, the next question will be what GOS to use. If you select VARIABLE, you will be asked for the GOS for each subsequent calculation. If a fixed GOS is selected, the next screen will be:

```
THIS DOES GOS CALCULATIONS

GOS is fixed at 0.01000

    traffic 4

  no. of circuits 10

        MAIN MENU
        SUB MENU
        *MORE CALCULATIONS
```

In this example the number of circuits required is 10. Try the same calculation again with ROUGH TRAFFIC and a variance of 25. This will first indicate that for variance-to-mean ratios of more than 5, there are some doubts about the efficacy of this method. However, if you continue by pressing ENTER, then it will return 27 circuits. This is considerably more than the 10 needed for the same level of smooth traffic. When rough traffic as above is being processed, you will notice that in the bottom left-hand corner of the screen the following will appear:

```
Equivalent traffic 123.4
Equivalent circuits 131.8
```

Overflow Parameters from Offered Traffic

Highlight OVERFLOW PARAMETERS... from the submenu and press ENTER. Sometimes it is necessary to determine the traffic carrying capacity of a fixed number of circuits. As with any situation where a traffic overflow occurs, the traffic that overflows will be rougher than normally distributed traffic and so the variance will be needed. As an example, suppose that 4 Erlangs of smooth traffic was offered to six circuits:

```
OFFERED TRAFFIC

ROUGH TRAFFIC
*SMOOTH TRAFFIC
```

Because the traffic is smooth, you press enter. The next screen will show:

```
THIS PROGRAM RETURNS THE MEAN AND VARIANCE OF THE OVERFLOW
TRAFFIC FOR A GIVEN NUMBER OF CIRCUITS AND TRAFFIC

     offered traffic        4

     circuits provided      6

     overflowed traffic     0.47
     overflow variance      0.79

     carried traffic        3.53


                         MAIN MENU
                         SUB MENU
                         *MORE CALCULATIONS
GOS = 0.117
```

This will tell you that the overflowed traffic was 0.47 Erlangs and that its variance is 0.79. The actual carried traffic will be 3.53 Erlangs.

Offered Traffic for Given Circuits and GOS

Highlight OFFERED TRAFFIC... from the submenu and press ENTER. Often it is necessary to determine the traffic capacity of a route when the number of circuits is known for a fixed GOS. This routine proceeds by iteration and uses an efficient binary chop method to converge on the required answer. The routine will return the answer promptly for circuits of a few hundred or less. The processing time increases in direct proportion to the number of circuits. Therefore, for very large numbers of circuits the procedure may seem somewhat slow. As an example, assume that you have 56 circuits and wish to use a GOS of 0.09. As illustrated below, the routine will indicate that the traffic is 55.055 Erlangs.

```
THIS ROUTINE RETURNS THE TRAFFIC THAT CAN BE OFFERED TO A
GIVEN NUMBER OF CIRCUITS FOR A GIVEN GOS

Number of circuits provided 56

GOS 0.09

            MAIN MENU
```

```
                        SUB MENU
                        *MORE CALCULATIONS

OFFERED TRAFFIC 55.055
```

Trunk radio systems often offer queuing capability. In order to best exploit this, the following routines can be used. If you select the queuing traffic routines, the following menu will appear:

```
QUEUING CALCULATIONS

QUEUING LENGTH FROM TRAFFIC AND CIRCUITS
CIRCUITS FOR GIVEN QUEUE AND TRAFFIC
TRAFFIC FOR GIVEN CIRCUITS AND QUEUE

*RETURN TO SUB-MENU

(Use arrow keys to select)
```

Selecting TRAFFIC FOR GIVEN CIRCUITS AND QUEUE will enable the calculation of the traffic that can be carried on a given number of circuits while keeping to a predefined queue length. As an example, assume that 45 circuits are provided with a required queue time of 3 s and an average call holding time of 150 s.

```
CALCULATE THE TRAFFIC CARRIED BY GIVEN CIRCUITS TRAFFIC, CALL
HOLDING AND QUEUE TIME

CIRCUITS PROVIDED          45
QUEUE TIME (secs)          3
CALL HOLDING TIME (secs)   150

THE TRAFFIC CARRIED IS     37.2 ERLANGS
```

The result is that the system can carry 37.2 Erlangs.

If it is required to calculate the number of circuits for a given queue and traffic chose the second option. Assuming that it is required to find the number of circuits needed to carry 20 Erlangs, for a queue of 3 s and a call holding time of 25 s the calculation will appear as follows:

```
CALCULATE THE NUMBER OF CIRCUITS TO CARRY A GIVEN TRAFFIC,
CALL HOLDING AND QUEUE TIME

TRAFFIC                     20
QUEUE TIME (secs)            3
CALL HOLDING TIME (secs)    25

THE NUMBER OF CIRCUITS REQUIRED   23
```

The other question that may be asked is, What is the queue length for a given traffic and number of circuits? For example, it may be required to calculate the queue length for 12 Erlangs offered to 15 circuits with a call holding time of 15 s. (Note that for queuing systems to work the traffic must always be less than the number of circuits or else the queue will become infinite.)

If this option is selected, then the following screen will appear:

```
CALCULATE QUEUE LENGTH FOR A GIVEN TRAFFIC AND NUMBER OF
CIRCUITS

TRAFFIC                  12
CIRCUITS                 15
CALL HOLDING TIME        15

QUEUE AVERAGE TIME       1.73 secs
AVERAGE NUMBER IN QUEUE  1.4
```

This tells us that the average queue length is 1.73 s and that, on average, 1.4 calls will be in the queue. This last figure may be significant if the queue length needs to be determined.

Dual-Mode Base-Station Circuits

Recently, dual-mode base stations have been introduced which have AMPS capacity and a second mode such as TDMA, E-TDMA, NAMPS, and so on. It has been traditional to fully provide each of these modes independently, so, for example, in an NAMPS system if 20% of the traffic is NAMPS-capable, then NAMPS channels to carry the whole of that traffic will be provided. In reality, unserved NAMPS mobiles can "overflow" to the AMPS circuits, and advantage can be taken of this to increase the base-station efficiency. These routines make no assumptions about which alternate system is used, and so it will always be necessary to specify the system gain. For example, TDMA offers three channels for one AMPS and so the gain is 3. NAMPS may have a gain of 2 or 3, depending on how it is implemented.

It needs to be understood that for dual-mode stations, there are two optimum configurations. One is a least-cost solution to carry the offered traffic, and the other is one that carries the maximum traffic, but not necessarily at the lowest cost per Erlang.

The program was written for AMPS dual mode but applies equally to trunked systems with wide area and local channels at a single site. In this case AMPS = local channels.

Simple calculations based on the assumption that the two modes independently carry their respective traffics (e.g., the AMPS channels only carry AMPS traffic, and the second mode carries only second-mode traffic) will always underestimate the actual traffic capacity. The true traffic capacity can be calculated using the dual-mode traffic routine, because this routine allows for overflow to the AMPS mode and calculates the optimum traffic for the configuration given.

For example, suppose we wanted to know the traffic capacity of a base station that has 12 AMPS channels, 9 NAMPS channels at a GOS of 0.05, and 60% of the phones NAMPS-capable. This calculation would appear as follows:

```
DUAL MODE TRAFFIC FOR GIVEN CHANNEL CONFIGURATION

      TOTAL AMPS CHANNELS              12
      TOTAL SECOND MODE CHANNELS       9
      GOS                             0.05
      PROPORTION OF DUAL MODE MOBILES


AMPS CHN TRAFFIC              8.0 ERLANGS
SECOND MODE CHN TRAFFIC       9.0 ERLANGS
SECOND MODE CHN GOS           0.222
DUAL MODE OVERALL GOS         0.0112
TOTAL TRAFFIC CARRIED         16.9 ERLANGS
```

Notice that two grades of service are given for the dual-mode system. The first is the channel GOS which at 0.222 would ordinarily be considered poor. However, the overall GOS of 0.0112 is achieved because of the ability of the dual-mode mobiles to overflow to the AMPS channel.

Generally, dual-mode operation will be progressively introduced into existing base stations that often will already be at maximum AMPS capacity. If it is desired to optimize an existing base station with a given number of existing AMPS channels, the second routine can be used. Two solutions are given: The first is the least cost per Erlang carried, and the second is the maximum traffic that can be carried regardless of cost. Suppose a base station has 56 AMPS channels, and it is desired to convert it to dual mode under the conditions that the GOS will be 0.05, the second mode channels cost 80% of the cost of an AMPS channel, 50% of the mobiles are dual-mode-capable, and the dual-mode system gives three channels for each original AMPS channel. This calculation would appear as follows:

```
DUAL MODE CHANNELS REQUIRED TO CARRY THE
MAX TRAFFIC or LEAST COST CONFIGURATION FOR N AMPS CHANNELS

      TOTAL AMPS CHANNELS AVAILABLE          56
      GOS                                    0.05
      RELATIVE COST OF SECOND MODE CHANNELS  0.8
      PROPORTION OF DUAL MODE PHONES         0.5
      DUAL MODE CAPACITY GAIN PER CHANNEL    3


DUAL MODE CIRCUITS COST MIN               30
TRAFFIC FOR MIN COST CIRCUITS             65.7 ERLANGS
DUAL MODE GOS FOR MIN COST                0.0100
DUAL MODE CIRCUITS TRAFFIC Max            42
MAX TRAFFIC                               68.6 ERLANGS
DUAL MODE GOS FOR MAX TRAFFIC             0.0019
```

Note also that two numbers will appear in brackets after the calculation is complete, which follows the "56" AMPS channels. In this case it is (46/42), which indicates the final number of AMPS channels for each solution.

It may be required to calculate the least-cost channel split for a base station for a given traffic. Assume that it was required to split an E-TDMA base station carrying 23 Erlangs at a GOS of 0.05, when 70% of the mobiles were dual-mode-capable and dual-mode channels cost 70% of the cost of an AMPS channel. This calculation would appear as follows:

```
CALCULATE DUAL MODE CIRCUITS MIX FOR
A GIVEN TRAFFIC, BASED ON LEAST COST

TRAFFIC (Erlangs)                              23
GOS                                            0.05
PROPORTION OF DUAL MODE MOBILES                0.7
DUAL MODE CHANNELS TO AMPS (capacity ratio)    6
PROPORTIONAL COST OF DUAL MODE CHANNELS TO AMPS  0.7
AMPS CHANNELS                                  13
SECOND MODE CHANNELS                           18
```

PART 2: MOBILE RF CALCULATIONS

The routine RANGE has a computerized algorithm that corresponds to the tables of Okumura for base heights in the range 30–1,000 m and uses the results of work done by the author for heights below 30 m. The tables can be used to find the field strength in the far field of any transmitter or the path loss. It can conversely be used to find the range of the transmitter for any given field strength. The cable loss for some common RF cables is also included. The routine allows the user to put in additional cables or to amend the original list. Finally, a routine that converts the most common units of field strength into one another is included. The accuracy of this RF prediction techniques is such that it is suitable for map studies and initial estimates only. It is not meant to replace detailed surveys.

Getting Started

Highlight MOBILE RF CALCULATIONS from the main menu, press ENTER, and the following screen will appear:

```
MOBILE RF CALCULATIONS A Selection of Path Loss Routines

RANGE FOR A GIVEN FIELD STRENGTH
RANGE FOR A GIVEN PATH LOSS
LOSS OVER A GIVEN DISTANCE
PHASED ANTENNA ARRAYS
CABLE LOSS
FREE SPACE CALCULATIONS
*EXIT
(Use arrow keys to select)
```

Highlight the desired function and press ENTER. When you choose any of the first three functions, the following table of assumptions will appear:

```
BASIC ASSUMPTIONS FOR CALCULATIONS
      Frequency in MHz                    = 850
      Base Station Height                 = 30
      % of Buildings to Land              = 15
      Vehicle Antenna Height              = 1.5
      Terrain Factor                      = 0
      Field Strength Reference (watts)    = 100

*Assumptions accepted
RETURN TO SUBMENU
```

Changes

You can change as many of these parameters as you wish by highlighting the appropriate line using the ARROW keys, pressing ENTER, inserting the change, and pressing ENTER again. Changes will be stored and become default values for subsequent calculations, until they are changed again. All heights are in meters. The percentage of buildings to land for the area considered is estimated as the area of land occupied by buildings to the area free of buildings. If it is zero, allow a figure of around 2% to 5% for foliage. Note that for values less than 2, the program will override the input and set it to 2. The terrain factor allows you to calibrate the calculation for your particular area. To do this, you need to do some local measurements and determine if there is a discrepancy. Where there is, you should estimate the average error (in decibels) and insert it here. This would apply particularly over unusual terrain such as swamps, deserts, and extremely flat or extremely undulating land. However, unless some feedback from surveys is available, assume a value of zero. Note that the factor may be either positive or negative.

The easiest way to estimate the terrain factor is to take the results of a survey (at least 200 readings should be used) and set all other parameters. Then for each measurement, compare the actual measurement with the result you get by running the LOSS routine. The AVERAGE value of the discrepancy for various ranges will be the appropriate value for the terrain factor. The reference field strength is the effective radiated power (ERP) of the transmitter. For trunked radio applications, this will usually be either 50 or 100 W (and for rural areas maybe 500 W). For the example you can accept all of the assumptions and simply press ENTER.

Range for a Given Field Strength

When you select this routine the screen will show the following:

```
RANGE IN KILOMETERS FOR A GIVEN FIELD STRENGTH

      Default input               = 0 dBuV/m
      Field Strength in dBuV/m     = 30
```

```
RANGE IN KILOMETERS IS 8.3

    MAIN MENU
    SUB MENU*
    MORE CALCULATIONS
```

For this example enter 30 (i.e., 30 dBuV/m) and the program will return a range of 8.3 km. Notice that the program will take a short while to arrive at the answer as it uses an iterative process to obtain the answer. By pressing ENTER you can calculate the range for another field strength. If you want to change any of the initial conditions, select return to the submenu and reopen the Table of Assumptions as before.

Range for a Given Path Loss

Select RANGE FOR A GIVEN PATH LOSS from the submenu and press ENTER.

Once again the Table of Assumptions will appear. Make any changes you choose to and press ENTER (Assumptions accepted) as above. The following screen will appear:

```
RANGE IN KILOMETERS FOR A GIVEN PATH LOSS

Default path loss    = 148 dB
Permissible path loss = ___
```

If you simply press ENTER in this mode, then the default value of 148 dB will be used. To use any other path loss, simply enter the value. Pressing ENTER with no change in the default value will return:

```
RANGE IN KILOMETERS IS 5.7
    MAIN MENU
    SUB MENU
    *MORE CALCULATIONS
```

Loss Over a Given Distance

An alternative mode is to find the loss for a given distance (range). The output will be both the loss in dBuV/m (referenced to the ERP) and the loss in dB. When the loss mode is selected, the following screen will appear:

```
LOSS CALCULATION FOR A GIVEN RANGE (in kilometers)

Loss calculation for what range? 67
Field strength -11.0 dBμV/m (referenced to 100.0 watts)

PATH LOSS IS 194.7
```

For this example, enter 67 (kilometers) and the program will return -11.0 dBμV/m and its reference in watts. Notice that the reading in dBμV/m will be dependent on the original ERP of the transmitter. Below this it will return the path loss.

Phased Antenna Arrays

This routine is designed to enable you to tailor design a radiation pattern using two phased antennas spaced at any distance. As it displays the result graphically, it will require a VGA- or EGA-compatible monitor. Most of the more interesting arrays involve spacings of less than 1 wavelength. For example, the well-known cardioid pattern occurs when two antennas spaced at 0.25 wavelengths are separated by 0.25 wavelengths. If phasing is specified as greater than one wavelength, the integer part is ignored. This is because a delay of 2.5 wavelengths is identical to a delay of 0.5 wavelengths in the way it might have any effect on the far-field pattern. To run this routine, select PHASED ANTENNA ARRAYS and a screen explaining how to implement the design comes up. To proceed press any key.

```
ANTENNA ARRAY PATTERN CALCULATOR


ANTENNA SEPARATION IN WAVELENGTHS   0.25
PHASE DELAY IN WAVELENGTHS          0.25
```

This input will generate a cardioid radiation pattern on the screen.

Cable Loss

The CABLE LOSS routine will calculate RF cable losses for the more common types of cable given the frequency in use and the cable length. Highlight CABLE LOSS on the submenu and press ENTER. The following screen will appear:

```
CABLE # CABLECO-AX CABLE LOSS CALCULATOR
1. RG58
2. RG213
3. 1/2 INCH        ADD NEW CABLES
4. 7/8 INCH        *CALCULATE LOSS
5. 1 1/4 INCH      SWITCH TO OTHER TABLE
6. 1 5/8 INCH      RETURN TO SUBMENU
7.
8.
9.


Press ENTER to continue ?

Frequency (MHz) 850 Cable Length (M)? 100

Loss is 4.5 dB
```

The default frequency is 850 MHz (i.e., if you press ENTER it will default to 850). Likewise the cable choice will default to 7/8 INCH, which is probably the most frequently used cable in cellular radio. The cable length should be in meters, and the loss will be given in decibels. You may find that the cable you most use is not on the list. If so, this presents no problem because you can add to the list or amend it. To do so, select the ADD NEW CABLES option. To add a new cable you will need access to the manufacturers loss graph for 100 m of cable versus frequency. The graph will appear in the following form.

Note that if your graph is for 100 ft, you should multiply the loss in dB by a factor of 3.3.

To calibrate the table, you will need to select two points on the graph as illustrated (note any two points will do). After selecting ADD NEW CABLES, the following screen will appear:

```
INPUT THE CABLE DATA

CABLE #___ CABLE NAME ___

1st  FREQUENCY ___
     LOSS ___

2nd  FREQUENCY ___
     LOSS ___

     ADD MORE
     *RETURN
```

The CABLE # is the position number of your new cable (from 1 to 30) that appears on the screen. You can use the spare numbers or you can overwrite your data on one of the existing cable types. The CABLE NAME is the name you give it (e.g., RG213) and can be any alphanumeric name up to 14 characters. From the loss graph depicted you may choose 100 MHz and 3 dB for the first frequency and loss and 1,000 MHz and 15 dB for the second. Note that there is no restriction on which two points are chosen, but it is best if they are quite far apart. After entering all the required data, you will be given the option to ADD MORE or RETURN (no more entries). The SWITCH TO OTHER TABLE option simply accesses more open cable numbers.

Free-Space Calculations

Free-space calculations involve RF propagation through an unobstructed medium such as a vacuum, or to a good approximation in air. Free-space path loss can take two forms. The familiar far-field losses in which the attenuation follows an inverse square law is widely understood. Near-field effects are significantly more complex, and they pertain to coupling of antennas separated by distances that are of the order of a wavelength (or even less). A method developed by P. S. Carter and published in the *Proceedings of the IREE* in 1932 calculates the mutual impedance of the transmit and receive antennas. The mutual impedance is defined as $Z_m = V_r/I_t$, where Z_m is the mutual impedance, V_r is the voltage induced into the receiving antenna, I_t is the current in the transmit antenna.

You will find that this routine returns the mutual impedance (which is complex) and also calculates the receive strength of the receive antenna and the free-space loss. Near-field calculations are of particular use for calculating the isolation offered between antennas for a given configuration on a tower. Caution should be exercised, however, because the localized effects of the tower itself cannot easily be accounted for, and its presence represents a deviation from true free-space conditions.

Far-Field Free-Space Path Loss

Highlight FAR-FIELD. . . and press ENTER. There are five available options:

```
FREE SPACE CALCULATIONS IN THE NEAR AND FAR FIELD
FAR-FIELD FREE-SPACE PATH LOSS
NEAR-FIELD PATH LOSS (half-wave dipole)
NEAR-FIELD RECEIVED POWER (half-wave dipole)
NEAR-FIELD PATH LOSS (any length dipole)
NEAR-FIELD PATH LOSS (any length dipole)
*RETURN TO SUB-MENU

FAR-FIELD FREE-SPACE PATH LOSS
```

Highlight the top option and press ENTER. In this example a range of 4 km and 6 MHz returns:

```
FAR-FIELD FREE-SPACE PATH LOSS of 60.1 dB.

CALCULATE FAR-FIELD FREE SPACE PATH LOSS (MORE THAN 300
WAVELENGTHS FROM THE ANTENNA)

Path distance in kilometers:    4

Frequency in MHz:               6
Path loss in dB:                60.1

MAIN MENU
RETURN TO SUB MENU
*MORE CALCULATIONS

NEAR-FIELD PATH LOSS (half-wave dipole)
```

This routine calculates the near-field path loss for distances between the send and receive antenna for separation from 0.01 wavelengths to 10 km.

In the following example a calculation is performed for the case of a horizontal separation of 1 m and a vertical separation of 1.5 m at 900 MHz:

```
CALCULATE NEAR-FIELD PATH LOSS (FROM 0.01 WAVELENGTHS TO 10
km FROM THE ANTENNA) DISTANCE IS IN METERS HORIZONTALLY AND
VERTICALLY
```

```
FREE-SPACE PATH CONDITIONS ARE ASSUMED

    Horizontal separation in meters:    1
    Vertical separation in meters:      1.5
    Frequency:                          900

    Mutual resistance      = 0.55111491
    Mutual reactance       = -0.65346256
    Coupling loss in dB    = 44.47

                        MAIN MENU
                        SUB MENU
                        *MORE CALCULATIONS
NEAR-FIELD RECEIVED POWER (half wavelength)
```

This routine calculates the near-field field strength for a given ERP at the transmit antenna. In the following example the horizontal separation is 1 m and the vertical separation is 1.5 m at 900 MHz, and the transmitted power is 100 W:

```
CALCULATE NEAR-FIELD STRENGTH (FROM 0.01 WAVELENGTHS TO 10 km
FROM THE ANTENNA). DISTANCE IS IN METERS HORIZONTALLY AND
VERTICALLY. FREE-SPACE PATH CONDITIONS ARE ASSUMED.

Horizontal separation in meters:    1
Vertical separation in meters:      1.5
Frequency:                          900
Transmitted power (ERP):            200
Gain Ant 1                          = 0
Gain Ant 2                          = 0
Mutual resistance                   = 0.55111491
Mutual reactance                    = -0.65346256
Field strength                      138.84 dBuV/m

                    MAIN MENU
                    SUB MENU
                    *MORE CALCULATIONS
```

This is essentially the same problem as in the prior example, but returning different parameters. The same types of calculations can be done for any length of dipole by choosing the appropriate menu item.

PART 3: MISCELLANEOUS CALCULATIONS

Highlight MISCELLANEOUS CALCULATIONS on the main menu and press ENTER for a selection of useful calculation routines.

```
MISCELLANEOUS CALCULATIONS
A Selection of Useful Routines

Units Conversion
```

```
Subscribers from Traffic and Calling Rate
Binary and HEX Numbers from Base 10
Frequencies from Channel Numbers (and Vice Versa)
dB from Forward and Reflected Power
VSWR Calculations
*Exit
```

Units Conversion

Highlight UNITS CONVERSION and press ENTER.
The units program facilitates easy conversion to and from dBm:µV:dBµV/m:dBµV/watts. Any impedance can be selected.

```
CONVERTS FIELD STRENGTH UNITS BASED ON 50 OHM TERMINATIONS

CONVERT         TO
*dBm            dBm
µV              *µV
dBµV/m          dBµV/m
µV/m            µV/m
dBµV            dBµV
watts           watts
imp             imp
Exit            Exit
```

After selecting an option in the CONVERT column, the TO column will appear. Select the TO choice. For example:

```
Convert dBm to µV           MAIN MENU
                            SUB MENU
dBm -122 = 0.18 µV          *MORE CALCULATIONS
```

In other words, −122 dBm converts to 0.18 µV. You can do any number of calculations of the same conversion option by pressing ENTER. To change the conversion option return to the SUBMENU and repeat the choice process.

Subscribers from Traffic and Calling Rate

Highlight this option on the MISCELLANEOUS CALCULATIONS submenu and press ENTER.

```
RETURNS SUBSCRIBERS FROM CALLING RATE (ERLANGS) AND TRAFFIC

Traffic                       78
Calling rate in Erlangs?      0.8
Subscribers supported         97
```

Entering a TRAFFIC value of 78 and a CALLING RATE of 0.8 will return the value of 97 SUBSCRIBERS SUPPORTED. It should be noted that the calling rate of 0.8 Erlangs is exceptionally high.

Binary and HEX Numbers from Base 10

To convert Binary/HEX numbers to Base 10 and vice versa, highlight this option and press ENTER. The following option menu will appear:

```
CONVERTS BINARY AND HEX TO BASE 10 NUMBERS AND VICE VERSA

Convert Binary/HEX to BASE 10
Convert BASE 10 to Binary/HEX
*EXIT
```

Let's assume that you want to convert a BASE 10 number to its HEX equivalent. Highlight the CONVERT BASE 10 TO BINARY/HEX option and press ENTER. The following screen will appear:

```
CALCULATES THE BINARY OR HEX EQUIVALENT OF A BASE 10 NUMBER

Convert to HEX
Convert to Binary
*EXIT
```

Highlight CONVERT TO HEX and press ENTER. You will be prompted to enter the Base 10 number. Let's assume you wish to convert the Base 10 number 67:

```
Value to convert 67        MAIN MENU
                           SUB MENU
                           *MORE CALCULATIONS
The HEX value is 43
```

Decibel Calculation

This calculation returns the power ratio in decibels of two power/current/voltage levels expressed as a gain. It also converts decibels into power or voltage/current gains.

For example, if an attenuator has 100 W at its input and 5.7 W at the output, you can use this routine to calculate the insertion loss in decibels.

```
CALCULATES dB FROM POWER LEVELS

POWER LEVEL IN 100
POWER LEVEL OUT 5.7        MAIN MENU
                           SUB MENU
```

```
                        *MORE CALCULATIONS
```

```
GAIN is -12.441 dBV
```

SWR Calculations

This selection leads to submenu of VSWR-related calculations.

```
DOES VSWR CALCULATIONS
```

```
LOSS CALCULATED FROM VSWR INTO S/C OR O/C
RETURN POWER LOSS (watts) TO VSWR
RETURN LOSS (dB) TO VSWR
VSWR CORRECTION FOR FEEDER LOSS
VSWR CORRECTION FOR FEEDER LOSS
*RETURN TO SUB-MENU
```

Loss Calculated from VSWR into S/C or O/C

The top selection enables you to calculate a feeder loss from a VSWR measurement done into an open or short circuit. This can be useful for tracking down feeder loss problems on installed systems. Notice that the calculation is preferably done for a short-circuited cable, because an open-circuit cable will disguise losses that occur near the end of the cable, such as a bad connector. This routine relies on the fact that an open- or short-circuit cable should reflect the power at the discontinuity 100%.

Assume that the cable has been short-circuited (it doesn't matter which because open and short circuits will give the same result) and a VSWR of 5 is measured.

```
CALCULATES CABLE LOSS FROM VSWR INTO SHORT OR OPEN CIRCUIT
CABLE
```

```
MEASURED VSWR 5        MAIN MENU
                       SUB MENU
                       *MORE CALCULATIONS
CABLE LOSS IS 1.8 dB
```

```
RETURN POWER LOSS TO VSWR
```

The next routine looks at the calculation of VSWR from a forward and reflected power reading, taking into account the cable loss, if any. This last factor is important because the reflected power will be reduced by any cable loss. Therefore, a measurement done at the base of a tower that does not properly account for the cable loss will always underestimate the true VSWR. In the case of a cable with a 100-W forward power, a measured 6-W return (reflected power), and a feeder loss of 3 dB, the calculation will return a VSWR of 1.649.

```
CALCULATES VSWR FROM FORWARD AND REFLECTED POWER
TAKING ACCOUNT OF FEEDER LOSS IF ANY


FORWARD POWER 100


REFLECTED POWER 6


FEEDER LOSS IN dB (ENTER = 0) 3      MAIN MENU
                                     SUB MENU
                                     *MORE CALCULATIONS
VSWR is 1.649
```

VSWR Correction for Feeder Loss

Given that a VSWR reading has been made (for example, with a VSWR meter) and that there is a correction to be made because of the feeder cable loss, then this routine can make this correction. Assuming that a VSWR of 1.5 was measured at the input of an antenna cable that had a loss of 3 dB, then the true VSWR can be calculated using this routine.

```
VSWR CORRECTION FOR FEEDER LOSS


MEASURED VSWR 1.5


FEEDER LOSS (dB) 3                    MAIN MENU
                                     SUB MENU
                                     *MORE CALCULATIONS
CORRECTED VSWR 2.33
```

PART 4: POWER CALCULATIONS

The power routines are designed to encompass the electrical power requirements of a trunked network. It should be noted that in most countries the final power design and implementation will require implementation by an approved electrical contractor or engineer. The power routines have been designed to enable the calculation of the power requirements and the selection of suitable cables. The cables can be specified either in standard metric sizes or in AWG gauge. It should be noted that even when AWG gauge is selected, metric units are still used for all calculations except the gauge. To begin the calculations, select POWER CALCULATIONS from the main menu. The following menu will appear:

```
POWER CALCULATIONS


WIRE Gauge METRIC CHANGE Metric/AWG Gauges
CABLE SELECTION CABLE
RESISTANCE CALCULATOR
```

```
AC POWER CALCULATOR
DC POWER CALCULATOR
*EXIT
```

DC Power Calculations

This routine calculates the current for a given power. Where repeated calculations at the same voltage are required, it is possible to fix the voltage for future calculations. For example, assuming a 125-W load running from a 24-V supply, you would proceed as follows:

```
DC POWER CALCULATOR

Voltage 24

    Fix the voltage for more calculations
         CONTINUE
Watts125           MAIN MENU
                      SUB MENU
                   *MORE CALCULATIONS
    DC CURRENT IS 5.21
```

To fix the voltage for more calculations, use the cursor to choose the fix voltage option and thus the previously selected voltage will be fixed for subsequent calculations. If on the CONTINUE prompt you press ENTER, the routine will proceed to ask for the wattage.

AC Power Calculations

This routine does a number of different calculations including calculation of current from watts, volt-amps, and power factor correction. It can do these calculations for single- and three-phase power. When the AC power window first appears, the default values for voltage will appear on the right-hand side of the screen. These values are permanently stored, but can be altered or bypassed if the user wants to freely input any voltage for the various calculations.

```
AC POWER CALCULATIONS          DEFAULT VALUES
                               CHANGE DEFAULT SETTINGS
CURRENT FROM POWER             DELTA VOLTAGE 240
CURRENT FROM VOLT AMPS         STAR VOLTAGE 415
POWER FACTOR CORRECTION        SINGLE PHASE VOLTAGE 240*
RETURN TO SUB MENU             BYPASS DEFAULTS NO
```

Power Factor Correction

Most large trunked installations, if not corrected for power factor, will present an inductive load to the supply. This will result in a higher current being drawn for a given power and often results in a penalty from the power company, which prefers to feed purely resistive loads. Power factor is corrected by placing capacitors across the feed line which are designed to cancel out the inductive component of the line. The power factor capacitors are rated in VARS (volt amps reactive) or KVARS (kilo VARS). To determine the VARS to be corrected for a 10,000-W load with a power factor of 0.85, choose POWER FACTOR CORRECTION.

```
CALCULATES THE VARS THAT MUST BE EQUALIZED FOR UNITY POWER
FACTOR, GIVEN POWER FACTOR AND POWER OR VA's

Power factor 0.85 Volt-Amps
*Watts
Power in watts 10,000        MAIN MENU
                            SUB MENU
                            *MORE CALCULATIONS

VARS to be compensated 6197.44
```

Notice that the calculation can also be done in volt-amps if desired.

Current from VOLT-AMPS

The current from volt-amps can be calculated from this routine starting with single-phase or three-phase (star or delta) power supplies. If you select this calculation, the screen will first bring up the options:

```
THREE PHASE*
SINGLE PHASE
```

By pressing ENTER, single-phase calculations will be assumed.

Alternatively, the cursor can be used to select three phase, which will bring up an additional prompt:

```
THREE PHASE
    STAR
*   SINGLE PHASE
*   DELTA
```

For example, assume that you want to calculate the current in a 10,000 volt-amp delta system:

```
CURRENT FROM THREE PHASE DELTA VOLT-AMPS

LINE VOLTAGE 240
Volt-Amps 10000              MAIN MENU
                            SUB MENU
                            *MORE CALCULATIONS
```

```
Current is 41.67
```

Notice that in this case the default line voltage of 240 has been assumed.

Current from Power

The current from power calculation proceeds in the same manner as the current from volt-amps calculation, first asking you to choose from single phase, or three phase (delta or star). When the main screen appears it asks additionally for the power factor.

Assume that the system consumes 10,000 W at a power factor of 0.7:

```
SINGLE PHASE POWER CALCULATOR

VOLTAGE IS 240
Power Factor 0.7
Power required 10000        MAIN MENU
                           SUB MENU
                           *MORE CALCULATIONS
LINE CURRENT IS 59.52 AMPS
```

```
It is worth noting how much greater the current required is
for 10,000 W at this power factor when compared to the
current for 10,000 volt-amps.
```

Change Default Settings

This menu option enables you to alter any or all of the default values. Note that selecting BYPASS DEFAULTS will inform the routine that you wish to enter user supplied values of voltage with each calculation. When all the changes have been made, move the cursor to NO MORE CHANGES and the cursor will return to the submenu, saving all the changes.

Cable Resistance Calculator

Highlight CABLE RESISTANCE CALCULATOR in the POWER CALCULATIONS submenu. This routine returns the cable resistance for a number of cables and for a given cable length. The operation of the routine is almost the same for metric and AWG gauges, but for demonstration sake we will consider an example of each.

Metric Gauges   Assume you have selected metric and want to find the resistance of 150 m of 10 mm^2 cable:

```
CABLE sq mm RETURNS RESISTANCE FOR A GIVEN DISTANCE IN METERS
```

```
1.0
```

```
1.5
2.5
4.0
6.0
10.0*      Cable length (meters) 150
16.0
25.
35.0       Temperature Correction 50.0
           *40 degree C OK
70.0
95.0       Use arrow keys to select
120.0
150.0
185.0
240.0      Resistance 0.2940     MAIN MENU 300.0
                                 SUB MENU 400.0
                                 *MORE CALCULATIONS 500.0
630.0
```

Note that initially only the table will show. You highlight 10 mm^2 using the ARROW keys, then press ENTER to make the selection.

AWG Gauges   First highlight CHANGE METRIC/AWG GAUGES in the POWER CALCULATIONS submenu and press ENTER. This will change the WIRE GAUGE default to AWG. For AWG gauges there is a wider selection of cables and, initially, an AWG selection will appear. If you attempt to scroll the arrow key beyond the end of the AWG table, a new table will appear. In this example, assume a selection of AWG gauge 22 for 67 m of cable.

```
AWG Gauge
32
30
28
26
24
22*        Cable length 67
20         (meters)
18
16         Temperature Correction
14
           *40 degree C OK
12
10         Use cursor to select
8
6
4          Resistance 4.8374      MAIN MENU
2                                 SUB MENU
1                                 *MORE CALCULATIONS
```

If the second table is selected, you can proceed as before, but in this example we have changed the default temperature value of 40°C to 55°C.

```
AWG Gauge
0
00
000
0000            Cable length 67
Cir. Mils       (meters)
250000
300000          Temperature Correction
350000          *40 degree C OK
400000
450000          Temperature in degree C 55
500000
600000
700000          Resistance 0.0029          MAIN MENU
800000                                     SUB MENU
900000                                     *MORE CALCULATIONS
1000000
```

```
CABLE SELECTION
```

The CABLE SELECTION routine enables you to choose a cable capable of carrying a specified current and (optionally) a specified maximum voltage drop. In the following example a single cable has been selected to carry 100 amps with no voltage drop restrictions:

```
CABLE LAYOUT

Select with arrow keys      DETERMINES METRIC CABLE
                            FOR A GIVEN CURRENT

*SINGLE CABLES              Cables loose on cable trays
                            Directly buried cable
                                Current to be carried 100
                            Individual conduit       (amps)
                            Voltage drop limited
Cable pair touching on cable tray   * Continue
Cable pair in conduit in air
Cable pair in conduit, insulating surround
Cable pair in buried conduit
Cable pair buried in unprotected ground
                                    MAIN MENU
                                    SUB MENU
                                    *MORE CALCULATIONS


    Cable sq mm 25.0
```

If the same calculation is repeated with the restriction that the voltage drop will not exceed 2 volts over a distance of 125 meters, the routine returns that a much heavier cable of 150 sq mm is required.

```
CABLE LAYOUT

Select with cursor      DETERMINES METRIC CABLE
                        FOR A GIVEN CURRENT
*SINGLE CABLES Cables loose on cable trays
    Directly buried cable
                          Current to be carried 100
Cable in individual conduit       (amps)

                                  * Voltage drop limited
Cable pair touching on cable tray  Continue

                                  Voltage drop 2
Cable pair in conduit in air

                                  Cable length 125
Cable pair in conduit, insulating surround
Cable pair in buried conduit
Cable pair buried in unprotected ground

                                  MAIN MENU
                                  SUB MENU
                                  *MORE CALCULATIONS
Cable sq mm 150.0
*VOLTAGE LIMITED
```

PART 5: BATCH PROCESSING

Batch Traffic Processing

Most of the programs so far mentioned are for individual calculations. The batch programs, on the other hand, are meant to process whole networks in a single run.

If "BATCH TRAFFIC PROCESSING" is selected, the following menu will appear:

```
BATCH TRAFFIC COMPUTATIONS
WRITE TRAFFIC FILES
DISPERSION CALCULATIONS
* RETURN TO SUB-MENU
```

Batch Traffic Computations.   This routine is designed to calculate the number of channels needed in a network of cellular or trunked radio bases stations, given as an input a set of traffic figures for each base station. The program will not only calculate the number of circuits needed on the basis of the traffic report, but will also predict future needs for any given subscriber capacity. Should the future traffic be such that the existing base stations are insufficient to carry the traffic load, the program will recognize this and add new sites as needed.

In order to be able to design the network, the program will first need to know the nature of the system being dimensioned, and the following screen will appear:

```
QUEUED TRUNK RADIO TRAFFIC
BASE STATIONS USING ERLANG B
BASE STATIONS USING ERLANG C
```

Once a selection has been made, the program treats the systems in much the same way. As an example, take the "Queued trunk radio traffic."

First the program will ask for the traffic file extension to be used. It is assumed that the traffic has already been written into a DOS file. The form of this file is

```
{string character, header}, {base station name}, {traffic},
{base station name}, {traffic}..........EOF
```

or

```
"Base station name"
"base 1", "45.1"
"base 2", "2.56"
"base 3", "5.5"
EOF
```

It is suggested that most users would eventually write a program to produce this file automatically from the traffic processing program. However, in the interim the selection "WRITE TRAFFIC FILES" will enable you to manually write the files.

It is suggested that a common extension be ascribed to all files, because this program will enable you to select from a menu. The program comes with some sample files with the DOS extension *.TRF. At the question prompt "TRAFFIC FILE EXTENSION"; in this case you input the letters TRF. {Note you may use any extension you prefer for you own files.} Once you have done this, a menu of all files with the extension .TRF will appear and you can scroll these, selecting the one you want by pressing ENTER.

In order to be able to design the network, the following questions need to be asked and answered.

1. Maximum Number of Channels per Site.   This is the physical upper limit of channels that can be operated from one site. For example, for a trunked radio system it may be 20, for cellular if the site is sectored it will be the maximum number of channels per sector (it might be 77, for example).
2. Minimum Number of Channels per Site.   This is the smallest number of channels that a site (or sector) will be equipped with. For small analog systems

it may be company policy to have at least two channels, and for GSM a minimum of six or seven channels may apply.

3. Existing Customer Base.   This is the number of customers at the time of the traffic reading. For the test data a customer base of 200 would be appropriate.

Call Holding Time

This will only be asked for a trunked system and is the average call holding time. A typical holding time would be 50 s.

Design Queue Length

This is also a parameter that only applies to trunked radio and may typically be 1, 3, or 10 s.

This will all be processed, and the prompt

```
EXIT PROGRAM
* MORE FORECASTS
```

will appear. If you choose more forecasts, you will be asked for the new customer forecast. Once you enter the number, the new channel requirements will be calculated. You will then be prompted to select one of the following:

```
PRINTER OUTPUT
* SCREEN OUTPUT
```

You can continue to look at new customer forecasts until you decide to EXIT.

Write Traffic Files

This routine enables you to write a traffic file for processing. Initially it will ask

```
FILE NAME
```

This can be up to 50 characters and is not the DOS file name. It is any heading you want to use.

The next screen will ask

```
BASE STATION ... here you input the base station/cell name
```

then

```
TRAFFIC
```

This is the measured traffic. This continues until all traffic sources are entered.

Next you will be prompted for the file name by "Save As". This is requesting the DOS file name.

Finally you will be asked if you want a printout. Input either a "Y" or "N" (note the case does not matter).

After this the file will be saved.

APPENDIX 7

ERLANG TABLES

The following tables compute the traffic carried by a given number of circuits for grades of service from 0.001 to 0.02 for both the Erlang B and Erlang C conditions. Although the Erlang B equation is the most widely used, its assumption that there are no repeated attempts on failed calls does not correspond well with the real-world situation of today. The Erlang C relationship does assume that the caller makes repeated call attempts until successful and so is more realistic.

Moreover, the Erlang B equations are often used as first approximations for trunk radio system capacities. Because most trunked systems have queuing of some kind (and hence calls encountering congestion are retried by the network), Erlang B is a very poor approximation and Erlang C would be better. Of course the correct queuing relationships should be used ultimately, and these appear as tables in this book and can be calculated with the software provided.

ERLANG B TRAFFIC TABLE

GOS Circuits	Traffic (Erlangs)					
	0.001	0.002	0.01	0.05	0.1	0.2
1	.001	.002	0.01	.052	0.11	0.25
2	0.458	.065	0.15	0.38	0.59	1
3	0.193	0.248	0.45	0.89	1.27	1.93
4	0.439	0.534	0.86	1.52	2.04	2.94
5	0.762	0.899	1.36	2.21	2.88	4.01
6	1.146	1.325	1.90	2.96	3.75	5.10
7	1.578	1.798	2.50	3.73	4.66	6.22

407

GOS Circuits	Traffic (Erlangs)					
	0.001	0.002	0.01	0.05	0.1	0.2
8	2.051	2.31	3.128	4.543	5.596	7.368
9	2.558	2.855	3.783	5.369	6.544	8.521
10	3.092	3.427	4.461	6.216	7.512	9.685
11	3.651	4.021	5.16	7.076	8.488	10.85
12	4.232	4.637	5.876	7.951	9.473	12.04
13	4.83	5.269	6.607	8.837	10.47	13.22
14	5.446	5.919	7.352	9.729	11.47	14.42
15	6.077	6.582	8.108	10.63	12.48	15.61
16	6.722	7.258	8.876	11.55	13.5	16.8
17	7.379	7.945	9.651	12.46	14.52	18.01
18	8.046	8.643	10.44	13.38	15.55	19.22
19	8.724	9.351	11.23	14.32	16.58	20.42
20	9.411	10.07	12.03	15.25	17.62	21.63
21	10.11	10.79	12.84	16.19	18.65	22.85
22	10.81	11.53	13.65	17.13	19.69	24.06
23	11.52	12.27	14.47	18.08	20.74	25.29
24	12.24	13.01	15.29	19.03	21.78	26.51
25	12.97	13.76	16.12	19.99	22.84	27.72
26	13.7	14.52	16.96	20.94	23.89	28.93
27	14.44	15.29	17.8	21.91	24.94	30.16
28	15.18	16.05	18.64	22.86	25.99	31.39
29	15.93	16.83	19.49	23.83	27.05	32.6
30	16.68	17.61	20.34	24.8	28.11	33.85
31	17.44	18.39	21.19	25.77	29.17	35.08
32	18.21	19.18	22.05	26.75	30.24	36.31
33	18.97	19.97	22.91	27.72	31.29	37.53
34	19.74	20.76	23.77	28.7	32.37	38.75
35	20.52	21.56	24.64	29.68	33.43	39.98
36	21.3	22.36	25.51	30.66	34.5	41.22
37	22.08	23.17	26.38	31.64	35.57	42.45
38	22.86	23.97	27.25	32.62	36.65	43.68
39	23.65	24.78	28.13	33.61	37.71	44.92
40	24.44	25.6	29.01	34.6	38.78	46.14
41	25.24	26.42	29.89	35.58	39.87	47.39
42	26.04	27.24	30.77	36.57	40.94	48.62
43	26.84	28.06	31.66	37.57	42.01	49.84
44	27.64	28.88	32.54	38.56	43.09	51.1
45	28.45	29.71	33.43	39.55	44.16	52.33
46	29.26	30.54	34.32	40.54	45.24	53.55
47	30.07	31.37	35.21	41.54	46.32	54.81

GOS Circuits	Traffic (Erlangs)					
	0.001	0.002	0.01	0.05	0.1	0.2
48	30.88	32.2	36.11	42.54	47.39	56.04
49	31.69	33.04	37	43.53	48.48	57.28
50	32.51	33.87	37.9	44.53	49.56	58.5
51	33.33	34.72	38.8	45.53	50.63	59.75
52	34.15	35.56	39.7	46.53	51.73	60.99
53	34.98	36.4	40.6	47.53	52.82	62.21
54	35.8	37.25	41.5	48.53	53.89	63.46
55	36.63	38.09	42.41	49.54	54.97	64.71
56	37.46	38.94	43.32	50.54	56.07	65.93
57	38.29	39.79	44.22	51.55	57.15	67.18
58	39.12	40.64	45.13	52.55	58.23	68.42
59	39.96	41.5	46.04	53.56	59.31	69.66
60	40.79	42.36	46.95	54.57	60.41	70.9
61	41.63	43.21	47.86	55.57	61.48	72.14
62	42.47	44.07	48.77	56.58	62.58	73.38
63	43.31	44.93	49.69	57.59	63.67	74.63
64	44.15	45.79	50.6	58.6	64.76	75.86
65	45	46.65	51.52	59.62	65.84	77.12
66	45.85	47.51	52.44	60.62	66.92	78.37
67	46.69	48.38	53.35	61.63	68.02	79.59
68	47.54	49.24	54.27	62.64	69.09	80.82
69	48.39	50.11	55.19	63.65	70.2	82.07
70	49.24	50.98	56.11	64.67	71.29	83.32
71	50.09	51.85	57.04	65.68	72.37	84.57
72	50.95	52.72	57.96	66.7	73.47	85.81
73	51.8	53.59	58.88	67.7	74.55	87.05
74	52.65	54.46	59.8	68.72	75.65	88.29
75	53.51	55.34	60.73	69.74	76.75	89.56
76	54.37	56.21	61.65	70.75	77.83	90.78
77	55.23	57.09	62.58	71.77	78.91	92.01
78	56.09	57.97	63.51	72.79	80.01	93.27
79	56.95	58.84	64.43	73.8	81.12	94.52
80	57.81	59.72	65.36	74.82	82.21	95.74
81	58.67	60.6	66.29	75.84	83.29	97
82	59.54	61.48	67.22	76.86	84.4	98.26
83	60.4	62.36	68.15	77.88	85.47	99.47
84	61.27	63.24	69.09	78.9	86.58	100.7
85	62.13	64.13	70.02	79.91	87.68	102
86	63	65.01	70.95	80.93	88.76	103.2
87	63.87	65.89	71.88	81.96	89.86	104.5

GOS Circuits	Traffic (Erlangs)					
	0.001	0.002	0.01	0.05	0.1	0.2
88	64.74	66.78	72.82	82.97	90.96	105.7
89	65.61	67.67	73.75	83.99	92.06	106.9
90	66.48	68.56	74.69	85	93.15	108.2
91	67.36	69.44	75.62	86.03	94.24	109.4
92	68.23	70.33	76.56	87.06	95.33	110.7
93	69.1	71.22	77.49	88.08	96.45	111.9
94	69.98	72.11	78.43	89.1	97.53	113.2
95	70.86	73	79.37	90.12	98.62	114.4
96	71.73	73.89	80.31	91.15	99.73	115.6
97	72.61	74.79	81.25	92.17	100.8	116.9
98	73.49	75.68	82.18	93.2	101.9	118.2
99	74.36	76.57	83.12	94.21	103	119.4
100	75.24	77.47	84.07	95.24	104.1	120.6
101	76.12	78.36	85	96.27	105.2	121.9
102	77	79.26	85.95	97.28	106.3	123.1
103	77.88	80.16	86.88	98.31	107.4	124.4
104	78.77	81.06	87.83	99.35	108.5	125.6
105	79.65	81.95	88.77	100.4	109.6	126.9
106	80.53	82.85	89.71	101.4	110.7	128.1
107	81.42	83.75	90.66	102.4	111.8	129.4
108	82.3	84.65	91.61	103.4	112.9	130.6
109	83.19	85.55	92.55	104.5	114	131.8
110	84.07	86.45	93.5	105.5	115.1	133.1
111	84.96	87.35	94.44	106.5	116.2	134.4
112	85.85	88.25	95.38	107.5	117.3	135.6
113	86.73	89.15	96.33	108.6	118.4	136.8
114	87.62	90.06	97.27	109.6	119.5	138.1
115	88.51	90.96	98.22	110.6	120.6	139.3
116	89.4	91.86	99.17	111.6	121.7	140.6
117	90.29	92.77	100.1	112.7	122.8	141.8
118	91.18	93.67	101.1	113.7	123.9	143.1
119	92.07	94.58	102	114.7	125	144.3
120	92.96	95.49	103	115.8	126.1	145.6
121	93.86	96.39	103.9	116.8	127.2	146.8
122	94.75	97.3	104.9	117.8	128.3	148
123	95.64	98.2	105.8	118.9	129.4	149.3
124	96.54	99.11	106.8	119.9	130.5	150.6
125	97.43	100	107.7	120.9	131.6	151.8
126	98.33	100.9	108.7	121.9	132.7	153
127	99.22	101.8	109.6	123	133.8	154.3

GOS	Traffic (Erlangs)					
Circuits	0.001	0.002	0.01	0.05	0.1	0.2
128	100.1	102.8	110.6	124	134.9	155.6
129	101	103.7	111.5	125	136	156.8
130	101.9	104.6	112.5	126.1	137.1	158
131	102.8	105.5	113.4	127.1	138.2	159.3
132	103.7	106.4	114.4	128.1	139.3	160.5
133	104.6	107.3	115.3	129.2	140.4	161.8
134	105.5	108.2	116.3	130.2	141.5	163
135	106.4	109.1	117.2	131.2	142.6	164.2
136	107.3	110	118.2	132.3	143.7	165.5
137	108.2	111	119.1	133.3	144.8	166.8
138	109.1	111.9	120.1	134.3	145.9	168
139	110	112.8	121	135.4	147	169.3
140	110.9	113.7	122	136.4	148.1	170.5
141	111.8	114.6	123	137.4	149.2	171.7
142	112.7	115.5	123.9	138.4	150.3	173
143	113.6	116.4	124.9	139.5	151.4	174.3
144	114.5	117.4	125.8	140.5	152.5	175.5
145	115.4	118.3	126.8	141.5	153.6	176.8
146	116.3	119.2	127.7	142.6	154.7	178
147	117.2	120.1	128.7	143.6	155.8	179.3
148	118.1	121	129.7	144.6	156.9	180.5
149	119	121.9	130.6	145.7	158	181.7
150	119.9	122.9	131.6	146.7	159.1	183
151	120.8	123.8	132.5	147.7	160.2	184.2
152	121.8	124.7	133.5	148.8	161.3	185.5
153	122.7	125.6	134.5	149.8	162.4	186.7
154	123.6	126.5	135.4	150.8	163.5	188
155	124.5	127.5	136.4	151.9	164.6	189.2
156	125.4	128.4	137.3	152.9	165.7	190.5
157	126.3	129.3	138.3	153.9	166.9	191.7
158	127.2	130.2	139.2	155	167.9	193
159	128.1	131.1	140.2	156	169	194.2
160	129	132.1	141.2	157.1	170.2	195.4
161	129.9	133	142.1	158.1	171.3	196.7
162	130.8	133.9	143.1	159.1	172.4	197.9
163	131.7	134.8	144.1	160.1	173.4	199.2
164	132.7	135.8	145	161.2	174.6	200.4
165	133.6	136.7	146	162.2	175.7	201.7
166	134.5	137.6	146.9	163.3	176.8	202.9
167	135.4	138.5	147.9	164.3	177.9	204.2

GOS Circuits	Traffic (Erlangs)					
	0.001	0.002	0.01	0.05	0.1	0.2
168	136.3	139.4	148.9	165.3	179	205.4
169	137.2	140.4	149.8	166.4	180.1	206.7
170	138.1	141.3	150.8	167.4	181.2	207.9
171	139	142.2	151.7	168.4	182.3	209.2
172	139.9	143.2	152.7	169.5	183.4	210.4
173	140.9	144.1	153.7	170.5	184.5	211.7
174	141.8	145	154.6	171.5	185.6	212.9
175	142.7	145.9	155.6	172.6	186.7	214.2
176	143.6	146.8	156.6	173.6	187.8	215.4
177	144.5	147.8	157.5	174.7	188.9	216.7
178	145.4	148.7	158.5	175.7	190	218
179	146.3	149.6	159.5	176.7	191.1	219.2
180	147.3	150.6	160.4	177.8	192.2	220.5
181	148.2	151.5	161.4	178.8	193.3	221.7
182	149.1	152.4	162.4	179.8	194.4	223
183	150	153.3	163.3	180.9	195.5	224.1
184	150.9	154.3	164.3	181.9	196.6	225.4
185	151.8	155.2	165.2	182.9	197.8	226.6
186	152.8	156.1	166.2	184	198.9	227.9
187	153.7	157.1	167.2	185	199.9	229.2
188	154.6	158	168.1	186.1	201.1	230.4
189	155.5	158.9	169.1	187.1	202.2	231.7
190	156.4	159.8	170.1	188.1	203.3	232.9
191	157.3	160.8	171	189.2	204.4	234.1
192	158.3	161.7	172	190.2	205.5	235.3
193	159.2	162.6	173	191.3	206.6	236.6
194	160.1	163.6	173.9	192.3	207.7	237.9
195	161	164.5	174.9	193.3	208.8	239.1
196	161.9	165.4	175.9	194.3	209.9	240.4
197	162.9	166.4	176.8	195.4	211	241.6
198	163.8	167.3	177.8	196.4	212.1	242.9
199	164.7	168.2	178.8	197.5	213.2	244.2
200	165.6	169.2	179.7	198.5	214.4	245.4
201	166.5	170.1	180.7	199.5	215.5	246.7
202	167.5	171	181.7	200.6	216.6	247.9
203	168.4	172	182.6	201.6	217.7	249.2
204	169.3	172.9	183.6	202.6	218.8	250.4
205	170.2	173.8	184.6	203.7	219.9	251.6
206	171.2	174.8	185.5	204.7	221	252.9
207	172.1	175.7	186.5	205.8	222.1	254.1

GOS Circuits	Traffic (Erlangs)					
	0.001	0.002	0.01	0.05	0.1	0.2
208	173	176.6	187.5	206.8	223.2	255.4
209	173.9	177.6	188.5	207.9	224.3	256.6
210	174.8	178.5	189.4	208.9	225.4	257.8
211	175.8	179.4	190.4	209.9	226.5	259.1
212	176.7	180.4	191.4	211	227.6	260.3
213	177.6	181.3	192.3	212	228.7	261.6
214	178.5	182.2	193.3	213.1	229.8	262.9
215	179.5	183.2	194.3	214.1	230.9	264.1
216	180.4	184.1	195.2	215.1	232	265.4
217	181.3	185	196.2	216.2	233.1	266.7
218	182.2	186	197.2	217.2	234.2	267.9
219	183.2	186.9	198.2	218.2	235.3	269.1
220	184.1	187.8	199.1	219.3	236.4	270.3
221	185	188.8	200.1	220.3	237.5	271.6
222	185.9	189.7	201.1	221.4	238.6	272.9
223	186.9	190.7	202	222.4	239.7	274
224	187.8	191.6	203	223.4	240.8	275.3
225	188.7	192.5	204	224.5	241.9	276.6
226	189.6	193.5	205	225.5	243.1	277.8
227	190.6	194.4	205.9	226.6	244.2	279.1
228	191.5	195.3	206.9	227.6	245.3	280.4
229	192.4	196.3	207.9	228.6	246.4	281.6
230	193.3	197.2	208.8	229.7	247.5	282.9
231	194.3	198.1	209.8	230.7	248.6	284.2
232	195.2	199.1	210.8	231.8	249.7	285.3
233	196.1	200	211.8	232.8	250.8	286.6
234	197.1	201	212.7	233.9	251.9	287.8
235	198	201.9	213.7	234.9	253	289.1
236	198.9	202.8	214.7	235.9	254.1	290.3
237	199.8	203.8	215.6	237	255.3	291.5
238	200.8	204.7	216.6	238	256.4	292.8
239	201.7	205.7	217.6	239.1	257.5	294.1
240	202.6	206.6	218.6	240.1	258.6	295.3
241	203.6	207.5	219.5	241.1	259.7	296.6
242	204.5	208.5	220.5	242.2	260.8	297.9
243	205.4	209.4	221.5	243.2	261.9	299.2
244	206.3	210.4	222.4	244.3	263	300.3
245	207.3	211.3	223.4	245.3	264.1	301.6
246	208.2	212.2	224.4	246.3	265.2	302.8
247	209.1	213.2	225.4	247.4	266.3	304

GOS	Traffic (Erlangs)					
Circuits	0.001	0.002	0.01	0.05	0.1	0.2
248	210.1	214.1	226.4	248.4	267.4	305.3
249	211	215.1	227.3	249.5	268.5	306.5
250	211.9	216	228.3	250.5	269.6	307.8
251	212.9	216.9	229.3	251.6	270.7	309.1
252	213.8	217.9	230.2	252.6	271.8	310.4
253	214.7	218.8	231.2	253.6	272.9	311.6
254	215.6	219.8	232.2	254.7	274.1	312.9
255	216.6	220.7	233.2	255.7	275.2	314
256	217.5	221.7	234.1	256.7	276.3	315.3

ERLANG *C* TRAFFIC TABLE

GOS	Traffic (Erlangs)					
Circuits	0.001	0.002	0.01	0.05	0.1	0.2
1	0.0009	0.0019	0.01	0.05	0.1	0.200
2	0.04	0.0642	0.146	0.342	0.500	0.740
3	0.189	0.2413	0.429	0.787	1.039	1.393
4	0.425	0.5142	0.81	1.319	1.653	2.103
5	0.734	0.8596	1.259	1.905	2.313	2.847
6	1.1	1.261	1.758	2.531	3.006	3.617
7	1.51	1.705	2.297	3.188	3.726	4.406
8	1.958	2.186	2.865	3.888	4.464	5.211
9	2.436	2.697	3.46	4.568	5.219	6.025
10	2.942	3.232	4.077	5.284	5.985	6.853
11	3.47	3.79	4.711	6.015	6.766	7.688
12	4.018	4.366	5.363	6.758	7.554	8.529
13	4.584	4.959	6.028	7.51	8.352	9.379
14	5.166	5.568	6.704	8.273	9.159	10.23
15	5.762	6.189	7.393	9.043	9.97	11.09
16	6.371	6.822	8.093	9.821	10.79	11.96
17	6.991	7.467	8.801	10.61	11.61	12.83
18	7.622	8.123	9.517	11.4	12.44	13.7
19	8.263	8.787	10.24	12.2	13.28	14.57
20	8.913	9.459	10.97	13	14.11	15.46
21	9.573	10.14	11.71	13.81	14.96	16.34
22	10.24	10.83	12.46	14.62	15.8	17.22
23	10.91	11.52	13.2	15.43	16.65	18.11
24	11.59	12.22	13.96	16.25	17.51	19

GOS Circuits	Traffic (Erlangs)					
	0.001	0.002	0.01	0.05	0.1	0.2
25	12.28	12.93	14.72	17.08	18.36	19.89
26	12.97	13.65	15.48	17.91	19.22	20.79
27	13.67	14.36	16.25	18.74	20.08	21.68
28	14.38	15.09	17.03	19.57	20.95	22.58
29	15.09	15.82	17.81	20.41	21.81	23.48
30	15.8	16.55	18.59	21.25	22.69	24.39
31	16.52	17.29	19.37	22.09	23.55	25.29
32	17.24	18.03	20.16	22.93	24.43	26.19
33	17.97	18.78	20.95	23.78	25.3	27.1
34	18.71	19.52	21.75	24.63	26.18	28.01
35	19.44	20.28	22.55	25.48	27.06	28.92
36	20.18	21.03	23.35	26.33	27.94	29.83
37	20.93	21.8	24.15	27.19	28.82	30.74
38	21.67	22.56	24.96	28.05	29.7	31.66
39	22.42	23.32	25.76	28.91	30.59	32.56
40	23.17	24.09	26.58	29.77	31.48	33.48
41	23.93	24.87	27.39	30.63	32.36	34.4
42	24.69	25.64	28.21	31.5	33.26	35.32
43	25.45	26.42	29.02	32.36	34.15	36.23
44	26.21	27.2	29.84	33.23	35.04	37.16
45	26.98	27.98	30.67	34.1	35.93	38.08
46	27.75	28.76	31.49	34.97	36.82	39
47	28.52	29.55	32.32	35.84	37.72	39.92
48	29.3	30.34	33.14	36.72	38.62	40.84
49	30.08	31.13	33.97	37.59	39.51	41.76
50	30.85	31.93	34.8	38.47	40.41	42.69
51	31.64	32.72	35.64	39.35	41.31	43.61
52	32.42	33.52	36.47	40.22	42.22	44.54
53	33.21	34.32	37.31	41.1	43.12	45.46
54	33.99	35.12	38.15	41.99	44.02	46.39
55	34.78	35.92	38.98	42.87	44.93	47.32
56	35.57	36.73	39.82	43.75	45.83	48.25
57	36.37	37.54	40.67	44.63	46.73	49.18
58	37.16	38.35	41.51	45.52	47.64	50.11
59	37.96	39.16	42.35	46.41	48.55	51.04
60	38.76	39.97	43.2	47.29	49.46	51.97
61	39.56	40.78	44.05	48.18	50.37	52.9
62	40.36	41.6	44.9	49.07	51.27	53.83
63	41.16	42.41	45.75	49.96	52.18	54.77
64	41.97	43.23	46.6	50.85	53.1	55.7

GOS Circuits	Traffic (Erlangs)					
	0.001	0.002	0.01	0.05	0.1	0.2
65	42.77	44.05	47.45	51.74	54.01	56.63
66	43.58	44.87	48.3	52.63	54.91	57.57
67	44.39	45.69	49.16	53.53	55.83	58.5
68	45.2	46.52	50.01	54.42	56.74	59.44
69	46.01	47.34	50.87	55.32	57.66	60.37
70	46.83	48.17	51.73	56.21	58.58	61.31
71	47.64	48.99	52.59	57.11	59.49	62.25
72	48.46	49.82	53.45	58.01	60.41	63.18
73	49.28	50.65	54.31	58.91	61.32	64.12
74	50.1	51.48	55.17	59.8	62.24	65.06
75	50.92	52.31	56.03	60.7	63.16	65.99
76	51.74	53.14	56.89	61.6	64.08	66.94
77	52.56	53.98	57.76	62.5	64.99	67.88
78	53.38	54.81	58.62	63.4	65.91	68.82
79	54.21	55.65	59.49	64.3	66.83	69.76
80	55.03	56.49	60.35	65.21	67.75	70.69
81	55.86	57.33	61.22	66.11	68.68	71.64
82	56.69	58.17	62.09	67.01	69.6	72.58
83	57.52	59	62.96	67.92	70.52	73.52
84	58.34	59.85	63.83	68.82	71.44	74.46
85	59.18	60.69	64.7	69.72	72.36	75.41
86	60.01	61.53	65.57	70.63	73.28	76.35
87	60.84	62.38	66.45	71.54	74.21	77.29
88	61.67	63.22	67.32	72.45	75.13	78.23
89	62.51	64.07	68.19	73.36	76.06	79.18
90	63.35	64.91	69.07	74.26	76.98	80.12
91	64.18	65.76	69.94	75.17	77.91	81.07
92	65.02	66.61	70.82	76.08	78.83	82.01
93	65.86	67.45	71.7	76.99	79.76	82.95
94	66.7	68.31	72.57	77.9	80.68	83.9
95	67.54	69.16	73.45	78.81	81.61	84.85
96	68.37	70.01	74.33	79.72	82.54	85.79
97	69.22	70.86	75.2	80.63	83.47	86.74
98	70.06	71.71	76.09	81.54	84.39	87.68
99	70.9	72.57	76.97	82.46	85.32	88.63
100	71.75	73.42	77.85	83.37	86.25	89.57
101	72.59	74.28	78.73	84.28	87.18	90.52
102	73.44	75.13	79.61	85.2	88.11	91.47
103	74.28	75.99	80.49	86.11	89.04	92.42
104	75.13	76.85	81.38	87.03	89.97	93.36

GOS Circuits	Traffic (Erlangs)					
	0.001	0.002	0.01	0.05	0.1	0.2
105	75.98	77.7	82.26	87.94	90.9	94.31
106	76.83	78.56	83.15	88.86	91.84	95.26
107	77.68	79.42	84.03	89.77	92.77	96.21
108	78.53	80.28	84.92	90.69	93.69	97.16
109	79.38	81.14	85.8	91.6	94.63	98.1
110	80.23	82	86.69	92.52	95.56	99.06
111	81.08	82.86	87.58	93.43	96.49	100
112	81.93	83.73	88.47	94.36	97.43	101
113	82.78	84.59	89.35	95.27	98.35	101.9
114	83.64	85.45	90.24	96.19	99.29	102.9
115	84.49	86.32	91.13	97.1	100.2	103.8
116	85.35	87.18	92.02	98.03	101.2	104.8
117	86.21	88.05	92.91	98.94	102.1	105.7
118	87.06	88.91	93.8	99.87	103	106.7
119	87.92	89.78	94.69	100.8	104	107.6
120	88.78	90.65	95.58	101.7	104.9	108.6
121	89.63	91.52	96.48	102.6	105.8	109.5
122	90.49	92.38	97.36	103.5	106.8	110.5
123	91.35	93.25	98.26	104.5	107.7	111.4
124	92.21	94.12	99.15	105.4	108.6	112.4
125	93.07	94.99	100	106.3	109.6	113.3
126	93.94	95.86	100.9	107.2	110.5	114.3
127	94.8	96.73	101.8	108.2	111.4	115.2
128	95.66	97.6	102.7	109.1	112.4	116.2
129	96.52	98.47	103.6	110	113.3	117.1
130	97.38	99.35	104.5	110.9	114.3	118.1
131	98.25	100.2	105.4	111.9	115.2	119
132	99.11	101.1	106.3	112.8	116.1	120
133	99.98	102	107.2	113.7	117.1	120.9
134	100.8	102.8	108.1	114.6	118	121.9
135	101.7	103.7	109	115.5	118.9	122.9
136	102.6	104.6	109.9	116.5	119.9	123.8
137	103.4	105.5	110.8	117.4	120.8	124.8
138	104.3	106.3	111.7	118.3	121.8	125.7
139	105.2	107.2	112.6	119.3	122.7	126.7
140	106	108.1	113.5	120.2	123.6	127.6
141	106.9	109	114.4	121.1	124.6	128.6
142	107.8	109.9	115.3	122	125.5	129.5
143	108.7	110.7	116.2	123	126.5	130.5
144	109.5	111.6	111.7	123.9	127.4	131.5

GOS	Traffic (Erlangs)					
Circuits	0.001	0.002	0.01	0.05	0.1	0.2
145	110.4	112.5	118	124.8	128.3	132.4
146	111.3	113.4	118.9	125.7	129.3	133.4
147	112.1	114.2	119.8	126.7	130.2	134.3
148	113	115.1	120.7	127.6	131.2	135.3
149	113.9	116	121.6	128.5	132.1	136.2
150	114.8	116.9	122.5	129.5	133.1	137.2
151	115.6	117.8	123.4	130.4	134	138.1
152	116.5	118.7	124.3	131.3	134.9	139.1
153	117.4	119.5	125.2	132.2	135.9	140.1
154	118.2	120.4	126.1	133.2	136.8	141
155	119.1	121.3	127	134.1	137.8	142
156	120	122.2	127.9	135	138.7	142.9
157	120.9	123.1	128.8	136	139.7	143.9
158	121.7	124	129.7	136.9	140.6	144.8
159	122.6	124.8	130.6	137.8	141.5	145.8
160	123.5	125.7	131.6	138.8	142.5	146.8
161	124.4	126.6	132.5	139.7	143.4	147.7
162	125.3	127.5	133.4	140.6	144.4	148.7
163	126.1	128.4	134.3	141.5	145.3	149.6
164	127	129.3	135.2	142.5	146.3	150.6
165	127.9	130.2	136.1	143.4	147.2	151.6
166	128.8	131	137	144.3	148.2	152.5
167	129.6	131.9	137.9	145.3	149.1	153.5
168	130.5	132.8	138.8	146.2	150	154.4
169	131.4	133.7	139.7	147.1	151	155.4
170	132.3	134.6	140.6	148.1	151.9	156.3
171	133.2	135.5	141.5	149	152.9	157.3
172	134.1	136.4	142.4	149.9	153.8	158.3
173	134.9	137.3	143.4	150.9	154.8	159.2
174	135.8	138.1	144.3	151.8	155.7	160.2
175	136.7	139	145.2	152.7	156.7	161.1
176	137.6	139.9	146.1	153.7	157.6	162.1
177	138.5	140.8	147	154.6	158.6	163.1
178	139.3	141.7	147.9	155.6	159.5	164
179	140.2	142.6	148.8	156.5	160.4	165
180	141.1	143.5	149.7	157.4	161.4	165.9
181	142	144.4	150.7	158.4	162.3	166.9
182	142.9	145.3	151.6	159.3	163.3	167.9
183	143.8	146.2	152.5	160.2	164.2	168.8
184	144.7	147.1	153.4	161.2	165.2	169.8

GOS Circuits	Traffic (Erlangs)					
	0.001	0.002	0.01	0.05	0.1	0.2
185	145.5	148	154.3	162.1	166.1	170.8
186	146.4	148.8	155.2	163	167.1	171.7
187	147.3	149.7	156.1	164	168	172.7
188	148.2	150.6	157	164.9	169	173.6
189	149.1	151.5	158	165.8	169.9	174.6
190	150	152.4	158.9	166.8	170.9	175.6
191	150.9	153.3	159.8	167.7	171.8	176.5
192	151.7	154.2	160.7	168.7	172.8	177.5
193	152.6	155.1	161.6	169.6	173.7	178.4
194	153.5	156	162.5	170.5	174.7	179.4
195	154.4	156.9	163.4	171.5	175.6	180.4
196	155.3	157.8	164.3	172.4	176.6	181.3
197	156.2	158.7	165.3	173.3	177.5	182.3
198	157.1	159.6	166.2	174.3	178.5	183.3
199	158	160.5	167.1	175.2	179.4	184.2
200	158.9	161.4	168	176.2	180.4	185.2
201	159.7	162.3	168.9	177.1	181.3	186.1
202	160.6	163.2	169.8	178	182.3	187.1
203	161.5	164.1	170.8	179	183.2	188.1
204	162.4	165	171.7	179.9	184.2	189
205	163.3	165.9	172.6	180.9	185.1	190
206	164.2	166.8	173.5	181.8	186.1	191
207	165.1	167.7	174.4	182.7	187	191.9
208	166	168.6	175.3	183.7	188	192.9
209	166.9	169.5	176.3	184.6	188.9	193.8
210	167.8	170.4	177.2	185.5	189.9	194.8
211	168.7	171.3	178.1	186.5	190.8	195.8
212	169.6	172.2	179	187.4	191.8	196.7
213	170.4	173.1	179.9	188.4	192.7	197.7
214	171.3	174	180.9	189.3	193.7	198.7
215	172.2	174.9	181.8	190.2	194.6	199.6
216	173.1	175.8	182.7	191.2	195.6	200.6
217	174	176.7	183.6	192.1	196.5	201.6
218	174.9	177.6	184.5	193.1	197.5	202.5
219	175.8	178.5	185.5	194	198.4	203.5
220	176.7	179.4	186.4	195	199.4	204.4
221	177.6	180.3	187.3	195.9	200.3	205.4
222	178.5	181.2	188.2	196.8	201.3	206.4
223	179.4	182.1	189.1	197.8	202.2	207.3
224	180.3	183	190.1	198.7	203.2	208.3

GOS Circuits	Traffic (Erlangs)					
	0.001	0.002	0.01	0.05	0.1	0.2
225	181.2	183.9	191	199.7	204.1	209.3
226	182.1	184.8	191.9	200.6	205.1	210.2
227	183	185.7	192.8	201.5	206	211.2
228	183.9	186.6	193.7	202.5	207	212.2
229	184.8	187.5	194.7	203.4	208	213.1
230	185.7	188.4	195.6	204.4	208.9	214.1
231	186.6	189.3	196.5	205.3	209.9	215.1
232	187.5	190.2	197.4	206.3	210.8	216
233	188.4	191.1	198.3	207.2	211.8	217
234	189.3	192	199.3	208.1	212.7	218
235	190.2	192.9	200.2	209.1	213.7	218.9
236	191.1	193.8	201.1	210	214.6	219.9
237	192	194.7	202	211	215.6	220.8
238	192.9	195.6	203	211.9	216.5	221.8
239	193.8	196.6	203.9	212.9	217.5	222.8
240	194.7	197.5	204.8	213.8	218.4	223.7
241	195.5	198.4	205.7	214.8	219.4	224.7
242	196.4	199.3	206.7	215.7	220.4	225.7
243	197.3	200.2	207.6	216.6	221.3	226.6
244	198.3	201.1	208.5	217.6	222.3	227.6
245	199.2	202	209.4	218.5	223.2	228.6
246	200.1	202.9	210.4	219.5	224.2	229.5
247	201	203.8	211.3	220.4	225.1	230.5
248	201.9	204.7	212.2	221.4	226.1	231.5
249	202.8	205.6	213.1	222.3	227	232.4
250	203.7	206.5	214	223.3	228	233.4
251	204.6	207.4	215	224.2	229	234.4
252	205.5	208.3	215.9	225.1	229.9	235.3
253	206.4	209.3	216.8	226.1	230.9	236.3
254	207.3	210.2	217.8	227	231.8	237.3
255	208.2	211.1	218.7	228	232.8	238.2
256	209.1	212	219.6	228.9	233.7	239.2

FIELD-STRENGTH UNITS
CONVERSION TABLE

This useful table provides conversion for all the commonly used units of field strength and over most of the ranges of values ordinarily encountered in the trunked radio industry. The units of dBm, μV, dBμV, and dBμV/m are all considered, the latter for the most commonly encountered frequencies.

Measuring equipment may come calibrated in any of these units; likewise, hardware specifications are often given in diverse units. In practice a faulty conversion can cause a lot of wasted time, and it is advisable to always use either a table or a computerized conversion. The software that comes with this book also has the same conversion capabilities.

			dBμV/m			
dBm	μV	dBμV	150 MHz	400 MHz	900 MHz	2,000 MHz
−130	0.07	−23	−11.6	−3	4	10.9
−129	0.08	−22	−10.6	−2	5	11.9
−128	0.09	−21	−9.6	−1	6	12.9
−127	0.1	−20	−8.6	0	7	13.9
−126	0.11	−19	−7.6	1	8	14.9
−125	0.13	−18	−6.6	2	9	15.9
−124	0.14	−17	−5.6	3	10	16.9
−123	0.16	−16	−4.6	4	11	17.9
−122	0.18	−15	−3.6	5	12	18.9
−121	0.2	−14	−2.6	6	13	19.9

			dBμV/m			
dBm	μV	dBμV	150 MHz	400 MHz	900 MHz	2,000 MHz
−120	0.22	−13	−1.6	7	14	20.9
−119	0.25	−12	−0.6	8	15	21.9
−118	0.28	−11	0.4	9	16	22.9
−117	0.32	−10	1.4	10	17	23.9
−116	0.35	−9	2.4	11	18	24.9
−115	0.4	−8	3.4	12	19	25.9
−114	0.45	−7	4.4	13	20	26.9
−113	0.5	−6	5.4	14	21	27.9
−112	0.56	−5	6.4	15	22	28.9
−111	0.63	−4	7.4	16	23	29.9
−110	0.71	−3	8.4	17	24	30.9
−109	0.79	−2	9.4	18	25	31.9
−108	0.89	−1	10.4	19	26	32.9
−107	1	0	11.4	20	27	33.9
−106	1.1	1	12.4	21	28	34.9
−105	1.3	2	13.4	22	29	35.9
−104	1.4	3	14.4	23	30	36.9
−103	1.6	4	15.4	24	31	37.9
−102	1.8	5	16.4	25	32	38.9
−101	2	6	17.4	26	33	39.9
−100	2.2	7	18.4	27	34	40.9
−99	2.5	8	19.4	28	35	41.9
−98	2.8	9	20.4	29	36	42.9
−97	3.2	10	21.4	30	37	43.9
−96	3.5	11	22.4	31	38	44.9
−95	4	12	23.4	32	39	45.9
−94	4.5	13	24.4	33	40	46.9
−93	5	14	25.4	34	41	47.9
−92	5.6	15	26.4	35	42	48.9
−91	6.3	16	27.4	36	43	49.9
−90	7.1	17	28.4	37	44	50.9
−89	7.9	18	29.4	38	45	51.9
−88	8.9	19	30.4	39	46	52.9
−87	10	20	31.4	40	47	53.9
−86	11.2	21	32.4	41	48	54.9
−85	12.6	22	33.4	42	49	55.9
−84	14.1	23	34.4	43	50	56.9
−83	15.8	24	35.4	44	51	57.9
−82	17.8	25	36.4	45	52	58.9
−81	19.9	26	37.4	46	53	59.9
−80	22.4	27	38.4	47	54	60.9

dBm	µV	dBµV	dBµV/m			
			150 MHz	400 MHz	900 MHz	2,000 MHz
−79	25.1	28	39.4	48	55	61.9
−78	28.2	29	40.4	49	56	62.9
−77	31.6	30	41.4	50	57	63.9
−76	35.4	31	42.4	51	58	64.9
−75	39.8	32	43.4	52	59	65.9
−74	44.6	33	44.4	53	60	66.9
−73	50.1	34	45.4	54	61	67.9
−72	56.2	35	46.4	55	62	68.9
−71	63	36	47.4	56	63	69.9
−70	70.7	37	48.4	57	64	70.9
−69	79.3	38	49.4	58	65	71.9
−68	89	39	50.4	59	66	72.9
−67	99.9	40	51.4	60	67	73.9
−66	112.1	41	52.4	61	68	74.9
−65	125.7	42	53.4	62	69	75.9
−64	141.1	43	54.4	63	70	76.9
−63	158.3	44	55.4	64	71	77.9
−62	177.6	45	56.4	65	72	78.9
−61	199.3	46	57.4	66	73	79.9
−60	223.6	47	58.4	67	74	80.9
−59	250.9	48	59.4	68	75	81.9
−58	281.5	49	60.4	69	76	82.9
−57	315.9	50	61.4	70	77	83.9
−56	354.4	51	62.4	71	78	84.9
−55	397.6	52	63.4	72	79	85.9
−54	446.2	53	64.4	73	80	86.9
−53	501	54	65.4	74	81	87.9
−52	562	55	66.4	75	82	88.9
−51	630	56	67.4	76	83	89.9
−50	707	57	68.4	77	84	90.9
−49	793	58	69.4	78	85	91.9
−48	890	59	70.4	79	86	92.9
−47	999	60	71.4	80	87	93.9
−46	1,121	61	72.4	81	88	94.9
−45	1,257	62	73.4	82	89	95.9
−44	1,411	63	74.4	83	90	96.9
−43	1,583	64	75.4	84	91	97.9
−42	1,776	65	76.4	85	92	98.9
−41	1,993	66	77.4	86	93	99.9
−40	2,236	67	78.4	87	94	100.9
−39	2,509	68	79.4	88	95	101.9

			dBμV/m			
dBm	μV	dBμV	150 MHz	400 MHz	900 MHz	2,000 MHz
−38	2,815	69	80.4	89	96	102.9
−37	3,159	70	81.4	90	97	103.9
−36	3,544	71	82.4	91	98	104.9
−35	3,976	72	83.4	92	99	105.9
−34	4,462	73	84.4	93	100	106.9
−33	5,006	74	85.4	94	101	107.9
−32	5,617	75	86.4	95	102	108.9
−31	6,302	76	87.4	96	103	109.9
−30	7,071	77	88.4	97	104	110.9
−29	7,934	78	89.4	98	105	111.9
−28	8,902	79	90.4	99	106	112.9
−27	9,988	80	91.4	100	107	113.9
−26	11,207	81	92.4	101	108	114.9
−25	12,574	82	93.4	102	109	115.9
−24	14,109	83	94.4	103	110	116.9
−23	15,830	84	95.4	104	111	117.9
−22	17,762	85	96.4	105	112	118.9
−21	19,929	86	97.4	106	113	119.9
−20	22,361	87	98.4	107	114	120.9
−19	25,089	88	99.4	108	115	121.9
−18	28,150	89	100.4	109	116	122.9
−17	31,585	90	101.4	110	117	123.9
−16	35,439	91	102.4	111	118	124.9
−15	39,764	92	103.4	112	119	125.9
−14	44,615	93	104.4	113	120	126.9
−13	50,059	94	105.4	114	121	127.9
−12	56,167	95	106.4	115	122	128.9
−11	63,021	96	107.4	116	123	129.9
−10	70,711	97	108.4	117	124	130.9
−9	79,339	98	109.4	118	125	131.9
−8	89,019	99	110.4	119	126	132.9
−7	99,881	100	111.4	120	127	133.9
−6	112,069	101	112.4	121	128	134.9
−5	125,743	102	113.4	122	129	135.9
−4	141,086	103	114.4	123	130	136.9
−3	158,301	104	115.4	124	131	137.9
−2	177,617	105	116.4	125	132	138.9
−1	199,290	106	117.4	126	133	139.9
0	223,607	107	118.4	127	134	140.9
1	250,891	108	119.4	128	135	141.9
2	281,504	109	120.4	129	136	142.9

dBm	μV	dBμV	dBμV/m 150 MHz	400 MHz	900 MHz	2,000 MHz
3	315,853	110	121.4	130	137	143.9
4	354,393	111	122.4	131	138	144.9
5	397,635	112	123.4	132	139	145.9
6	446,154	113	124.4	133	140	146.9
7	500,593	114	125.4	134	141	147.9
8	561,675	115	126.4	135	142	148.9
9	630,210	116	127.4	136	143	149.9
10	707,107	117	128.4	137	144	150.9
11	793,387	118	129.4	138	145	151.9
12	890,195	119	130.4	139	146	152.9
13	998,815	120	131.4	140	147	153.9
14	1,120,689	121	132.4	141	148	154.9
15	1,257,433	122	133.4	142	149	155.9
16	1,410,864	123	134.4	143	150	156.9
17	1,583,015	124	135.4	144	151	157.9
18	1,776,172	125	136.4	145	152	158.9
19	1,992,898	126	137.4	146	153	159.9
20	2,236,068	127	138.4	147	154	160.9
21	2,508,910	128	139.4	148	155	161.9
22	2,815,043	129	140.4	149	156	162.9
23	3,158,530	130	141.4	150	157	163.9
24	3,543,929	131	142.4	151	158	164.9
25	3,976,354	132	143.4	152	159	165.9
26	4,461,542	133	144.4	153	160	166.9
27	5,005,932	134	145.4	154	161	167.9
28	5,616,749	135	146.4	155	162	168.9
29	6,302,096	136	147.4	156	163	169.9
30	7,071,068	137	148.4	157	164	170.9
31	7,933,868	138	149.4	158	165	171.9
32	8,901,947	139	150.4	159	166	172.9
33	9,988,149	140	151.4	160	167	173.9
34	11,206,887	141	152.4	161	168	174.9
35	12,574,334	142	153.4	162	169	175.9
36	14,108,635	143	154.4	163	170	176.9
37	15,830,149	144	155.4	164	171	177.9
38	17,761,720	145	156.4	165	172	178.9
39	19,928,976	146	157.4	166	173	179.9
40	22,360,680	147	158.4	167	174	180.9
41	25,089,096	148	159.4	168	175	181.9
42	28,150,428	149	160.4	169	176	182.9

dBm	μV	dBμV	dBμV/m			
			150 MHz	400 MHz	900 MHz	2,000 MHz
43	31,585,300	150	161.4	170	177	183.9
44	35,439,288	151	162.4	171	178	184.9
45	39,763,536	152	163.4	172	179	185.9

APPENDIX 9

QUEUING TABLES

Not many books have attempted to include queuing tables, mainly because there are too many variables involved. Queuing calculations need to account of call holding time, average queue length, and circuits. To simplify the presentation the traffic capacity for a given average call holding time is presented in separate tables for each of a number of different call holding times. If the exact conditions that the system is operating under cannot be found in these tables, a good approximation can be had by interpolating between the conditions either side of the actual conditions. Alternatively, use the software provided if more precision is required.

Queuing Table 1   Average Call Holding Time = 15 seconds

	Average Time in Q in seconds					
Circuits	1	3	5	10	20	30
1	0.032	0.091	0.1434	0.2499	0.4001	0.5
2	0.4195	0.6492	0.7847	0.9997	1.243	1.39
3	0.9589	1.315	1.511	1.812	2.135	2.32
4	1.575	2.044	2.301	2.677	3.071	3.283
5	2.272	2.837	3.139	3.616	4.039	4.245
6	2.957	3.693	4.004	4.52	4.999	5.253
7	3.77	4.522	4.902	5.478	5.998	6.242
8	4.533	5.383	5.836	6.457	6.998	7.21
9	5.341	6.278	6.737	7.379	7.918	8.24
10	6.192	7.207	7.657	8.384	8.908	9.223
11	7.021	8.087	8.681	9.316	9.898	10.2
12	7.881	8.988	9.549	10.35	10.89	11.22
13	8.772	10	10.52	11.29	11.88	12.18
14	9.601	10.84	11.52	12.24	12.87	13.19
15	10.55	11.79	12.4	13.24	13.86	14.21
16	11.42	12.76	13.42	14.26	14.85	15.18
17	12.3	13.75	14.32	15.21	15.92	16.16
18	13.2	14.61	15.37	16.16	16.91	17.17
19	14.12	15.63	16.27	17.11	17.9	18.18
20	14.91	16.49	17.26	18.15	18.9	19.19
21	15.85	17.54	18.26	19.11	19.89	20.2
22	16.82	18.41	19.17	20.17	20.89	21.15
23	17.7	19.48	20.18	21.13	21.88	22.16
24	18.6	20.37	21.21	22.09	22.88	23.17
25	19.61	21.26	22.14	23.05	23.87	24.17
26	20.43	22.26	23.06	24.13	24.87	25.18
27	21.46	23.26	24.1	25.09	25.86	26.19
28	22.28	24.16	25.03	26.06	26.86	27.13
29	23.22	25.18	26.08	27.02	27.85	28.13
30	24.18	26.21	27.02	27.99	28.85	29.14
31	25.14	27.12	27.95	29.03	29.84	30.14
32	26.1	28.02	29.02	30.07	30.84	31.15
33	26.95	29.07	29.96	31.04	31.83	32.15
34	27.93	29.98	30.89	32.01	32.83	33.16
35	28.93	31.04	31.99	32.98	33.82	34.16
36	29.78	31.95	32.94	33.95	34.81	35.16
37	30.78	32.87	33.88	34.92	35.81	36.17
38	31.63	33.95	34.82	35.98	36.8	37.17
39	32.65	34.86	35.77	36.95	37.8	38.13
40	33.51	35.78	36.89	38.02	38.79	39.13

Queuing Table 2   Average Call Holding Time = 20 seconds

	Average Time in Q in seconds					
Circuits	1	3	5	10	20	30
1	0.0477	0.131	0.2002	0.3331	0.5	0.6003
2	0.4955	0.7555	0.9061	1.14	1.39	1.527
3	1.08	1.469	1.684	1.999	2.32	2.487
4	1.738	2.243	2.52	2.91	3.283	3.467
5	2.463	3.076	3.403	3.842	4.245	4.46
6	3.206	3.924	4.341	4.802	5.253	5.465
7	4.006	4.804	5.209	5.762	6.242	6.43
8	4.866	5.72	6.201	6.723	7.21	7.428
9	5.675	6.67	7.087	7.76	8.24	8.447
10	6.514	7.579	8.135	8.73	9.223	9.455
11	7.386	8.508	9.039	9.7	10.2	10.45
12	8.29	9.452	10.04	10.67	11.22	11.44
13	9.228	10.42	10.96	11.64	12.18	12.42
14	10.1	11.4	11.99	12.61	13.19	13.42
15	10.99	12.28	12.91	13.65	14.21	14.42
16	11.89	13.29	13.84	14.62	15.18	15.41
17	12.81	14.17	14.91	15.6	16.16	16.4
18	13.75	15.21	15.84	16.65	17.17	17.42
19	14.56	16.1	16.77	17.64	18.18	18.4
20	15.52	17.17	17.79	18.62	19.19	19.42
21	16.5	18.08	18.82	19.6	20.2	20.4
22	17.33	18.98	19.77	20.58	21.15	21.42
23	18.34	20.09	20.71	21.56	22.16	22.41
24	19.27	21	21.75	22.54	23.17	23.4
25	20.21	21.91	22.7	23.52	24.17	24.42
26	21.16	22.94	23.65	24.5	25.18	25.43
27	22.12	23.98	24.72	25.54	26.19	26.39
28	22.97	24.9	25.67	26.52	27.13	27.4
29	23.94	25.83	26.62	27.58	28.13	28.42
30	24.92	26.88	27.57	28.56	29.14	29.43
31	25.78	27.81	28.66	29.55	30.14	30.41
32	26.77	28.73	29.61	30.53	31.15	31.38
33	27.78	29.81	30.57	31.52	32.15	32.39
34	28.65	30.74	31.53	32.5	33.16	33.41
35	29.66	31.67	32.64	33.48	34.16	34.42
36	30.69	32.77	33.61	34.47	35.16	35.39
37	31.57	33.7	34.57	35.45	36.17	36.4
38	32.44	34.64	35.53	36.44	37.17	37.41
39	33.48	35.58	36.5	37.52	38.13	38.37
40	34.36	36.71	37.46	38.51	39.13	39.38

Queuing Table 3   Average Call Holding Time = 30 seconds

	Average Time in Q in seconds					
Circuits	1	3	5	10	20	30
1	0.032	0.091	0.1434	0.2499	0.4001	0.5
2	0.4195	0.6492	0.7847	0.9997	1.243	1.39
3	0.9589	1.315	1.511	1.812	2.135	2.32
4	1.575	2.044	2.301	2.677	3.071	3.283
5	2.272	2.837	3.139	3.616	4.039	4.245
6	2.957	3.693	4.004	4.52	4.999	5.253
7	3.77	4.522	4.902	5.478	5.998	6.242
8	4.533	5.383	5.836	6.457	6.998	7.21
9	5.341	6.278	6.737	7.379	7.918	8.24
10	6.192	7.207	7.657	8.384	8.908	9.223
11	7.021	8.087	8.681	9.316	9.898	10.2
12	7.881	8.988	9.549	10.35	10.89	11.22
13	8.772	10	10.52	11.29	11.88	12.18
14	9.601	10.84	11.52	12.24	12.87	13.19
15	10.55	11.79	12.4	13.24	13.86	14.21
16	11.42	12.76	13.42	14.26	14.85	15.18
17	12.3	13.75	14.32	15.21	15.92	16.16
18	13.2	14.61	15.37	16.16	16.91	17.17
19	14.12	15.63	16.27	17.11	17.9	18.18
20	14.91	16.49	17.26	18.15	18.9	19.19
21	15.85	17.54	18.26	19.11	19.89	20.2
22	16.82	18.41	19.17	20.17	20.89	21.15
23	17.7	19.48	20.18	21.13	21.88	22.16
24	18.6	20.37	21.21	22.09	22.88	23.17
25	19.61	21.26	22.14	23.05	23.87	24.17
26	20.43	22.26	23.06	24.13	24.87	25.18
27	21.46	23.26	24.1	25.09	25.86	26.19
28	22.28	24.16	25.03	26.06	26.86	27.13
29	23.22	25.18	26.08	27.02	27.85	28.13
30	24.18	26.21	27.02	27.99	28.85	29.14
31	25.14	27.12	27.95	29.03	29.84	30.14
32	26.1	28.02	29.02	30.07	30.84	31.15
33	26.95	29.07	29.96	31.04	31.83	32.15
34	27.93	29.98	30.89	32.01	32.83	33.16
35	28.93	31.04	31.99	32.98	33.82	34.16
36	29.78	31.95	32.94	33.95	34.81	35.16
37	30.78	32.87	33.88	34.92	35.81	36.17
38	31.63	33.95	34.82	35.98	36.8	37.17
39	32.65	34.86	35.77	36.95	37.8	38.13
40	33.51	35.78	36.89	38.02	38.79	39.13

Queuing Table 4   Average Call Holding Time = 60 seconds

Circuits	Average Time in Q in seconds					
	1	3	5	10	20	30
1	0.0163	0.0477	0.076	0.1434	0.2499	0.3331
2	0.3129	0.4955	0.6065	0.7847	0.9997	1.14
3	0.7796	1.08	1.25	1.511	1.812	1.999
4	1.33	1.738	1.96	2.301	2.677	2.91
5	1.933	2.463	2.725	3.139	3.616	3.842
6	2.619	3.206	3.547	4.004	4.52	4.802
7	3.34	4.006	4.343	4.902	5.478	5.762
8	4.057	4.866	5.275	5.836	6.457	6.723
9	4.828	5.675	6.152	6.737	7.379	7.76
10	5.597	6.514	7.062	7.657	8.384	8.73
11	6.412	7.386	7.926	8.681	9.316	9.7
12	7.197	8.29	8.808	9.549	10.35	10.67
13	8.011	9.228	9.705	10.52	11.29	11.64
14	8.856	10.1	10.62	11.52	12.24	12.61
15	9.732	10.99	11.55	12.4	13.24	13.65
16	10.53	11.89	12.51	13.42	14.26	14.62
17	11.46	12.81	13.47	14.32	15.21	15.6
18	12.3	13.75	14.46	15.37	16.16	16.65
19	13.16	14.56	15.31	16.27	17.11	17.64
20	14.03	15.52	16.33	17.26	18.15	18.62
21	14.92	16.5	17.19	18.26	19.11	19.6
22	15.83	17.33	18.23	19.17	20.17	20.58
23	16.75	18.34	19.1	20.18	21.13	21.56
24	17.51	19.27	20.07	21.21	22.09	22.54
25	18.46	20.21	21.04	22.14	23.05	23.52
26	19.42	21.16	21.92	23.06	24.13	24.5
27	20.19	22.12	23.02	24.1	25.09	25.54
28	21.19	22.97	23.92	25.03	26.06	26.52
29	22.08	23.94	24.8	26.08	27.02	27.58
30	22.98	24.92	25.82	27.02	27.99	28.56
31	23.89	25.78	26.84	27.95	29.03	29.55
32	24.82	26.77	27.74	29.02	30.07	30.53
33	25.62	27.78	28.63	29.96	31.04	31.52
34	26.69	28.65	29.67	30.89	32.01	32.5
35	27.5	29.66	30.73	31.99	32.98	33.48
36	28.45	30.69	31.64	32.94	33.95	34.47
37	29.41	31.57	32.54	33.88	34.92	35.45
38	30.23	32.44	33.45	34.82	35.98	36.44
39	31.2	33.48	34.52	35.77	36.95	37.52
40	32.18	34.36	35.43	36.89	38.02	38.51

Queuing Table 5   Average Call Holding Time = 120 seconds

Circuits	Average Time in Q in seconds					
	1	3	5	10	20	30
1	0.008	0.0244	0.044	0.076	0.1434	0.2002
2	0.23	0.3716	0.4593	0.6065	0.7847	0.9061
3	0.6305	0.8801	1.024	1.25	1.511	1.684
4	1.123	1.471	1.661	1.96	2.301	2.52
5	1.678	2.138	2.366	2.725	3.139	3.403
6	2.274	2.84	3.079	3.547	4.004	4.341
7	2.958	3.548	3.885	4.343	4.902	5.209
8	3.63	4.31	4.673	5.275	5.836	6.201
9	4.364	5.129	5.561	6.152	6.737	7.087
10	5.112	5.947	6.384	7.062	7.657	8.135
11	5.854	6.743	7.238	7.926	8.681	9.039
12	6.638	7.568	8.124	8.808	9.549	10.04
13	7.389	8.425	8.951	9.705	10.52	10.96
14	8.169	9.313	9.895	10.62	11.52	11.99
15	8.976	10.13	10.77	11.55	12.4	12.91
16	9.814	11.08	11.65	12.51	13.42	13.84
17	10.68	11.94	12.56	13.47	14.32	14.91
18	11.46	12.81	13.47	14.46	15.37	15.84
19	12.26	13.7	14.41	15.31	16.27	16.77
20	13.21	14.61	15.29	16.33	17.26	17.79
21	14.04	15.53	16.17	17.19	18.26	18.82
22	14.9	16.31	17.16	18.23	19.17	19.77
23	15.76	17.26	17.98	19.1	20.18	20.71
24	16.56	18.23	18.98	20.07	21.21	21.75
25	17.37	19.02	19.9	21.04	22.14	22.7
26	18.28	20.02	20.84	21.92	23.06	23.65
27	19.2	20.92	21.78	23.02	24.1	24.72
28	20.04	21.84	22.74	23.92	25.03	25.67
29	20.89	22.76	23.58	24.8	26.08	26.62
30	21.86	23.7	24.66	25.82	27.02	27.57
31	22.72	24.51	25.52	26.84	27.95	28.66
32	23.6	25.58	26.37	27.74	29.02	29.61
33	24.48	26.41	27.5	28.63	29.96	30.57
34	25.38	27.37	28.36	29.67	30.89	31.53
35	26.28	28.35	29.22	30.73	31.99	32.64
36	27.18	29.18	30.22	31.64	32.94	33.61
37	27.96	30.16	31.25	32.54	33.88	34.57
38	29.03	31.16	32.12	33.45	34.82	35.53
39	29.82	32	32.99	34.52	35.77	36.5
40	30.76	33	34.02	35.43	36.89	37.46

GLOSSARY

Airtime. The total time that a channel is occupied, including call time, call setup, and cleardown time.

AM. Amplitude modulation. An analog modulation system that modulates (varies) the transmitter power level in proportion to the signal to be transmitted.

Antenna gain. The gain of an antenna expressed relative to either the gain of a dipole (dBd) or the gain of an omnidirectional antenna (dBi).

ASCII. American Standard Code for Information Interchange. An 8-bit code with 1 bit for parity checking that is used to send 127 characters, consisting of numbers, letters, and symbols commonly found on a computer keyboard.

Base station. A base station is the transmitter site including all transceivers, the tower, and buildings that house the equipment.

Baseband. The frequency, or band of frequencies, that are transmitted over a carrier system.

Bit error rate. The fraction number of error bits in a digital code.

Blocking. A call lost due to lack of resources either at the switch or at the base station.

Carrier. The radio signal that transmits a signal.

Cavity. A tuned circuit, which, as the name implies, is in the form of a resonant cavity.

CDMA. Code division multiple access. A modulation system that sends multiple signals on the same channel at the same time, but distinguishes the desired signal by a digital code that identifies each signal. In this case the unwanted signals are seen only as background noise. This is usually associated with spread-spectrum systems.

Channel. A pair of frequencies used by a mobile or base site to transmit one conversation or transaction.

Collinear antenna. A vertically stacked array of dipoles. Can usually be identified by the tube encasing that gives it the form of a pipe.

Coaxial cable. A pair of conductors consisting of a central conductor surrounded by the outer conductor. These cables are used because of their low loss and relative immunity from interference.

Combiner. A device that combines a number of RF channels. It may combine channels at the transmit level, but it may also refer to a device that combines audio channels (for example, there may be a number of dispatch operators who need to connect to a single audio line via a combiner.

Coupler. A device for connecting two or more sources of RF energy into a single cable or port.

Coverage. The area over which the mobile service meets a predefined standard.

CTCSS. Continuous-tone-controlled squelch systems. This may be used in either trunked or conventional mobile systems and refers to a system that transmits a (usually) subaudible tone, that operates the squelch on the receivers. Sometimes an audible tone is used for this purpose, and it is notch-filtered out of the audio path. It is used in conventional PMR so that mobiles in different groups can share common channels without intruding on each other.

dBd. The gain of an antenna compared to the gain of a dipole.

dBi. The gain of an antenna compared to the gain of an omnidirectional (isotropic) radiator (one that radiates equally in all directions).

Deviation. The amount of change in frequency, from the center frequency that occurs in FM modulation.

DID. Direct in dial. This is used to describe any system that allows direct indialing to a mobile (or conversely to a landline).

Diffraction. Propagation around an obstructing obstacle.

Digital. A processing system whereby a signal is represented as a discrete code. Generally the signal is processed in binary representation, meaning that it is coded as "1s" and "0s."

DOD. Direct out dialing. This is used to describe any system that allows automatic out dialing.

DTMF. Dual-tone multifrequency. DTMF is the tone set that is widely used in the POT (plain old telephone) for tone dialing. DTMF can be used for signaling over a mobile path, but there are problems. First, the DTMF detectors require a minimum decode time that is of the order of 50 ms (the actual time is decoder dependent). Secondly, by its nature, mobile radio transmissions can be sporadic and a tone that is sent for too long may be double-decoded if it is interrupted. For example, if a "6" is held down for too long and a break occurs, it may be interpreted as "66." Nevertheless, DTMF is widely used in both conventional and trunked radio systems and is often used for telephone interconnect.

Dual mode. Refers to any system that can operate in two different ways. In trunk radio, this mostly means operation as a trunked radio as well as conventional PMR.

It may refer, however, to a mobile that can operate in digital and analog modes, or even to one that can operate in two different trunked modes. The choice of operation can be either automatic or manual.

Duplex. A system that allows simultaneous transmission and reception (as distinguished from PTT operations where the operator can either transmit or receive but not both together). Also known as diplex.

EMI. Electromagnetic interference.

Erlang. This is a unit of circuit traffic that is defined in a number of ways. It can be defined as follows: One Erlang is an equivalent circuit in use at the time of observation. More usually it is defined to be the average occupancy of a circuit over an hour (or some other period). Thus a channel which is in use 51% of the time over the period of measurement would be said to be carrying 0.51 Erlangs. If more than one circuit in a group is in use, the traffic in Erlangs is additive.

ERP. Effective radiated power. This figure gives a measure of the gain of the antenna system. An isotropic radiator (an almost hypothetical radiator that radiates equally in all directions) or a dipole antenna is used as a reference. Thus a high-gain antenna may (in its direction of optimal radiation) have a gain that is 10 dB higher than a dipole. This antenna would be said to have a 10 dBd gain (the last "d" referring to a dipole). When comparing antenna gains, it is important to be clear whether the units of gain are dBd or dBi (referred to an omnidirectional radiator). The conversion factor is dBd = dBi + 2.1.

FCC. The Federal Communications Commission, the regulatory body in the United States.

FDMA. Frequency division multiple access. In FDMA systems, individual traffic channels are separated by being transmitted on separate frequencies.

Feeder. The cable that connects the antenna to the transmitter.

Feedline. See Feeder.

FM. Frequency modulation. An analog modulation system that modulates (varies) the frequency of the transmitter, in response to the signal to be transmitted.

Frequency. The rate at which the electric and magnetic fields vibrate per second. The units are hertz, usually abbreviated as Hz.

Full-duplex. A system that allows simultaneous talk and listen capabilities. Examples are a telephone and a mobile phone. The opposite is simplex, which requires push-to-talk operation.

Gain. The factor (usually expressed in decibels) by which a signal is amplified.

GOS. Grade of service. The probability that a call will fail. The lower the grade of service, the lower the probability of call failure. Typical GOS for a trunked system would be around 0.05.

GOS 2. When referred to the Erlang *C* table or to a queuing system, the GOS refers to the probability that a call will retry or have to queue.

Ground plane. The plane directly below a quarter wave of other unbalanced antenna. It is put there to balance the radiation path.

Group call. A call to multiple users simultaneously. This approximately simulates the "all stations call" characteristic of conventional land mobile radio.

Hands-free. A voice-operated system that allows the user to talk and listen without the need to hold the microphone or the speaker.

IC. An integrated circuit, the building blocks of modern telecommunications equipment.

Interference. The reception of unwanted signals from any source.

Intermodulation interference. Interference generated by the interaction between two or more carriers at a point of nonlinearity. When two signals interact at a nonlinear interface (like a rusty fence), they produce signals equal to their sums and differences together with the sums and differences of integer multiples of the carriers.

Isolator. A device that allows a signal to pass in one direction only.

Isotropic. Something that is the same in all directions. Referred to RF, this usually means a signal radiator that is uniform in all directions (e.g., a point source).

kHz. Kilohertz, or 1,000 Hz.

Leaky cable. See Lossy cable.

Links. Refers to the circuits connecting various parts of the network together (such as the switch to the base station).

LNA. Low-noise amplifier. An amplifier designed to amplify weak signals without adding too much internally generated noise into the system. Often used on towers (connected at the antenna) or before the RF input of the transceiver. The noise performance of a system with an LNA will be determined largely by the noise figure of the LNA.

Lossy cable. Cable designed to be deliberately high loss, in a way that the losses are caused by radiation leaks in the cable outer. This cable is often used as a distributed antenna in tunnels and buildings.

Mast. A structure for supporting antennas that relies on guys (cables) for support.

Modem. A modulating/demodulation system that converts digital signals to analog (for send down an analog line) and then back to digital again.

Modulation. The method by which the transmitted signal is impressed on the carrier.

MTBF. Mean time between failure. A measure of the average reliability of the components of a network, usually quoted in average years (or hours) of service that can be expected. Because there are no standardized ways of measurement, these figures can be difficult to interpret.

Multicoupler. A device that couples two or more transmitters into one antenna.

Multipath fading. Fading caused by the (self-) interference of the signal when it has more than one direct path to the receiver.

NiCad or NICAD. A rechargeable nickel cadmium battery.

Omnidirectional antenna. An antenna that transmits a uniform signal in all directions (as opposed to a directional antenna like a yagi).

Overdialing. A connection that requires the user to dial an initial access number, wait for an answer and response, and then dial further digits.

PABX. Private automatic branch exchange. An in-house switch to connect local calls.

Path budget. The losses that can be incurred over the radio path and still allow reception within the specified performance figures.

Phase-locked loop. A circuit that can lock onto a desired frequency using a frequency reference and a phase comparator.

PMR. Public mobile radio.

Portable. A handheld trunk or conventional radio.

POTS. Plain old telephones.

Preselector. A filter that is placed between the antenna and the receiver to reduce the effect of out-of-band signals.

PSTN. Public switched telephone network. Refers to the telephone network providing public access.

PTT. Press to talk; the usual way of communicating from mobiles.

Queuing. In trunked radio, this refers to holding call requests in a queue until a free channel becomes available. The alternative to a queue is to send a busy tone when the system is not available and require the caller to manually redial.

Rayleigh fading. Another term for multipath fading. Rayleigh was the first to seriously study interference patterns in light (for which the mathematics is essentially the same).

Refraction. Propagation in other than a straight line by bending around an obstacle.

Repeater. A device that can be either active or passive and that receives a signal and then relays it on.

RF. Radio frequencies that vary from 10 kHz to 300 GHz.

SINAD. Signal-to-noise-plus-distortion ratio.

Standby time. The number of hours that a handheld mobile can operate in the receive mode only.

Spread spectrum. A spread-spectrum system is one that has a wide bandwidth relative to the transmitted signal, and the transmitted bandwidth is independent of the modulation.

Talk time. The time that a fully charged battery can hold the transmitter on for a connected call.

TDMA. Time division multiple access. A digital modulation system which transmits a number of signals over the same frequency by allocating different time slots to each of the signals.

Tower. A self-supporting structure for holding antennas.

Traffic. Calls in progress, usually measured in Erlangs.

Transceiver. A combined transmitter/receiver that usually defines one RF channel (digital systems may have multiple channels on each transceiver).

Trunk radio. A multichannel mobile radio system with automatic frequency assignment for the mobiles across the frequencies. Distinguished from cellular radio by the ability to call and talk with groups of mobiles simultaneously.

UHF. The frequency range from 300 to 3,000 MHz.

VLSI. Very large scale integrated circuit. These are at the heart of modern communications systems.

VSWR. Voltage standing wave ratio. The ratio of the maximum voltage to the minimum voltage on the line. Ideally, a system will have a VSWR of unity, with any other (higher) value indicating some impedance mismatch and thus some power losses. Ordinarily, an SWR of around 1.5 is acceptable for most RF applications.

Wavelength. The distance from a point on a wave to the next corresponding point; for example, it may be the distance between two consecutive troughs or between two consecutive peaks.

Wide area. Wide-area trunked radio systems operate over the coverage area of two or more base stations.

INDEX